THE HISTORICAL RECORDS OF THE FIFTH ROYAL IRISH LANCERS

HISTORICAL RECORDS OF

THE FIFTH (ROYAL IRISH) LANCERS

FROM THEIR FOUNDATION AS

WYNNE'S DRAGOONS

(IN 1689) TO THE PRESENT DAY

BY

WALTER TEMPLE WILLCOX

MAJOR THIRD HUSSARS

LATE CAPTAIN FIFTH (ROYAL IRISH) LANCERS

DEDICATED

TO THE PAST & PRESENT

OFFICERS

NON-COMMISSIONED

OFFICERS AND MEN

OF THE

FIFTH LANCERS

TITLES OF THE REGIMENT

Wynne's (Enniskillen) Dragoons
1689.

Ross' Dragoons
1695.

The Royal Dragoons of Ireland
1704.

The 5th (or Royal Irish) Dragoons
about 1752.

The 5th (or Royal Irish) Regiment of (Light) Dragoons (Lancers)
1858.

The 5th (Royal Irish) Lancers
1861.

HONOURS

BORNE BY THE 5TH LANCERS ON THE APPOINTMENTS

"BLENHEIM" "RAMILLIES"
"OUDENARDE" "MALPLAQUET" "SUAKIN, 1885"
"SOUTH AFRICA, 1899-1902"
"DEFENCE OF LADYSMITH"

INTRODUCTION

Shortly after my appointment as Honorary Colonel of the 5th Royal Irish Lancers, it was a source of satisfaction and pleasure to me to learn that the Historical Records of the Regiment were being compiled, for I consider it essential that every Officer, non-commissioned Officer, and man should be conversant with the history of his regiment, in order that he may gain that "esprit de corps" with which it is so important that all ranks should be imbued—an element which forms, I am glad to say, one of the strongest features of the British Army.

The study of such Historical Records, recounting the deeds of those who have helped to make their Country's History, fosters and maintains this spirit, and is an incentive to all ranks to endeavour to sustain the good name and high state of efficiency for which their corps were celebrated in the past.

Surely, therefore, the present record of the exploits of past members of the 5th Royal Irish Lancers must prove most interesting and instructive to those now serving in the Regiment.

The author tells me that his aim has been to recount facts of interest to other Cavalry Regiments, as well as to his own, for he thinks that, while "esprit de corps" should be nurtured in each individual Regiment it should also be cherished throughout the British Army as a whole.

I cannot speak too highly of the care and trouble which Captain Willcox (recently promoted to a majority in the 3rd Hussars) has taken in the production of this record.

His self-imposed task has occupied him for three years, and it has been most admirably performed. Everyone who has served or is still serving in the Regiment must owe him a deep debt of gratitude.

T. A. COOKE, MAJOR-GENERAL.
HON. COL. 5th (ROYAL IRISH) LANCERS.

September, 1906.

PREFACE

It may be a popular belief that a soldier as a rule has but little inclination towards a literary effort. The belief, if such there be, is possibly justifiable, for a soldier should undoubtedly live by his sword rather than by his pen. On the other hand, now-a-days the profession of arms entails a large amount of study and some literary effort on the part of her sons.

Such study, outside purely professional subjects, should, we are told, take the form of History in its connection with War. I would, however, venture to add that the first historical study to which a young Officer should devote himself on first obtaining the honour of His Majesty's Commission is the history of his own regiment; and not only the Officer but the recruit should surely know of the actions of his predecessors.

To recall, however, a possibly forgotten past is quite a waste of time if no useful lessons or fruitful results are forthcoming from the labours which the effort entails. As regards the lessons to be learnt much depends upon the ability of the would be historian, but fruitful results in the furtherance of that *esprit de corps* for which British regiments are so justly famed, will always follow in the course of any effort in the publishing of a regimental history.

To this end some three years ago the following story of the Regiment was considered, thought out and commenced.

One may be allowed to say that the difficulties of the undertaking were great. Our Regiment is now the 5th Lancers.

**

of changes of stations and mere matters of detail are dull reading, and also for want of space, that period has been but lightly touched upon.

The account of the great Rebellion of 1799 in Ireland tells its own tale, and in its relation my object is " neither to vindicate, nor to set down aught in malice. "

I regret exceedingly that a sudden departure for India prevents my obtaining the biographies of certain Officers who have been at the head of the Regiment, more especially that of Lord Rossmore and of his Lieutenant-Colonel, Stewart, who later, in the Peninsula wars, rose to great distinction. For the same reason I am unable correctly to check the list of Officers who have served in the Old and Present Regiment, and I must tender my apologies for any mistakes in the Roll.

I have met with the greatest kindness and help on all sides ; I would especially thank Brigadier-General Allenby C.B., Commanding the 4th Cavalry Brigade, for his encouragement and help at the commencement of my labours, and also my thanks are due to Lieutenant-Colonel H. Graham D.S.O. and the Officers of the 5th Lancers for coming forward and taking over the final work of publication, which, owing to the exigencies of the Service, I am prevented from doing. With them, too, I have to thank my step-Father, Mr James Dodds, for his promise of co-operation. To Mrs King I tender my thanks for her kindness in promising the use of a painting of the 5th Lancers. I have made much use of the Departments at the British Museum and the Public Record Office, and have experienced the greatest kindness from the officials at both those institutions. The advice of the Librarian at the War Office was most valuable, while the Royal United Service Institution and the Prince Consort's Library at Aldershot were of the greatest help. My thanks are also due to Mr W. Morris Colles, of the Authors' Syndicate, for his sound advice and practical help.

To Major-General Cooke C.V.O., Colonel of the Regiment, my thanks are due, for his symphathy in my undertaking.

That this story might have have better told I well know, but should it prove of interest to my Comrades the Officers, Non-Commissioned officers and Men of the 5th Lancers, I should feel that the labours of the past three years have not been in vain.

I have only to add that my effort is but a small thank offering for the extremely happy thirteen years which I have passed in the Regiment, and which end today.

W. T. W.

West Cavalry Barracks
Aldershot.
17th December, 1906.

PUBLISHER'S NOTE.

Although this work was completed nearly eighteen months ago, arrangements for publication were not made until March of this year [1908]. The conditions under which the volume is published did not allow sufficient time for sending proofs to the author, who is at a serious disadvantage in being unable to supervise the printing of his book, which had to be seen through the press by the publishers.

While taking full responsibility for any imperfections the volume contains, the publishers crave the indulgence of the author and of the public on the score of the circumstances in which it had to be produced.

LIST OF AUTHORITIES

AMERY, Leopold Charles Maurice Stennett.
The Times History of the War in South Africa, 1899-1900. Edited by L. S. Amery. London, 1900 etc.

ARCHER, Thomas.
The War in Egypt and the Soudan : an episode in the history of the British Empire. 4 vols. London, 1885-87.

ARMY LISTS. (1859 to date) London.

BOYER, Abel.
The History of Queen Anne, wherein all the civil and military Transactions of that reign are impartially related. London, 1735.

British Military Library, or Journal : comprehending a complete body of military knowledge, etc., 2 vols. London, 1798-1801.

CHURCHILL, John, Duke of Marlborough.
The Letters and Despatches of John Churchill, Duke oj Marlborough. 1702 to 1712. 5 vols. London, 1845.

COXE, William. Archdeacon of Wilts.
Memoirs of John, Duke of Marlborough with his original correspondence collected from the family records . . . and other authentic sources. 3 vols. London, 1847.

DALTON, Charles.
English Army Lists and Commission Registers 1661-1714. 6 vols. London, 1892-1904.

D'ALTON, John.
Illustrations, historical ana genealogical, of King James's Irish Army List. Dublin, 1855.

D'AUVERGNE, Edward.
The History of the Campaign in Flanders for the year 1691. *Compleating the history of the seven Campaigns of his Majesty to the Treaty of Ryswick*, etc., London, 1735.

DAVIS, John. Colonel.
The History of the Second Queen's Royal Regiment, now the Queen's Royal West Surrey Regiment. 6 vols. London, 1887-1906.

DAVITT, Michael.
The Boer Fight for Freedom. London.

FORTESCUE, Hon. John William.
History of the British Army. London, 1899.

GALLOWAY, William.
The Battle of Tofrek fought near Suakin,... 1885 *under... Sir J. C. McNeill, in its relation to the Mahdist insurrection in the Eastern Sudan.* London, 1887.

GLEICHEN, Albert Edward Wilfrid. Count.
With the Camel Corps up the Nile. London, 1888.

GRAHAM, John.
Derriana, consisting of a history of the Siege of Londonderry and defence of Enniskillen, in 1688 *and* 1689. Londonderry. 1823. 2nd edition. Dublin, 1829.

GRAHAM, John.
A History of Ireland, from the relief of Londonderry in 1689, *to the surrender of Limerick in* 1691. Dublin, 1839.

GRANT, James.
British Battles on Land and Sea. London, 1899.

GROSE, Francis.
Military Antiquities respecting a history of the English Army from the Conquest to the present time. 2 vols. London, 1801.

HAMILTON, Andrew.
A true relation of the Actions of the Inniskilling Men, from the first taking up of arms in December 1688 *for the defence of the Protestant religion.* London, 1690.

KNIGHT, Henry Raleigh. Captain.
Historical Records of the Buffs, East Kent Regiment, 3rd Foot,

formerly designated the Holland Regiment and Prince of Denmark's Regiment. London, 1905.

MACAULAY, Thomas Babington. Baron.
History of England, from the accession of James II. London, 1849-61.

The Field of Mars: being an alphabetical digestion of the principal naval and military engagements in Europe, Asia, and America, particularly of Great Britain and her Allies, from the ninth century to the Peace of 1801.... *Selected from the best Historians and Journalists.* 2 vols. London, 1801.

MUSGRAVE, Sir Richard.
Memoirs of the different rebellions in Ireland from the arrival of the English, with a particular detail of that which broke out the XXIIId of May, MDCCXCVIII. Dublin, 1801.

SMET, John Francis. Surgeon.
Historical Records of the 8*th King's Royal Irish Hussars.* To 1803. London, 1874.

STEEVENS, George Warrington.
From Capetown to Ladysmith. An unfinished record of the South African War. Edited (with a concluding chapter) by Vernon Blackburn. London, 1900.

STORY, George Warter.
A True and Impartial History of the most material occurrences in the Kingdom of Ireland during the Two last years with the Present State of Both Armies. London, 1691.

STORY, George Warter.
An Impartial History of the Wars of Ireland (from March 1689 *to March* 1692*). With some remarks upon the present state of that Kingdom.* 2 parts. London, 1693.

WALTON, Clifford.
History of the British Standing Army, A.D. 1660 *to* 1700. London, 1894.

CONTENTS.

CHAPTER VI.

CHAPTER VII.

CHAPTER VIII.

CHAPTER IX.

CHAPTER X.

CHAPTER XI.

CHAPTER XII.

CHAPTER XIII.

CHAPTER XIV.

CHAPTER XV.

CHAPTER XVI.

CHAPTER XVII.

CHAPTER XVIII.

CHAPTER XIX.

APPENDICES.

LIST OF ILLUSTRATIONS

The Fifth Lancers on the Long Valley.

From a Painting in possession of Mrs. Augustus King.

QUIS SEPARABIT
5TH
ROYAL IRISH

UNIFORMS OF THE REGIMENT AT DIFFERENT PERIODS OF ITS HISTORY.

FIFTH (or Royal Irish) DRAGOONS.

To GENERAL LORD ROSSMORE, this representation of the Uniform of the FIFTH (or Royal Irish) REGIMENT of DRAGOONS, is respectfully dedicated. by his Lordship's obedient Servants.

London: March: 1800.

THE EDITORS.

Fifth (Royal Irish) Dragoons about 1780.

From a Water-colour Painting in the Officers' Mess.

THE CHARGE OF THE FIFTH LANCERS AT ELANDSLAAGTE.

FROM AN OIL PAINTING IN THE OFFICERS' MESS.

A Mounted Sergeant, 1905, Fifth Lancers.

THE DRUM BANNERS.

THE DRUM HORSE.

LIST OF MAPS AND PLANS

A HISTORY OF

THE FIFTH LANCERS

CHAPTER I.

The Irish Revolution 1688-1691.—*The raising of Wynne's Dragoons.—Siege of Derry and Defence of Enniskillen.—Dundalk Camp.—The Boyne.—Aughrim.—Limerick.*

At the close of 1688 the Earl of Tyrconnel foresaw the possibility of a struggle in Ireland between William of Orange and King James and set himself to organise a Jacobite army in Ireland. His efforts were so successful that he was able to attach to the Jacobite cause no less than 19 regiments of infantry, 4 or 5 of cavalry, 20,000 stands of arms and several guns.

Tyrconnel's successful efforts in organising this Jacobite army had the result of arousing the suspicions of the North of Ireland, and the people of the North immediately began to organise on their part on behalf of King William the Third.

There were a number of minor engagements in the North in which the people of the North were mostly defeated, although they were more or less successful in some. King William's adherents however, seeing it was useless to carry on these detached actions in the open, proceeded to concentrate their forces, chiefly at Derry and Enniskillen.

Andrew Hamilton, in his pamphlet entitled *The actions of the Enniskillen Men*, writes that the inhabitants of Inniskilling and of the country round that town received orders early in December 1688 from Lord Tyrconnel to provide quarters for two Foot Companies of Irish which he proposed to send to the town, and on Friday the 14th December, news arrived that the two companies of foot whose presence they so much feared, were within 18 miles of Enniskillen. On the 15th December, the men of Enniskillen wrote the following letter to the Officers commanding in Londonderry:—

Gentlemen,

The frequent intelligence we have from all parts of this Kingdom, of a general massacre of the Protestants, and two companies of foot of Sir Thomas

Newcomen's regiment.... being on their march to garrison here, and now within 10 *miles, hath put upon us the resolution of refusing them entrance; our design being only to preserve our own lives, and the lives of our neighbours, this being the most considerable pass between Connaught and Ulster; and hearing of your resolutions, we thought it convenient to impart this to you, as likewise to beg your assistance both in your advice and relief, especially in helping us with some powder, and in carrying on a correspondance with us hereafter, as we shall with God's assistance, do with you, which is all at present, Gentlemen, from your faithful friends and fellow Christians.*

The Inhabitants of Enniskillen.

On Sunday the 16th at 10 a.m., word reached the town that the two companies they so much feared had arrived at Lisbellaw. Most of the inhabitants were in church at the time, but soon got under arms, resolved to advance and meet the enemy. On being drawn out they were found to amount to some 200 infantry and 150 horse, and the sight of the Enniskillen men was sufficient to drive the two companies, headed by their officers, to Cavan, where they remained until they received orders from Tyrconnel to march to other quarters.

On the 18th December Gustavus Hamilton was appointed Governor of the Town, and under him was formed a garrison drawn from the town itself and the surrounding districts.

This garrison soon consisted of 12 Companies of Horse and Foot, "upon which they thought fit to Regiment themselves under Gustavus Hamilton and Thomas Lloyd, and to send to England for arms and ammunition, meanwhile causing a great number of pikes to be made, and beating out old scythes, they fixing them on poles."

In the news of the accession of William and Mary in March 1689, Andrew Hamilton states "we rejoiced unspeakably."

When James II landed at Kinsale in March 1689 with some 1800 followers from France, the situation stood briefly thus:—

The Earl of Tyrconnel, James' Lord Lieutenant, had been stirring to prevent the Protestants from "concentralizing" so as to become formidable, and he had occupied in King James' name as many stations as he could. The Protestants, on the other hand, were endeavouring to hold their own until support should arrive from England. Both parties were very determined and rancorous, because their quarrel was religious as well as political and national.

The Irish had it all their own way in the South, and were also predominant in Connaught and Leinster, but in the North the Protestants were more numerous, and were banding themselves together for resistance.

James arrived in the North of Ireland in April, and appearing before Derry, demanded the surrender of the Town. The answer he received was: "No surrender" accompanied by the discharge of every description of fire-arm, whereupon the forces of King James laid siege unsuccessfully to the place for 105 days, until it was relieved by a force from England.

Meanwhile the people of Inniskilling, in spite of the manifestations of Lord Gilmoy, the threats of King James and of the non-arrival of arms from England, remained firm in their defence of the city.

On the 6th May 1689, Lieut.-Colonel Lloyd with 12 Foot Companies and some horse went towards Ballyshannon, and meeting the enemy's horse near Beleek, soon put them to rout; killing about 120 of them and taking prisoners and 2 small pieces of cannon, several horses and some good arms, without the loss of a single man. "This was the first time our men encountered the enemy in the field and having had so great success in the beginning it encouraged our men very much."[1] Redhills and Ballinacarg in the county of Cavan, were taken by Lloyd with 1500 Horse and Foot. The taking of these places spread consternation amongst the Irish, and Lloyd returned to Enniskillen with 3000 cows and oxen, 2000 sheep and some horses, again without losing a man.

On June 4th, the Governor of Enniskillen, hearing that the Irish Army besieging Derry had sent a great many of their horses to graze near Omagh, despatched 2 troops of Dragoons under Captains Gore and Crosby to the parish of Kilskerry, where they stayed two days, and picked up a troop of Horse and 2 Companies of Foot that were quartered there. In the evening, at about sunset, they proceeded together towards Omagh, and before eight o'clock the next morning they returned with 80 good horses, and nearly as many more smaller and inferior horses fit for labour, and 300 cows. By this enterprise they dismounted about 3 troops of the enemy's horse, and would have surprised their foot at Omagh, if notice of their coming had not been sent to the enemy, which gave the latter time to secure their position, but not to save their cattle.

1. Andrew Hamilton.

On the 10th of June, the Governor of Enniskillen, hearing of the dreadful state of the Protestants in Londonderry, who, it was generally thought, would be obliged to surrender in a few days if not relieved, marched with 2000 of his men towards that city. At Omagh he possessed himself of the whole of the town except the fort, which he invested. In a few hours, however, urgent messages were received from Enniskillen that Colonel Sarsfield with 5 or 6000 men had advanced and laid siege to Ballyshannon, and that Colonel Sutherland had appeared with another army before Belturbet. Each of these places being within 20 miles of Enniskillen, the danger of an attack from these armies appeared imminent. A consultation was at once held, when it was agreed that it was their imperative duty to return to the protection of their own town, and not to proceed, according to their previous purpose, to the relief of another. The next day, therefore, the whole party returned to Enniskillen.

On the 15th of June, the Enniskilleners received intelligence that Colonel Sutherland's force at Belturbet was daily increasing, as the Irish were flocking to him from all parts of the country. Colonel Lloyd was ordered to march against Sutherland with the greatest strength of Horse and Foot he could collect, and on the 17th he arrived at Maguire's bridge, half way between Enniskillen and Belturbet, with his little army, which Irish rumours had swelled to the imaginary number of 15,000. From McGuire's bridge a spy fled on Lloyd's approach, and informed Sutherland that all the forces of Enniskillen were in full march to attack him. Sutherland had with him at Belturbet only 2 regiments of Foot, a regiment of Dragoons, and a few troops of Horse. He gave credit to the exaggerated numbers of his opponents, and no longer considered it safe to remain in Belturbet. There was no place of any strength there but the church and the graveyard about it, the latter being but weakly fortified, and not nearly large enough to contain the men he had with him. He therefore retreated towards Monaghan, leaving a detachment of 80 Dragoons and 200 foot under the command of Lieut. Col. Edward Scott to defend Belturbet. The next day happening to be remarkably wet, Lloyd's army could not march from their quarters, and so the retreat of Sutherland was effected without a pursuit; but on the succeeding day the Enniskilleners appeared before Belturbet. Colonel Lloyd, advancing at the head of his men against the town, ordered Captain

Robert Vaughan and Capt. Hugh Galbraith, with their two troops of Dragoons, on the forlorn hope. When within two miles of the town, they were fired on by a troop of dragoons, upon which they alighted from their horses and lined the ditches upon both sides of the road, " which unusual manœuvre, " together with the appearance of the main body of their army coming up at the moment, caused the enemy to retreat to Belturbet. Here, with the rest of their party, they took post in and about the church, and in the Archbishop of Dublin's house adjoining it ; but after two hours skirmishing, they held out a white flag and surrendered upon the condition that their lives should be spared, but that the common soldiers should be stripped of their red coats. The officers were not included in this stipulation, and had all their money, under ten pounds each, returned to them. The prisoners taken numbered 300, including Colonel Scott and 13 other officers. Two hundred of the meanest prisoners were next morning discharged, the victors being unwilling to have the trouble of maintaining them, and the rest, with their officers, were brought to Enniskillen, together with about 700 muskets, some gunpowder, 80 dragoon horses, with all the accoutrements belonging to them, 20 horse loads of biscuit, 50 barrels of flour, 120 barrels of wheat, and as many red coats as served two companies of men, who were in much need of such clothing.

Early in July, General Kirk, who had arrived from England with a fleet for the relief of Londonderry, sent a ship round from Lough Swilly to Ballyshannon, for the purpose of ascertaining the wants of the Enniskillen garrison and offering a supply of ammunition and other necessaries. Whereupon on the 4th July an expedition was sent to inform him of their wants, and to bring back arms and ammunition. With the expedition was sent a statement of their strength, as follows :— " 17 troops, 30 Foot Companies and some few troops of Dragoons ; our Foot are indifferently well armed, but our Horse and Dragoons not so well. " [1]

Kirk had but few arms fit for horsemen, but, on July the 12th he gave the Enniskilleners 20 barrels of gunpowder, 600 firelocks, and a thousand muskets, together with bullets and match, 8 small pieces of cannon and a few hand grenades.

" Major General Kirke, gave us commissions for a Regiment of Horse consisting of 16 troops, and 50 Private men in each troop

1. Andrew Hamilton.

besides officers, for a Regiment of Dragoons consisting of 12 troops and the like number of private men in each troop, and for 3 Regiments of Foot each consisting of 18 Companies of 60 private men in each. He told us he could spare none of his private men, but gave us some very good officers, viz.—Col. Will Wolseley to be our Commander-in-chief and Colonel of Horse, Colonel Will Berry to be Lieutenant-Colonel to our Horse: Captain James WYNNE, a gentleman of Ireland, but then a Captain in Col. Stuart's Regiment, to be Colonel of our Dragoons: and for our 3 Regiments of Foot, Governor Gustavus Hamilton, Lieut. Col. Lloyd and Major Tiffan were made Colonels."[1] These Regiments were brought on to the establishment of the Regular Army on 1st January 1689 by the following Royal Warrant[2] :—

W. R.

"Whereas we have thought fitt to forme a Regiment of Horse together with Two Regiments of Dragoons and Three Regiments of Foot out of our Inniskilling Forces, and to take them into our pay and entertainment. We do hereby make and pass this our establishment for the said Forces to commence the First day of January 1689, in the first year of our Reign."

Then follow the establishment charges, those for a Regiment of Dragoons being:

FIELD AND STAFF OFFICERS

	£.	s.	d.	
Colonel as Colonel		15.	0	per diem.
Lieut.-Colonel as Lt.-Col.		9.	0	"
Major, who has no troop for himself and horses	1.	0.	0	"
Chaplain		6.	8	"
Adjutant		5.	0	"
Chirugion		6.	0	"
Gun Smith 4s—and 1s for his servent.		5.	0	"

ONE TROOP

Captain 8s and 3 horses 3s	11.	0	"
Lieutenant 4s 2 " 2s	6.	0	"
Cornet 3s and 2 " 2s	5.	0	"
Quartermaster for self & horse.	4.	0	"

1. Andrew Hamilton.
2. Harl. MSS. no. 4847.

3 Sergeants, each 18d & 3s horses . .		7.	6	„
3 Corporals, each 12d & 3s „ . .		6.	0	„
2 Drummers, each 12d & 2s „ . .		4.	0	„
60 Dragoons, 18d a day for man and horse.	4.	10.	0	„

A regiment of Dragoons to consist of 8 troops.

The above mentioned Royal Warrant called into existence

Wolseley's Inniskilling Horse—disbanded in 1697.

Wynne's Inniskilling Dragoons—afterwards the 5th Royal Irish Dragoons, disbanded in 1799, and re-embodied as the 5th Royal Irish Lancers in 1858.

Cunningham's Inniskilling Dragoons—now the 6th Inniskilling Dragoons.

Hamilton's Inniskilling Foot
Lloyd's „ „
Tiffan's „ „ now the 27th Inniskilling Fusiliers.

The following officers were appointed to the Regiment on formation, commissions bearing date 20th June 1689.

CAPTAINS

Jas. Wynne — Colonel
.... — L[t] Colonel
.... — Major
Hugh Galbraith
Percy Gethin
Chas. Newcomen
Hugh Caldwell
Chas. Ross

LIEUTENANTS

Robt. Drury

CORNETS

Mat. Watts
Mat. Wells

Although it has no immediate connection with the Dragoons of Enniskillen, it may not be out of place here to turn for a moment to the famous siege of Derry. The following market prices, taken from Walker's Diary, testify to the extent of the sufferings from famine of the garrison, and the degree of heroism which animated them in their refusals to surrender.—Horse flesh, each pound, one shilling & eight pence. A quarter of a dog, fattened by eating dead bodies, five shillings and six pence. A dog's head, two shillings and six pence. A Cat, four shillings and six pence. A rat, fattened by eating human flesh, one shilling. A mouse, six pence.

Walker, Bishop of Derry, Governor of that city and garrison, Colonel of volunteers and the last warrior Bishop of English history, says that by the 27th July, the defenders of Londonderry had no prospect of subsistence otherwise than by eating the bodies of the dead ; and he mentions a fat gentleman of his acquaintance who hid himself for several days, because he imagined that some of the soldiers who were perishing by hunger, looked at him with a greedy eye.

The regiments of horse in the place had been reduced to foot; having in their extremity been forced to kill and eat their horses.

At this time the Irish General McCarthy was besieging the Castle of Crom on Lough Erne ; and another Irish force under Colonel Sarsfield was at the same time close to Ballyshannon, the design being to effect a junction before Enniskillen and to crush that stronghold. In order to prevent this junction, and to force a separate action with one or other of these leaders, Colonel Wolseley at Enniskillen detached Lt.-Col. Berry on the 30th July with some four troops of horse, two of Dragoons, and three companies of foot, to turn McCarthy.

On the 31st of July Berry came up with the Enemy about four miles beyond Lisnaskea. Finding they far outnumbered his own detachment, he despatched an orderly to Wolseley at Enniskillen asking for assistance ; he meantime retreated past Lisnaskea so as to take up a position a mile from the town, where he would have a bog to cover his front. The causeway over the bog was scarcely wide enough to allow of two horsemen riding abreast. He placed his dragoons and infantry in a thicket of underwood at the end of the causeway, drawing a body of horse a little further off as a reserve, with which he proposed to support the other, and he gave the word " Oxford. "

The pursuing Irish soon came into view. McCarthy had ordered Colonel Anthony Hamilton with a force to attack Berry at once.

On arriving at the bog, Hamilton dismounted his dragoons and gallantly led them along the causeway on foot. Both parties opened fire. The Enniskilleners were concealed in the thicket, while Hamilton's men were in full view on the narrow causeway. Hamilton was very soon wounded, and his successor in the command was immediately after shot dead. Deprived of their leaders, and with their comrades dropping around them, the Irish began to retreat, while the Enniskilleners "raising a shout, and crying out that the rogues were running," took the bog on each side of the narrow road over which their reserve of Horse galloped, and quickly turned the retreat into a flight, and chased the fugitives through the streets of Lisnaskea, killing some 200 and taking 30 prisoners, and capturing a large number of arms. It was still but nine o'clock, and Berry returned to his ground and rested his men until a message arrived at about noon from Wolseley, ordering him to effect a junction with him at Lisnaskea. On the junction of the two bodies, Wolseley called the officers together, and also consulted the men, and explained the risk they ran of being caught between McCarthy and Sarsfield, and of their shortness of provisions, and that their only alternative to fighting McCarthy at once was to return home. They were unanimous in giving the preference to "fighting their way to the enemy's provision carts, rather than return to Enniskillen for their dinner."[1]

McCarthy's force consisted of some 6000 men, horse, foot and dragoons, while with Wolseley were 16 troops of horse, 3 of dragoons, and 21 companies of foot, in all barely 2000 men.

Accordingly, having given the word "No Popery," this gallant little band marched towards the enemy.

McCarthy had raised the siege of Crom and was advancing towards the Enniskilleners, but on coming in sight of them between Donagh and Newtown-Butler, he retreated on Newtown-Butler, and took up a position about half a mile from that place, on a steep hill with a bog in front of it.

The forlorn hope of the Enniskilleners led their advance by half a mile, and on reaching the enemy's position, Colonel Tiffan with his battalion of foot entered the bog on the right of the causeway, while Colonel Lloyd's battalion moved on the left, and Colonel WYNNE's Dragoons, divided into two parts, supported Tiffan and Lloyd on foot. Berry, with the horse, advanced along the causeway, and Wolseley came on in the rear with the main body.

McCarthy allowed the van of the Enniskilleners to cross the bog before opening fire, but before they could climb the hill he retired through Newtown-Butler (which was set fire to) to a similar position about a mile beyond the town. McCarthy posted

1. *Siege of Derry.*

his guns so as to cover the causeway across the bog, and checked the advance of the Enniskillen horse along the causeway. Wolseley recalled the horse, and ordered his infantry across the bog in extended order, Tiffan on the right, and Lloyd on the left, with WYNNE'S Dragoons (dismounted) again supporting either wing. The Enniskilleners crossed the bog with some loss, captured the cannon and "killed the cannoneers," who had courageously maintained their fire to the last moment. The horse then came along the causeway in support.

The brunt of the Enniskillen attack was borne by the Irish right. McCarthy, seeing this, sent orders to a regiment on the left to move to the right. There was some mistake in the delivery of the order, and the regiment, instead of facing to the right, went to the right about. The troops in rear of this regiment imagined they were about to retreat. The Irish cavalry promptly turned and galloped off, leaving the infantry to their fate. The regiment that had faced about, seeing the cavalry go off, ran too, whereupon the right threw down their arms and fled also. By this time the Enniskillen horse had reached the spot, and there ensued a most bloody rout. The majority of the fugitives made towards Crom, and crossing a bog took refuge in a wood near Lough Erne, where they were followed by the Enniskillen foot, who gave no quarter to any but the officers. Rendered desperate, 500 of the fugitives took to the waters of the Lough, and only one escaped drowning.

Of McCarthy's 6000 men, 400 were captured, 2000 killed and 500 drowned. McCarthy, who had striven in vain to rally his troops, was captured, and all the Irish guns, ammunition and colours were taken. The loss amongst the Enniskilleners did not amount to a hundred.

"There was a very remarkable stroke given by Captain William Smith in the battle: with one blow of his sword, he cut off the upper part of a man's skull, just under the hat. As much of the skull as was within the hat, with all the brains it contained, was struck away from the under part of it, and not so much as a fibre of skin remained to keep them together." [1]

"After this victory, the name of the Enniskillen men became a terror to the Irish." [2]

1. *Siege of Derry.*
2. Andrew Hamilton.

"The Enniskilleners having made their name a terror to the Irish, had only to rest and refresh themselves, and regiment themselves according to the commissions which General Kirk had sent them."[1]

Londonderry was relieved by Kirk on the 4th of August, and on the 7th, that General sent orders to Wolseley at Enniskillen to send him 500 horse and 200 dragoons to escort his infantry to meet the Duke Schomberg, who was on his way from England to Belfast Lough with an army which King William had raised in England.

According to these orders our Horse and Dragoons came to Derry, and "marched before the Major-General till his party joined Duke Schomberg at Carrickfergus," in other words, the Horse and Dragoons acted as Kirk's advanced guard.

Schomberg landed near Bangor on the 13th of August with 10,000 men, and on the following day started for Carrickfergus, arriving at that place on the 15th August; and, besieging it for eleven days, captured it on the 26th.

The arrival of the Inniskilling horse, dragoons, and foot, and 3000 more men from England, made up Duke Schomberg's army to 15,000 men.

There are two incidents recorded in the Siege of Carrickfergus which are of interest.

The first is that its defenders discovered the Duke Schomberg's tent, and waiting until he was at dinner, they fired two or three 6 *lb.* shot at the tent. A writer points out that this was a gross breach of manners and of military etiquette. The other incident is curious. A breach was made in the wall of Carrickfergus, and was about to be rushed, when the inhabitants of the town drove a large herd of bullocks towards the breach. The guns were turned on the bullocks, killing them, and earth being thrown over the beasts they formed a rampart which was defended for some considerable time.

By the terms of the surrender of Carrickfergus, the garrison was permitted to march out to Newry, to the great chagrin of the Enniskilleners.

On the surrender of Carrickfergus, the English Army marched to Belfast, and on the 2nd of September continued the march to Newry *via* Dromore and Lough Wickland.

1. *History of Ireland.*

The Enniskillen Horse and 5th and 6th Dragoons formed the advanced guard.

These regiments, whose exploits had often been talked over by the English soldiers, were the subject of great curiosity when they first joined Schomberg's force. Every one expected to see a perfectly equipped and admirably drilled body of men. Instead of this, there rode into camp three regiments of irregulars without uniforms, mounted on all sorts of horses. Some men had holsters, while others carried their pistols stuck into their belts; and the majority of the privates had their servants riding behind them on small country ponies called "garrons."

Another eye-witness writes:—"The Inniskilling Dragoons came there to us. They are but middle sized men, but they are, nevertheless, brave fellows. I have seen them like masty (mastiff) dogs run against bullets."

Story, an Army Chaplain, in his *Impartial History of the War*, gives us the following anecdote in connection with that advanced guard of Enniskilleners. He met the Dragoons, who had received orders to find the enemy's outposts and to establish contact with them and to report to Schomberg. To use his own words, "They showed me the enemy's scouts upon a hill before us; I wisht them to go and beat them off, and they answered 'with all their hearts; but they had orders to go no further than where they saw the enemy's scouts;' though they seemed to be dissatisfied with it, and added, they should never thrive so long as they were under orders."

Schomberg had formed a high opinion of the Enniskillens, and reposed more confidence in them than in the freshly raised English regulars. He had sent to England for uniforms for them.

On the approach of the English Army, the Duke of Berwick abandoned Newry after setting fire to the town, and retreated to Drogheda, where he joined King James. Schomberg sent "a trumpet" to Berwick, to say if any more towns were thus wantonly burnt, his army would give no quarter for the future; a timely hint of which Berwick availed himself. The word trumpet, as used here, meant a message sent by a cavalry trumpeter; a message sent by a drummer was termed a "drum."

After two days rest at Newry, Schomberg's army proceeded to Dundalk, where they arrived on the 7th September, and pitched their camp about a mile North West of the town, on extremely damp, low-lying and marshy ground.

The Enniskillen horse, foot and dragoons, together with the Guards and some artillery, were posted on the Northern side of the town to protect it from attack.

Meanwhile, King James had collected an army at Drogheda, and on the 21st September he appeared before the camp at Dundalk, but Schomberg refused to entertain any idea of fighting. Fever had broken out in his camp, and he was doubtful of the loyalty of the French troops with him. He found out afterwards that there was a conspiracy amongst them.

Diversions were made by King James all along Schomberg's right towards Sligo, and it was during one of these, on the 27th September, that a force of Enniskillen Horse and Dragoons about 1000 strong under Colonel Lloyd, succeeded in utterly routing an Irish force of some 5000 men near Boyle and Sligo.

Duke Schomberg was so delighted to hear of the gallantry and success of the men from Enniskillen, that, "having ordered all the Enniskillen horse, foot and dragoons in his camp to be drawn out, he rode along the line with his hat off, and caused the Dutch guards and the Enniskillen foot to make three running fires, which were answered by the Enniskillen horse and by the cannon upon the works, as also from the ships that lay at the mouth of the river, as an honourable mark of his approbation."[1]

The camp at Dundalk was now getting into a most insanitary condition ; a raging sickness of fever, occasioned by the unwholesome situation of the camp, a wet season, bad provisions and want of medicines, clothes and comfortable bedding, swept away and disabled a great number of Schomberg's officers and soldiers. King William wrote several letters to Schomberg, pressing him to move on Dublin, but Schomberg alleged many good reasons for not doing so.

In addition to the sickness, his English troops were in a bad state of disorganisation and training. The confidential Report on the state of the army throws some light on the deplorable condition of some of the regiments. The report on one reads : "the Colonel ill, and as incapable as all the other officers, who are usually absent ; as bad a regiment as possible, except...... which is worse." Another, "the Major very assiduous, but the Colonel

[1] *Impartial History of the War.*

neglects the regiment." One regiment is accused of much debauchery and drinking; while another "has hardly any good officers, an entire absence of good order, and badly clothed." Another regiment has "a very assiduous Colonel, but he is too easy to the Officers, who are the most negligent that can be imagined. Often he is the only officer present with the regiment, which he never quits; yet the regiment is in a bad condition; arms almost useless." One regiment has "a Colonel who has a good opinion of himself, but is not really efficient."

The question of the payment of the army was another of Schomberg's difficulties. Not only did the military officers rob their men, but the Commander-in-Chief could get but little money. For this, the Treasurer of the Army, a most shameless individual, was responsible. The only corps in the force which was regularly paid, was an independent troop of cavalry, which this Official, by name William Harbord, had, by some jobbery, contrived to "raise" and command. The troop in reality only consisted of himself, two clerks, whom he put down as officers, and a standard which he kept in his bedroom; and yet he managed to draw pay regularly for it as a complete troop of cavalry.

It is not to be wondered at that Schomberg considered his army unfit to advance, and that it shewed itself equally unfit for the monotony of a standing camp.

In the beginning of November, the Irish army went into winter quarters, and Schomberg lost no time in following their example, and on 5th November the camp was broken up.

On the 4th December, Belturbet was taken by Wolseley with a force of Enniskilleners, and the dragoons of Enniskillen went into Winter quarters there on the 12th of that month.

The loss of life in the camp at Dundalk stood thus :—

Total of Army in Camp		14.000
Losses — Died in Camp	1.700	
Died in course of removal to Belfast .	800	
Died in Hospital at Belfast . . .	3.800	6.300
	Survivors —	7.700

Story writes, "Death has become so familiar in the camp at Dundalk that men were only sorry when their comrades were carried away for burial, because their dead bodies had been used to stop the chinks in the huts."[1] The same author writes that during the march of a foot regiment to Winter quarters, some men were lodged one night in a stable, and the morning two of them were dead. The chaplain visited the stable, and found the survivors had arranged the bodies of their two dead comrades to serve as seats round a fire.

On the 12th February, 1690, the Duke of Berwick reinforced the Irish garrison at Cavan with some three to four thousand men, with the object of next day dislodging Wolseley's force from Belturbet. Wolseley, however, had received notice of this design, and marched from Belturbet in the same evening that Berwick entered Cavan, with 3 troops of horse, 2 of dragoons, and 700 foot, in all about 1000 men.

Wolseley's only chance of success lay in surprising his enemy; he accordingly made a détour so as to cross the river Annalee a couple of miles above Ballyhaise. This entailed a march of 14 miles over bad roads and a deep ford, "over which the horse were obliged to carry the foot," and it was daylight before he came within a mile of Cavan. The approach of the Enniskilleners was perceived, and the garrison beat to arms. The advanced guard of cavalry was checked in a narrow lane, and retreated on the infantry, some of whom, Story relates, were so furious at the idea of running from Irishmen, that they fired at the cavalry and killed several of them.

Wolseley now brought forward his Foot, and the enemy retired on the main body, which held a fortified position on a hill near the town. The Enniskilleners deployed to the attack. After a while the enemy, giving a yell of victory, fired a volley and charged, but the balls flew harmlessly overhead, and before they could re-load, the Enniskilleners poured in such a brisk fire that the Irish fled. The victors, instead of pursuing the enemy into the fort, began to loot the town, which gave the Irish an opportunity to sally out of the fort and attack them.

Wolseley's reserve of some 300 men kept the enemy at bay, while he and his officers endeavoured to rally their men in the town. They found this more difficult than they imagined, and they had to set fire to the town to compel the men to leave the burning houses.

The Irish infantry were again driven into the fort, while the horse fled and disappeared altogether. The Magazines were blown

1. *Impartial History of the War.*

up, and stores destroyed, and Wolseley returned to Belturbet; his men being too fatigued to attempt the capture of the fort, having been marching and fighting since 4 p.m. the previous day.

Wolseley captured 200 prisoners, and killed 10 officers and 300 men, the Enniskilleners only losing about 30 men.

The following order, issued by the Duke of Schomberg to the English army in Ireland, is interesting:

A PROCLAMATION

BY FREDERICK, DUKE OF SCHOMBERG

Lord General of Their Majesties forces &c.

"Whereas the horrid and detestable crimes of profane cursing, swearing, and taking God's Holy Name in vain, being sins of much guilt and little temptation, have, by all nations and people, and that in all ages, been punished with sharp and severe penalties, as great and grievous sins: And we, to our great grief and trouble, taking notice of the too frequent practice of these sins by several under our command; and that some have arrived to that height of impiety that they are heard more frequently to invoke God to Damn them than to Save them; and this notwithstanding the heavy and dreadful judgments of God upon us at this very time for these and our other sins, and notwithstanding the penalties enjoined by Their Majesties' Articles of War on these offenders; and we, justly fearing that their Majesties' Army may be more prejudiced by these sins than advantaged by the conduct and courage of those guilty of them, do think fit strictly to charge and command all officers and soldiers under our command that they and every one of them from henceforward do forbear all vain cursing, swearing, and taking God's Holy Name in vain, under the penalties enjoined by the aforesaid Articles, and our further displeasure; and that all officers take particular care to put the said Articles of War in execution on all under their respective commands guilty of the said offences, as they will answer the contrary at their utmost peril.

Given at our Head Quarters... the 18th January 1690.

SCHOMBERG.

From March to June 1690, the military operations consisted of the nature of raids and counter-raids, attended with considerable loss of life on both sides, and captures of cattle by Schomberg's forces.

In April, Schomberg besieged the fortress of Charlemont, but the Enniskilleners were not employed in the investment. The place fell to Schomberg, but some days before its capture, a "trumpet" was sent to summon it to surrender. "Tell your gineral," shouted O'Regan the governor, "that he is an owld knave, an' by Sin Patrick, he shall not have the town at all, at all."

This Teague O'Regan, an old gentleman, must have been somewhat of a character. Story writes that on the surrender of the town, O'Regan, riding out to meet the Duke, wore an old red coat, a worn out, long, ill dressed wig, on which was perched a narrow brimmed white beaver hat, much too small for him and cocked on one side; a yellow cravat round his scraggy throat, and a muff hanging round his neck, and he had evidently "drink taken." He was mounted on a very old and starved stallion, which was very lame, and had an unpleasant trick of kicking and squealing when anyone approached him. The two Generals, the Duke Schomberg and Teague O'Regan, met and began to exchange the usual compliments, which, however, were cut short by Teague's vicious stallion, which drowned every syllable with his squeals, clearing a ring round him by lashing out on all sides. As soon as the ridiculous scene was over, and he could control his laughter and speak, the Duke observed that if Teague's horse was very mad, Teague himself was very drunk. O'Regan was afterwards knighted by King James.

All this time, regiments, recruits, and reinforcements, and also an artillery train had been arriving at Belfast from England and the Continent for the English army.

King William himself landed at Carrickfergus on the 14th June, having left London on 4th June, and before a week was over, the Army was assembled between Armagh and Newry. The King occupied himself in making minute inspections of the different regiments, which he formed into brigades, and in gaining a knowledge of his generals and staff. By the 27th, his army was concentrated on Dundalk. The following list from Story gives the strength of the Army, and shows the English regiments that took part in the battle of The Boyne river. In giving this list, Story makes the following note. "Because several people may be curious to know what Number of Men we had at the Boyne, and also how many the Enemy were, I have inserted the Exact Number of our Horse and Foot, as it was taken at Finglas."

HORSE.

	Men.
Life Guards 1st & 3rd Troops & Horse Grenadiers	368
Oxford's Blues (Royal Horse Guards)	368
Lanier's (1st Dragoon Guards)	360
Villiers' (2nd " ")	245
Coy's (5th " ")	236
Byerley's (6th " ")	244
Schomberg's (7th " ")	242
Russell's	242
Langston's (Princess Anne's,	225
Wolseley's (Enniskillen)	423
Harbord's Troop	38
Total Horse.	2,991

DRAGOONS.

Hayford's Mathews' (1st Royal Dragoons)	406
Leveson's (3rd Dragoons, now 3rd Hussars)	246
WYNNE'S Enniskillen (5th Dragoons now 5th Royal Irish Lancers)	260
Cunningham's Enniskillen (6 Inniskilling Dragoons)	358
Total Dragoons.	1,270

FOOT REGIMENTS.

Douglas'	648
Kirke's (2nd Queen's)	666
Trelawney's or Queen's Regt. (4th King's Own)	553
Lloyd's (5th Northumberland Fus.)	652
Babington's (6th R. Warwickshire)	416
Beaumont's (8th King's Liverpool)	526
Stuart's (9th Norfolk)	660
Hammer's (11th Devonshire)	593
Brewer's Wharton's (12th Suffolk)	571
Hastings' (13th Somersetsh. L. I.)	606
Meath's (18th Royal Irish)	678
Gus. Hamilton's (20th Lancs. Fus.)	560
Bellasis' (22nd Cheshire)	628
Herbert's (23rd R. Welsh Fus.)	600
Deering's (24th S. Wales Borderers)	600
Tiffan's, Enniskillen (27th R. Inniskilling Fus.)	625
Fowke's	439
Lisburne's	611
Earle's	693
Mitchelburne's	664
S. John's	589
Drogheda's	660
Geo. Hamilton's	583
White's	600
Hamilton's	600
Total.	15,021

French Forces, Horse		395
Foot		2,231
	Total.	2,626

Dutch Forces, Horse		1,683
Dragoons		621
Foot		3,704
	Total.	6,008

Danish &c. Forces, Horse		812
Foot		4,581
	Total.	5,393

SUMMARY.

English Forces including 1700 for Officers & Sergeants and 300 killed at the Boyne	21,282
Foreign Forces including 1200 for Officers & Sergeants and 700 killed at the Boyne	15,927
Grand Total of K. William's Army	37,209

The list from Story was taken on 5th July after the battle of The Boyne, and, consequently, addition has been made for killed and wounded.

The strength of the Irish Army north of the Boyne at this time is doubtful; Walton puts it down as 45,260, (Foot 32,950, Horse 12,310), but other writers bring it nearer 30,000.

James retreated before the advance of William, who arrived at the Boyne on the 30th of June, and found James in position on its southern bank.

During the march of William's Army to the Boyne, the Enniskillen Horse and Dragoons (5th and 6th) were always in the advanced guard. The two regiments of Enniskillen Dragoons were at this time clothed in grey.

On the march, an Irish gossoon was hanged for kidnapping English soldiers, for whom he received half a crown a head. His practice was to promise to shew the soldiers where cattle and loot were to be found, and so induce them to follow him. He wore an English dragoon's hat and waistcoat, the owner of which he had stabbed in the back, while his worthy father held the man in conversation.

Just before being hanged, the boy, bent on a little business combined with amusement, offered to hang one of his own countrymen, a prisoner, for six pence.

At daybreak on the last day of June, William started his army in three columns, and himself rode forward to reconnoitre the Boyne. In their position on the southern bank of the river, the Irish army possessed a good front and a safe line of retreat.

On James' right lay Drogheda, held by his soldiers, and on his left was a very difficult bog. In his front was the river Boyne, with its rugged steep banks and difficult fords. The Irish position was on a series of terraces, on the summit of which was the village of Donore commanding the view to the North.

Three miles south of Donore is the village and pass of Duleek, through which lay the high road from Drogheda to Dublin. Some three miles higher up the river, and on the left of the Irish camp, is the bridge of Slane, and an extensive bog lay between the Camp and the bridge, and one narrow road crossed the bog. In the centre of the position, and close to the river, was the hamlet of Oldbridge, to which was a good ford.

During King William's reconnaissance, the enemy opened fire on him with two six-pounders from near Oldbridge. The incident is thus related in "The Field of Mars":—

"King William.... as he reconnoitred their situation, was exposed to the fire of some fieldpieces, which the enemy had purposely planted against his person. They killed a man and two horses close by him, and the second bullet rebounding from the earth, grazed upon his right shoulder, so as to carry off part of his clothes and skin, and produce a considerable contusion. This accident, which he bore without the least emotion, created some confusion amongst his attendants, which the enemy perceiving, concluded he was killed, and shouted aloud in token of their joy. Their whole camp resounded with acclamations. .....William rode along the line, to shew himself to his Army after his narrow escape."

William, meanwhile, ordered his escort to dismount and rest beside their horses in full view of the enemy. Some of Coy's and Byerley's Horse and Life Guards were there, and for some hours did the King make them remain on the spot exposed to the enemy's artillery fire. Eventually he ordered them to a more sheltered position, saying, "Now I see that my men will stand."

At 9 p.m., William held a Council of War, and declared to his generals his intention of forcing the river. Schomberg opposed the idea as being too hazardous, but was over-ruled. Schomberg then suggested that a strong force should be at once sent to occupy Slane Bridge, so that in the morning it might move along the Dublin road and cut off the enemy's line of retreat at the pass of Duleek. This was opposed, and William decided against it, whereupon the Duke retired in high dudgeon to his tent, and when, later, the order of battle was brought to him, he growled that it was the first he had ever had to receive since he commanded armies.

The road from Slane Bridge to the pass of Duleek was undoubtedly the key of the position, and, curiously enough, James was also advised by one of his Generals to send immediately a large force to secure the bridge, which advice was also over-ruled, and a weak force of a regiment of dragoons only was sent on this duty.

In the order of battle for the English Army, King William was to lead the left attack towards Drogheda in person, the Duke Schomberg the main or centre attack on Oldbridge, while his son, Count Schomberg, was to lead the right attack against Slane Bridge.

The two armies were dressed so much alike that William ordered his men to wear a green sprig in their hats, while James' troops wore a white cockade in theirs.

Before retiring to bed at midnight, King William rode through the camp by torchlight, and gave the word "Westminster" for the night.

The 1st of July, a lovely summer morning, "as if the Sun itself had a mind to see what would happen," was to witness the first pitched battle fought by the Standing Army of England. In the battle of the Boyne we shall chiefly confine ourselves to the doings of the dragoons of Enniskillen.

William's left attack was composed of all the Enniskillen Dragoons, (WYNNE's and Cunningham's), and Villiers', Mathews' and Leveson's regiments of horse.

It was to move towards Drogheda, and crossing at the fords, was to hold the Irish right, and, as the centre or main attack developed, to harass the enemy.

Early in the morning the right attack under Count Schomberg forced the river at Slane Bridge and by the ford of Rosnaree just below it, and at ten o'clock, William, learning of this success, ordered the advance of the old Duke Schomberg's main attack on Oldbridge, and the King himself rode off to lead the left wing.

A short account of Schomberg's main attack will not be out of place, as during it that fine old soldier met his death.

The Band of the Blue Dutch Guards struck up a march as the Dutch and French regiments marched into the river, supported by the Enniskillen foot and Hanmer's and Nassau's brigades. The Dutchmen climbed the opposite bank and engaged the Irish Foot Guards at Oldbridge. Schomberg's regiments pressed on out of the river and the Irish regiments retreated before them. The Irish Horse then made some gallant charges, cutting their way through various regiments, finally being repulsed by Hanmer's men. These several charges had told on the whole front; the Frenchmen were nearly broken ; the Dutch had received a check ; the Danish horse was routed ; the English brigade held its own, but made no progress. Considerable confusion arose, and Duke Schomberg, noticing the crisis and that Caillemotte had fallen, without waiting to put on his armour, rode into the river and placed himself at the head of the leaderless French Protestants, and urged them on. It was soon after this that Schomberg was killed. There were numerous reports as to how he was shot. In Graham's *History of Ireland*, we find that the French refugees, having allowed some of the Irish Horse through their lines, thinking them friends, but, perceiving their mistake, commenced firing rashly, by which means they shot the Duke in the neck. A report current in the army to this day was, that the Duke was shot by a trooper in King James' Guards who was a deserter from Schomberg's regiment a year before.

Walton, in his *History of the British Standing Army*, gives the following tradition he heard from an inhabitant of Oldbridge:—

"On the appearance of the English Army on the north bank of the Boyne, the inhabitants of Oldbridge abandoned their homes, many of the men joining James' army,—all except one poor deformed lame man or "bokkha." This bokkha refused to leave, and damning the heretics who were coming to lay waste his home, he swore he would have a shot with his old duck gun at the bloody-minded Prince of Orange. On the morning of the battle he concealed himself in a double ditch by the Oldbridge ford. Lying low in his hiding place, he made out the principal personages in the Enemy's army, and bided his time until he could get a shot at some important person. He watched the charges of the Irish horse, the wavering of the French regiments, and the fall of Caillemotte. He then perceived with joy a horseman ride down to the river, who was wearing a richly laced coat, an embroidered sword belt and plumes in his hat. This, he thought, must be William of Orange, and he was about to ride close by his hiding place. He examined the priming of his duck gun, and seeing the officer point the soldiers towards his empty home, decided, even if he were not the Prince of Orange, he was quite worth killing. Chance favoured the bokkha, for the officer turned in the saddle to speak to the soldiers, and his horse pulling at the reins stooped for a long drink. Taking a deliberate aim, he fired, and Schomberg fell, never to rise again." History does not relate what happened to the bokkha.

Duke Schomberg, the gallant old soldier, and one of the greatest captains of his time, was buried in St. Patrick's Cathedral, Dublin, and two hundred and fourteen years later, the successors of WYNNE's Enniskillen Dragoons, who fought under Schomberg, and under him shared the horrors of the pestilential camp at Dundalk, erected a monument close by his grave to their comrades in the 5th Royal Irish Lancers who fell in the South African War of 1899-1902.

To return to the Battle of the Boyne, at about 11 o'clock King William, with his left wing of cavalry, was putting into effect his movement against the right flank of the enemy. Placing himself at the head of the dragoons from Enniskillen, William told them that they should be his guards that day. "Gentlemen," he said, "I have heard much of your exploits, and now I shall myself witness them;" and he led his division down to the river side. Here a number of them, including the King, got bogged. William was obliged to dismount while Private David M'Kinley, of the Enniskillen dragoons, extricated his horse. A regiment of Irish dragoons on the other side of the river, howled in derision at the endeavours of the horsemen to reach the stream. "Pass if you can!" they shouted. Local tradition gives this as the reason for the passage being still called the "Pas-if-you-can."

At length they got into the water, and the Irish regiment retired upon its supports, when the Enniskilleners, deploying into line, charged with such fury, that they not only broke the enemy, but became so scattered in pursuit that they in turn were driven back on to the Danish Horse, by a fresh body of Irish Cavalry. William was now at the head of the Danes, who, overwhelmed by fleeing friends and pursuing foes, gave way, carrying the King with them, on to a British regiment whose immovable rock of seventeen-foot pikes turned the tide of battle : William was able to rally the flying troopers, and the Irish squadrons were cut to pieces.

The Irish foot now got the order to charge and drive the main body of William's infantry back into the river. They advanced, but only the Irish foot guards actually charged, for a murmur was passed along the ranks that the army was taken in flank by cavalry. The Irish foot-soldiers saw the grey uniforms of the relentless Enniskilleners on their right, and turning about, pressed back up the hill as fast as they could go.

In vain their officers endeavoured by example shewn in sacrificing their own lives, to rally them. "The Horse, the Enniskilleners!" was all that the panic-stricken infantry would reply.

The gallantry of the Irish Horse at this point saved a rout. With such effect did they charge that they gained time for their foot to re-form at Donore. But it was only for a short time ; a rapid and disorderly retreat on Duleek was commenced. Two regiments of Irish dragoons were ordered to the front to check the pursuit, but both turned and followed the foot. The Irish Horse regiments, however, behaved with the greatest gallantry. Time after time they charged down the grassy slopes endeavouring to check the pursuit.

At Platin House, halfway between Donore and Duleek, General Hamilton had drawn up a body of Irish cavalry in a field, into which a gap from a by-road was the only entrance. Eight troops of Enniskilleners under Wolseley came riding along this by-road in advance of the army, and two troops promptly entered the field. Wolseley ordered the men, by mistake, to form to the right, thus bringing the men with their backs to the enemy. Orders to wheel to the right were given, but the result was confusion. The Irish charged, and cutting down some fifty troopers, drove the others pell mell on to the crowded troops in the lane, who were quite incapable of resistance and were chased out of the lane. King William, coming up with some Dutch cavalry, rode up to the Enniskilleners and asked them "What they would do for him." The soldiers turned their faces to the enemy again, and, re-forming with the Dutch, charged their pursuers, who were now joined by more of the Irish cavalry.

The hardest fighting of the day took place now. Ten times did the gallant Irish horse charge to gain time for the retreat of their panic-stricken infantry, and ten times did the equally gallant Enniskilleners and Dutchmen beat them back. Numbers of the Irish Officers were killed, and the Irish Life guards and other troops were literally cut to pieces. Eventually they had to retire, but no troops could have done more than they did.

While this action was taking place, General de Ginckell, with Cunningham's Enniskillen Dragoons, Leveson's Dragoons and some Dutch Cavalry, had encountered the enemy still more to the left towards the Drogheda road. Here again the Irish Horse displayed their accustomed valour, and de Ginckell and his Dutch regiment were driven pell mell down a lane. The Enniskillens and Leveson's dragoons promptly dismounted and lined the hedge, besides manning a house that overlooked the lane, and by their unexpected and heavy fire drove back the Irishmen.

The whole of the Irish Cavalry at length retreated, after having thus fought for half an hour against superior numbers. This half hour saved King James' army from annihilation.

Count Schomberg's attack on the right joined the rest of the army in Duleek. The cavalry followed up the enemy for some three miles, but the retreat was so excellently covered by the French troops on the Irish side, that the enemy suffered little loss.

Almost all the baggage and stores were seized by the conquerors. The Enniskilleners had taken not less than three hundred cars, and found among the booty ten thousand pounds in money, and much plate, many valuable trinkets, and all the rich camp equipage of Tyrconnel and Lanzon.

King James II fled through Dublin to Waterford, where he took ship to France, his army retreating southwards through Dublin. Of the Irish army, the Irish Horse and the French regiments of Lanzon were the only troops who marched through Dublin with any sort of order, with their kettledrums and trumpets sounding a march.

On the 5th of July King William's army encamped at Finglass, a suburb of Dublin, and on the 7th and 8th he held a very particular muster, or review of his troops. On the 9th the army marched southwards, the main army with the King to Limerick, while a detached force under Lieutenant-General Douglas was sent to invest Athlone. Douglas' detached force consisted of Langston's, Russell's and Wolseley's (Enniskillen) Horse, WYNNE's and Cunningham's (Enniskillen) Dragoons and ten regiments of foot, in all some 7,500 men.

While the King was at Dublin, he published a proclamation offering protection to those of the Irish peasantry who should remain neutral. But the men of Douglas' division plundered indiscriminately on the march, and this violation of the King's protection papers had a most baneful effect on the people. It is told that on one occasion four soldiers were caught looting, and Douglas ordered their execution. He was asked to reduce the sentence, which he did by making the malefactors throw dice, and the holder of the lowest throw alone suffered the death penalty.

On the 17th of July Douglas arrived before Athlone and summoned the place to surrender. " These are my terms, " replied Governor Grace, firing a pistol at the messenger : " these only will I give or receive ; and after my provisions shall be consumed, I will defend Athlone until I eat my boots. "

Douglas proceeded to invest the place, but was not long in discovering that its strength had been underestimated. He had only a dozen guns, the heaviest of which were 12 pounders, and he was without sufficient powder for these. He also heard that Sarsfield, with 15,000 men, was on the march to cut off his communication with Dublin and the main army.

Douglas was considered one of Willliam's best officers, but that he was very illiterate, his letter to the Earl of Portland will show:—[1]

" My Lord,

I have done my best endeavours at Athlon. The necessaries which belong to our train wer so smal, that even pouder was scarce. At the beginning wee had bot eighteen barrill of pouder for tenn piece of cannon and two houitzers.

They mak head at Rachray, which is fourteen myles distant from my camp. All their passes on the Shannon and many more, the enemie had possessed before I came here, or at least the nixt day after, so that it is generally believed that they will endeavour to keep the Province of Connaught for their winter quarter, and

1 *History of Ireland.*

hereby prolong the war, or obteen terms for themselves—this reasons is lykways my not having pouder to mak a breach on their retrenchments makes me judge it absolutely necessary at present to reteir to Molengair....

James Douglas."

The besieging army had by this time grown dispirited from a continuance of fruitless efforts to take the town, and sickly from scanty provisions. At midnight of the 24th and the 25th of July Douglas raised the siege and marched towards Limerick to rejoin the King. Captain Rapin, Douglas' quartermaster-general, says that the valour displayed on both sides was admirable.

Douglas' retreat was not accomplished without difficulty. Walton writes, "Ignorant of the strength of the enemy in the vicinity, obliged to take precautions against possible pursuit by the garrison of Athlone or a surprise by Sarsfield's reported force, Douglas had to accept for his route those roads where he would be least likely to fall in with the enemy, and where he would have the advantage of passes capable of defence by so small a force. This restriction to by-roads, on which there were few farms and no towns, caused such a failure in the supply of provisions that for four days together, the army was without bread."

Douglas rejoined King William on the 27th of July, at Cahirconlish, a place about six miles from Limerick.

On the 8th of August William's advanced guard, 200 Horse and Dragoons in the Van, marched towards Limerick, and very early next morning came into contact with the Irish outposts. The army pressing forward gradually drove in the outposts, and the King called on Limerick to surrender. Major General Boisleau, who commanded the garrison, replied that he was surprised at the summons, and that he thought the way to gain the Prince of Orange's good opinion was by a vigorous defence of the City with which His Majesty King James II had entrusted him.

On the same day General de Ginckell, with a force of dragoons and foot, reconnoitred the ford of Annaghbeg, two miles above the town, and discovered it strongly held by the Irish. However, it was abandoned during the night, and in the morning de Ginckell secured the passage.

King William had brought with him nothing but field guns, but there was on the road from Dublin a train of ammunition wagons and siege guns, and a set of tin pontoon boats. The escort for this valuable convoy, moving through an enemy's country, consisted of only two troops of Villiers' Horse under Captain Poulteney. Naturally the Irish had information of the approach of the convoy.

On the morning of the 11th, Mr. Manus O'Brien, an Irish Country gentleman, arrived in camp with the news that Sarsfield from Limerick had passed the Shannon in the night by the bridge at Killaloe, with a body of five or six hundred horse and dragoons, "designing something extraordinary." The English Officers laughed at the news, and suggested Sarsfield was hunting mares' nests in the mountains; while one great officer commenced asking about a possible cattle raid in the country, to which O'Brien replied he was sorry to see King William's honour regarded less by General Officers than cattle were. He was, however, taken to the King, "who grasped the situation in a moment, understanding at once that the Irish must have had news of the convoy, and that Sarsfield had been sent to waylay it; he instantly ordered Sir John Lanier to take a body of five hundred horse to meet the convoy and escort it to camp." Sir John Lanier failed in his orders, for he did not start until about two o'clock in the morning, and then in a most leisurely fashion.

Sarsfield, meanwhile, left Limerick on the night of the 10th with eight hundred cavalry, and crossing the river at Killaloe, marched quietly towards Cashel until he received certain news that the train was at that town. "All the 11th he lurked in the mountains along the course to be taken by his prey. In the afternoon the train halted at Ballynedy Castle, seven miles from Limerick: the men of the escort turned their horses loose to graze, detailed the usual corporal's guard, ate their suppers, smoked their pipes and went to sleep; they had never marched alone in an enemy's country before, they had no practical military education, and therefore they merely complied with routine orders and trusted to luck. Not a sentry, not a vedette was posted beyond the precincts of the camp, not a word of notice of approach had been sent on to headquarters."

Sarsfield, arriving quietly at the spot, effected a complete surprise, the escort and wagoners being either cut to pieces or running away. The Irish drew the carriages and guns together, and filling the latter with gunpowder, fixed their mouths in the ground and set fire to the whole heap, which blew up with a " most astonishing explosion. "

Sir John Lanier viewed the flash of the explosion from a distance of three miles. He wheeled half his force to the left, and the other half under Villiers towards O'Brien's bridge, in order to intercept the Irish retreat over the Shannon, " but Sarsfield marched another way and got into Limerick. "

This well executed success greatly inspirited the Irish, and the want of guns greatly delayed the English works. Other guns were ordered up from Waterford, while a force of three hundred foot, mounted on ponies, kept the road.

On the night of the 17th of August the trenches were opened, and the Irish beaten out of an advanced redoubt, and the next night the besieged made a sally. The two English regiments, having relieved the right of the trenches, were ordered to lie down, and the majority of the officers and men falling asleep, the enemy were able to surprise them. The English soldiers, suddenly aroused, began firing at everything and everybody. Some " Danish troops, who manned the trenches to the left, finding themselves fired upon, took the English for the enemy and returned the fire, while the Irish fired on both. This confusion lasted for over an hour before the English and Danes discovered the true state of affairs, when they united in the charge and repulsed the sallying party. "

During the following days the siege gradually progressed, the semi-circle of guns and trenches surely closing in until they were within a range of eighty yards. For three days and nights before the 27th, a storm of shot, shell and red hot balls rained upon the town, and on this date a breach appeared in the walls near St. John's Gate. During this period of the siege, a heavy and continuous rain poured down. So heavy was it that " to mount fresh batteries was an impossibility : the trenches were knee deep in mud : the soldiers were never dry: sickness increased to a plague : the camp became a swamp, and those who could afford it burnt bowls of spirits in their tents to keep down the damp. " These facts and the near exhaustion of ammunition, rendered it imperative to carry the place at once or raise the siege.

An assault under General Douglas was determined upon, and at 3.30 p.m. on the 27th of August the signal was given by three guns. Five hundred grenadiers, drawn from every regiment, rushed to the counterscarp and " showered grenades and bullets on to the

Irish, who replied with cannon and musquet. Captain Carlisle, of Drogheda's regiment, who had been twice wounded before reaching the top of the glacis, was first into the ditch, and was instantly shot, and his subaltern took his place ; the covered way was quite in the hands of the besiegers, and the supporting regiments followed up the success. The Grenadiers then impetuously rushed on to the breach, while, unfortunately, the supporting regiments followed those of the Irish, who were retiring along the covered way to St. John's Gate, which they reached only to have the gate shut in their faces. By this time the grenadiers were well within the town, but the Irish, seeing how few they were, and unsupported, rallied and returned to the fight. The Brandenbergers were the first to support the grenadiers, and had actually taken the Black Battery, when, with a horrible explosion, the enemy's powder there blew up, killing and wounding many. The grenadiers had now to fall back and regain the covered way, and for three hours did a sharp fight continue, in which the Irish women boldly joined ; and when they failed to obtain more deadly missiles, threw stones and broken bottles. At length when the ammunition was spent, while the Irish fire increased, the troops were recalled from the covered way. In this assault of a little over three hours' fighting, one thousand five hundred men had fallen on the side of the besiegers alone, of whom five hundred were left dead on the spot.

On the 29th the rain came down again in torrents, with every appearance of continuing. A Council of War was held, and on the 31st of August the King raised the siege, and with a rear guard of five thousand men, marched to Cahircoulish, first blowing up all the shell etc. that could not be taken away.

The following description of the siege by Corporal Trim is related by Sterne, and is evidently taken from some old soldier who was present : —

" We were scarcely able to crawl out of our tents at the time the siege of Limerick was raised, and had it not been for the quantity of brandy we set fire to every night, and the claret and cinnamon and Geneva with which we plied ourselves off, we had both left our lives in the trenches.... The City of Limerick, the siege of which was begun under His Majesty King William himself, lies in the middle of a devilish wet swampy country ; it is surrounded with the Shannon, and is by its situation one of the strongest fortified places in Ireland ; it is all cut through with drains and bogs ; and besides, there was such a quantity of rain fell during the siege, the whole country was like a puddle ; 'twas that and nothing else brought on the flux. Now there was no such thing after the first ten days, as for a soldier to lie dry in his tent without cutting a ditch round it, to draw off the water : nor was that enough for those who could afford it, without setting fire every night to a pewter dish full of brandy, which took off the damp of the air, and made the inside of the tent as warm as a stove. "

A curse to William's army during this campaign of 1690, were the Rapparees, who were bodies of armed peasantry independent of the armies, and who hung around the flanks of the English forces, stripping the dead, murdering and looting the wounded, and pillaging when opportunity offered. " They were extremely cunning, and when they feared detection, the Rapparees would sink down into the long grass or other cover, and how they would dismount the locks of their pieces and stow them away in their clothes or some dry spot, how they would then stop the muzzles with corks and the touch holes with small quills and chuck away the piece confidently into a pond or other equally secure hiding place ; and then you may see an hundred of them without arms who look like the poorest humblest slaves in the world, and you may search without finding a gun ; and to do mischief, they can all be ready in an hour's warning. " [1]

The English Army now retired into Winter quarters.

King William having returned to England, and Marlborough having arrived in Ireland and captured Cork and Kinsale and returned to England, the Spring of 1691 found General de Ginckell in command of, and busily reorganising the English Army in Ireland, in preparation for the third and final year of the Irish War. It is with de Ginckell that we continue to follow the doings of WYNNE's Dragoons. Both armies were eagerly making preparations for the approaching campaign.

Tyrconnell, in January, offered every trooper or dragoon who would desert from the service of the " Prince of Orange and repair to Limerick, or any other Irish town, two pistoles of gold or silver, and every foot soldier a pistole in like coin. "

An attempt was made by the Irish to erect defences on the roads leading to Athlone ; but de Ginckell, hearing of this, ordered

1. Story.

out a body of troops to force them, which was successfully done. It is to be noted that with this force was a corps of wounded infantry called Monk's dragoons, composed of a few picked men taken from each company of Kirk's regiment of Foot, who were mounted, and ordered to do dragoon duty under Lieutenant Monk of the same regiment.

In May the campaign opened, and on the 6th of June de Ginckell marched from Mullingar to Rathcondra, sending forward a strong body of cavalry to Ballymore, to prevent any attempted relief of that fort.

Just before arriving at Rathcondra, Douglas with Wolseley's Horse, and WYNNE's and Cunningham's Dragoons, and eight regiments of foot, joined the main army.

The next day the united forces marched to Ballymore and invested the place.

When de Ginckell appeared at Ballymore, the wretched starving aged men and women and children, who had been expelled over the frontier by the Irish, came flocking round the troops and were to be seen eagerly picking up the offal and refuse of the camp, and devouring dead horses.

Story relates a little incident which occurred about this time. He narrates that his brother was killed in an attack upon a certain fort, and the Irish irregulars captured the body. The soldiers of the Company sent and asked for their Officer's body, that they might bury it. The request was refused, but the soldiers were told that "their Officer was a brave man and they, (the Irish Irregulars), would bury him themselves." They added that "his own company drummers were to beat the dead march before him, but that they would fire three volleys over his grave themselves to show their appreciation of his bravery."

De Ginckell's army arrived before Athlone on 19th of June, having besieged and captured the fort at Ballymore on the way.

Athlone stands on either side of the Shannon, and the two portions of the town are called English town and Irish town, the two being connected by a stone bridge. The Irish had considerably strengthened the defences of Irish town, but curiously enough, no great endeavour had been made in the defences of English town.

The enemy's outposts were met and driven in, and batteries were erected against English town on the rising ground to the North west, and, by noon next day, a wide breach appeared in the North west bastion. A Council of War resolved that English town should be stormed, and a force of foot, with a body of cavalry in readiness to support, advanced to the assault at 5 p.m. the same day. The advance party was led by a French subaltern, and after a short and severe fight, English town was in the hands of the attackers, and batteries were erected against Irish town.

"A desperate struggle now took place day by day for the possession of the bridge connecting the two towns, the Irish contesting it to the utmost, and the English gaining ground only inch by inch."

The English general having information of a ford higher up the river by which he might cross and take the enemy in rear, despatched a small party under a lieutenant of horse to reconnoitre, with orders to return directly he had tested the ford by crossing it. The young officer crossed the ford and went on after a herd of cattle he espied in the distance. The enemy consequently discovered the party's proceedings, and within a few hours the newly found ford was strongly defended by Irish earthworks. The lieutenant was tried and cashiered.

In the daily desperate fight for the Bridge of Athlone, the English were continuously gaining ground and repairing the captured arches as they were taken, and by the night of the 26th, with the exception of the broken arch on the Irish side, the bridge was in their possession. On the 27th, the breastwork erected by the Irish on their end of the bridge was burned by the English grenades, and on that they succeeded in laying planks over the gap of the ruined arch. Next morning the enemy observing this, a Sergeant and ten Scottish soldiers of Maxwell's regiment volunteered to wreck the English work over the gap. Putting on armour, they boldly set about their daring task in the face of an appalling fire, but every man of them was killed. Another party of ten equally gallant fellows took their place, and under the close and heavy fire of the English actually succeeded in throwing all the planks into the river, but only two of them returned.

On the 30th, at a Council of War, it was decided to carry the place by assault.

To ascertain the depth of a ford, three Danish soldiers under sentence of death were promised their pardon if they would try whether it was fordable or not. They readily consented, and putting on armour, entered the river at distances from each other, the English firing at them all the time as if they were deserters. The Irish, thinking such was the case, left them alone until the Danes facing about, the design became apparent, when a hot fire was opened on them from the other side of the river.

It is satisfactory to know that all three got back safely with a well earned pardon, two being slightly wounded. They had discovered that the ford did not reach their breasts. The river had not been known so low in the memory of living man.

A little before 6 o'clock on the evening of the 30th, " instead of the usual relief of the guards, " there marched down to the English trenches some two thousand men, every man carrying fifteen rounds of ball and having a green bough, the " sign of battle, " to place in his hat. [1] De Ginckell went to the trenches to encourage the soldiers, and " distributed a bag of guineas as some sort of acknowledgement among those who were thus foremost in danger. "

At length the signal rang out, — the toll of the bell from the church steeple — and instantly 60 men in body armour left the trenches and entered the river, and close upon them followed the remainder of the storming party.

So unexpected was the attack, that the Irish were driven through the town, and their own gates shut in their faces, and by half past six o'clock the whole place was in the hands of the English.

" St. Ruth, in command of the Irish forces, was astounded no less at the audacity, than at the success of de Ginckell's stroke; and early next morning he left, leaving the castle, which had not yet surrendered, to fall into the hands of the English. "

During the attack, the English found the streets encumbered by rubbish and masonry knocked down by their own cannon, and the bumps they received as they fell over the ruins " occasioned the soldiers to utter volleys of oaths, " which drew a memorable rebuke on them from Major-General Mackay, " a veteran officer, noted for religion and virtue, as well as for valour and conduct. " " Soldiers, " said he, " ye have more reason to fall upon your knees and thank God for your victory, than to blaspheme His Name ; you are brave men, and would be the best of men if you would swear less. "

1. It is interesting to note that the ancient practice of donning a badge as a "sign of battle" had not quite expired in King William's reign ; for throughout the wars in Ireland, and in the subsequent campaign in Flanders, it was the custom of the English soldiers on going into action to fasten a green bough in their hats, an emblem which originated at the battle of the Boyne.

St. Ruth crossed the Shannon from Athlone and went to Aughrim, while de Ginckell, having put the place in a state of defence, left there on the 10th of July for Limerick, which city he intended to capture. He arrived at Ballinasloe on the 11th, and found St. Ruth in position at Aughrim, distant 3½ miles.

De Ginckell reconnoitred the Irish position and found St. Ruth strongly posted on the top of Aughrim or Kilcommodon Hill, occupying a front of two miles. His left rested on the castle of Aughrim which was strongly held, while his right rested on Urachree. An extensive bog lay across the foot of the slope of the hill. The only passes suitable for cavalry were the Ballinasloe Road, which passed over the front of the Urachree ridge and dipped between the extensive bog and another bog between Aughrim and Urachree, and passed close under the castle ; and a road which diverged from the Ballinasloe Road along the Urachree ridge, and skirting the bog between the two ridges, ran along the rear of the Aughrim hill. There were several foot tracks by which infantry could avoid the worst places of the bog.

On the evening of the 11th de Ginckell issued orders for an advance against the enemy in the morning ; the baggage was to be left behind under a guard of two regiments. The force was to be early under arms without beat of drum, "the arms to be clean and bayonets fixed," and a good quantity of ammunition was to be taken ; the grenadiers were to be on the right and left of each regiment, with two grenades per man, and five pioneers were to march at the head of each regiment, to be ready to act when called for. The word that night was "Dublin."

At 6 a. m. on the 12th of July the troops marched out of Ballinasloe, and as they crossed the river Suck, they were formed in a double line of battle.

The English Cavalry formed the right wing, and the left wing cavalry consisted of French and Danish horse.

These, with twenty-eight foot regiments in the centre, gave the English strength as some 19,000 men.

The Irish army of some 20,000 men was drawn up in two lines, with the infantry in the centre and cavalry on both flanks. A reserve of cavalry was posted to the left rear, under the command of Sarsfield.

St. Ruth, seeing that the English were determined to give battle, ordered masses to be said in every part of his army, while

AUGHRIM

The Line of Battle 12th July 1691.

1st Line.

Major-Generals	Brigadiers	Regiments
Scravemore	Villiers	Leveson (3rd D. G.) WYNNE (5th Lan. 2 squad.) Oxford (Blues) Langston Ruvigny Villiers (2nd D. G.)
Mackay	Bellasyse	Kirk (2nd Foot) Gus. Hamilton (20th Foot) Herbert (23rd Foot) Lord G. Hamilton Ffoulks Bellasyse (22nd Foot) Breiver (12th Foot)
Tettan	La Melonière	La Melonière Du Cambon Belcastell Greben Danish Foot Do Do
La Forrest	Eppinger	La Forrest's Dragoons Schested Donep Boncour Montpouillon Eppinger

2nd Line.

Major-Generals	Brigadiers	Regiments
Ruvigny	Leveson	Cunningham (6th Drag.) WYNNE (5th Lan. 1 squad.) Lanier (2 D. G.) Wolseley Byerley (6th D. G.)
Talmack	Stuart	Stuart (9th Foot) Earle (19th Foot) Tiffan (27th Foot) Crighton St. Johns Lisburn Meath (18th Foot)
Nassau	Prince of Hesse	Nassau Regiment Lloyd (6th Foot) Cutts Danish Foot Do Do
Holstaple	Schack	Schack's Dragoons Nieuhense Do Zuhistern Do Reidesell Do Ginckell Do Eppinger Do

the priests obliged the soldiers to promise they would give no quarter.

A heavy fog hung over the swamp until about noon, when it lifted and discovered the two armies in battle array.

In the early afternoon de Ginckell detached a small force of Danish horse to force the pass of Urachree, who, however, were beaten back, and a party of two hundred Enniskillen Dragoons, (chiefly Cunningham's), were ordered down to prevent the Irish from crossing while the infantry of the centre advanced.

It was evident to de Ginckell that to render the centre attack of any avail, either the Ballinasloe or Urachree road must be forced, and the Enniskilleners were therefore ordered to force the ford and fall upon the right flank of the enemy. They forced the ford, and driving the enemy from the cover of a house, were in turn overwhelmed by his supports, and driven back on to Eppinger's Dragoons, who had been sent to reinforce them.

The Irish were reinforced, and Enniskilleners and Eppingers were now in so tight a corner that the Blues had to be sent to their assistance. Portland with the latter regiment charged the Irish, and the combined efforts of the three regiments drove the enemy back over the stream, which they did not cross again.

The English generals now met in a Council of War, and considered whether the attack should be delayed until next day, and orders were sent for the tents at Ballinasloe. Eventually, however, it was decided not to delay, but to bring the Irish to a decisive action.

The plan of action was that Aughrim, being the key of the position, was to be captured at all costs. The Irish right at Urachree was to be threatened, while an infantry holding attack was to be directed on the centre, and the main and decisive attack was to be delivered against the Irish left upon Aughrim.

At 4.30 p.m. the English line advanced, and at 5 p.m. the firing re-commenced at Urachree, and the cannon of both sides was playing on the main body of either army. The fighting was desperate, both sides displaying great courage, and it was not until 6.30 p. m. that the Irish right had sufficiently given way to admit of the centre or holding attack being developed.

The centre attack consisted of Wharton's, Herbert's, Creighton's, and Earle's foot regiments, supported by Stuart's and Ffoulks' foot. The men got so elated at their success at first that they could not be restrained, and the attack developed into a severe frontal one and was carried into the Irish position, when they found themselves confronted by a mass of Irish Cavalry, who came pouring down on them through the gaps in the fences, as had previously been arranged by St. Ruth. The English regiments, in spite of the greatest gallantry, were overwhelmed, and beaten back into the bog, leaving so many of their men mingled with the Irish dead, that the spot on the hillside was known afterwards as "The Bloody Hollow."

Meanwhile, the attack had been extended along the right by the Enniskillens, St. John's, Hamilton's, Meath's and the French foot. The Irish remained concealed and held their fire until the English were within twenty yards, when from the hedges came a terrible fire. In spite of frightful losses, the English and French steadily drove the Irish, who fought gallantly, from hedge to hedge, until at length the regiments became so intermingled and confused, that when the Irish foot cleared away for their cavalry to swoop down, the victorious but disordered regiments had to retire.

De Ginckell next sent Lanier's and also de Ruvigny's horse from his right to assist his hard-pressed infantry on the left, in their feint against the Irish right at Urachree. St. Ruth, seeing this, and thinking the English main attack was about to develop against Urachree, reinforced his right from his left at Aughrim.

De Ginckell, discovering this, ordered the centre attack forward again, and commenced his main or right attack on Aughrim. This consisted of WYNNE's Enniskillen dragoons, the Blues, Cunningham's Enniskillen dragoons, Leveson's dragoons, Villiers' and Byerley's horse, supported by Kirk's and Gus Hamilton's foot regiments.

The English cavalry rode steadily forward under an incessant fire from the hedges on the Irish position, and following a by-road which branched off from the Ballinasloe road to Aughrim, reached a very narrow passage across the bog. This pass, with slippery and boggy sides, would only permit of two horses crossing at a time, and with showers of bullets in front and a heavy press of men and horses behind, it required steady and brave troops to attempt it.

When St. Ruth saw the cavalry making for the crossing, he could not believe they were going to attempt it, and when he saw them actually reach the place, he cried, " By Heaven, they were gallant fellows, and it was a pity they should thus court death." The horse and dragoons successfully leapt or scrambled their horses over the stream, and led by the Blues, they charged along the firm ground that bordered the bog, under a close fire from Aughrim Castle ; and Kirk's and Hamilton's foot regiments managed to reach a large dry ditch close under, and more or less covered from, the fire of the castle.

Meanwhile, the desperate fight between the infantry continued on the bog, the Irish and English alternately gaining or giving ground.

St. Ruth, seeing the successful crossing of the English cavalry, rode off to his left and brought away some Irish horse to oppose the English at Aughrim, and on the hill side just above the Bloody Hollow, he was killed by a chain-shot or round-shot.

Walton, in his *History of the British Standing Army* says the " chain-shot asserted to be the fatal shot is preserved in St. Patrick's Cathedral. "

The English infantry on the right was meanwhile reinforced, and under General Talmach pressed forward, and an advance of the whole line of infantry taking place, a heavy enfilading fire was poured in on the Irish left.

St. Ruth had not communicated his plans to a successor, and the Irish Army without a leader, broke, and, pursued by the English, fled in utter confusion, until kindly night ended the horrid slaughter.

Walton relates that just before the Irish gave way, their ammunition ran short, and on a fresh supply being brought up, it was found that the bullets had been cast too small for the calibre of the fire arms; whereupon the Irish soldiers tore the buttons from their coats and tried to make them serve as bullets.

Historians do not agree in the losses incurred during this battle, but the Irish losses during the fight and pursuit are said to have amounted to 7000 killed, and the English to some 2000.

In the official lists furnished to the General two days after the battle, it is noticed that WYNNE's dragoons are shewn as not having lost a single officer, man, or horse, but as at this time "there were several pecuniary inducements to officers to conceal the vacancies in their regiments," it is quite possible that these lists were not to be trusted.

The cavalry regiments especially mentioned for gallantry at the battle of Aughrim were: the Blues, WYNNE's and Cunningham's Enniskillen Dragoons, and de Ruvigny's French Horse, which rather leads one to believe the foregoing statement as to the correctness of the official lists.

Mr. Story, the historian, writing of the bravery of English soldiers at this battle, says "that they marched boldly up to their old ground again from which they had lately been beat; which is only natural to Englishmen: for it is observable that they are commonly fiercer and bolder after being repulsed than before; and what blunts the courage of other nations commonly whets theirs, I mean the killing of their fellow soldiers before their faces."

Though using the word "Englishmen," he writes of the whole of the British troops.

The Irish too, horse and foot, fought gallantly at Aughrim. "Never," says Captain Parker, an eyewitness of the battle, "did the Irish fight so well in their own country as they did on this day."

Story relates a great many of the dead were left unburied on the ground, and large numbers of dogs fed on the carcases of horses and men, which made them so savage that people found it unsafe to pass that way alone. He tells a pretty story of a greyhound, the property of a fallen Irish Officer. This dog would not suffer his master's body to be touched, and when the other dogs left the place, this one remained guarding his master's bones, obtaining food at night from the houses about. For six months did he keep watch, until one day a soldier happened to pass close to the Irishman's remains, whereupon the dog flew at him, and the soldier, alarmed at the attack, shot the poor beast with his firelock.

The remains of the Irish Army, after their defeat at Aughrim, retreated as best they could to Limerick.

De Ginckell, four days after the battle, marched by way of Loughrea and Athenry on Galway, which surrendered with the honours of war, being permitted to march out of the town to Limerick "with their arms, six pieces of cannon, drums beating, colours flying, match lighted, and bullet in mouth." WYNNE'S Dragoons were employed in the reduction of Galway, and on the 20th July, as soon as it was dark, formed part of a force which were floated over the river on pontoons, about two miles north of the town, but met with no resistance except from a small force of dragoons "who fired on the first and then retreated." Upon the attack by de Ginckell, next morning, the city surrendered.

De Ginckell them moved slowly to Limerick with his cavalry, reconnoitring down the line of the Shannon.

The following account of the siege of Limerick is taken from *The Field of Mars* published in 1801 :—

"The General (de Ginckell) next marched to this town (Limerick), and while Captain Cole sailed up the Shannon with a squadron and some frigates, he drove the enemy from their advanced posts. The batteries were opened, and a line of contravallation was formed on the 25th August, 1691, while the Irish army lay encamped on the other side of the river Killala, and the fords were guarded with four regiments of their dragoons. After the town had been almost reduced to a heap of ruins, and large breaches made in the walls, the guns were dismounted, the outposts evacuated, and all such other motions were made as if the English were inclined to abandon the siege ; for on the 17th September, it was warmly debated in a Council of War, whether they should prosecute the siege, or turn it into a blockade, and in the meantime pass over the Shannon and destroy the enemy's forage in the county of Clare. It was so far carried for the latter, that an engineer was ordered to go with a guard to Kilmallock, and fortify that place, but was countermanded and a great number of pallisades brought to Mackay's fort...... as if the army intended to winter there. On the 19th a party was ordered to pass the river, either to prosecute the siege with vigour, or if not found feasible, to burn their forage and prevent all supplies.

"A battery was raised between Mackay's fort and the Old Church, to flank the Irish, in case of a sally from St. John's Gate, and four mortars were brought from the great battery to this fort, it being judged the best for bombarding, as the whole town lay in a line from it. When the ships that came up the river with Captain Cole had appeared, and fired into the Irish horse camp, the enemy could scarcely credit it, having been persuaded that either the English had no ships there, or else that the Dauphin, with a strong fleet from France, would soon destroy them; which was industriously propagated and hourly expected, so infatuated were the credulous Irish in the favorite opinion of succours from the French King.

On the 20th most of the heavy cannon were sent on shipboard, which caused it to be generally believed that Limerick would not be taken this year. But Ginckell, far from desponding, exerted all his military capacity on this arduous occasion; for in two days after, he pressed with most of the horse and dragoons, over a bridge of boats they had laid, into the county of Clare, on the other side of the river, leaving Mackay and Talmash to the command on this side, while the enemy fired on them all the morning from several batteries without any effect. In the afternoon a party of English troops was attacked by a superior force of the enemy, till, being sustained by four regiments of foot, they drove and obliged them to retire under their cannon. Then the English were commanded to advance and attack the works that covered St. Thomond's Gate, and other smaller fortifications, from which the Irish played their cannon, and made such discharges with their small shot, that the English were ordered not to approach so near: the grenadiers, notwithstanding, pressed so hotly on the re-inforced detachments, that they drove them to the draw-bridge, which the officers, fearing the English would enter instantly, drew up, and thus sacrificed the Irish who were on the outside, to the fury of the English, who killed or drowned most of them.

The English then lodged themselves within ten yards of the bridge; and the Irish, finding all communication cut off between them and their horse, and despairing of the French succours, began to open their eyes and think seriously of capitulation. In the meantime the siege was carried on with great fury, and the fire was excessive hot on both sides the next day, though it rained and blew incessantly; but towards night, on the 22nd, the rain began to cease, and both storms ended together, for about six o'clock the Irish beat a parley. The next day Sarsfield and Wachope desired a cessation of arms for three days, till they could send to Lieutenant-Colonel Shelden, who lay with 1,600 horse at Six mile Bridge, in order that he and his men should be included in the articles of capitulation, which was granted, and the English prisoners in the town were released. On the 26th

Sarsfield and Wachope dined with General de Ginckell and it was agreed to exchange hostages. The next day the Irish sent out their proposals, which were so high, that de Ginckell sent word, that though he was a stranger to the laws of England (being a Dutchman), yet from the face of the demands, he was sure they could not be listened to; and so ordered a new battery to be raised: but upon the request of the Irish, he sent twelve articles, which in the end proved to be the sum of the capitulation; which put an end to the French intrigues for holding out the siege, and thereby the war. This step was also mortifying to some English, who hoped this war, if continued, would end in the total destruction of the Irish interest, whereby they might enjoy the forfeited lands.

On the 1st October, the Lords Justices came into the camp, and after some conferences with the commissioners, on the part of the garrison, and their troops in the County of Clare, the articles for the surrender of the City of Limerick, with the castles of Ross, Clare, and other places, were finally concluded on the 3rd. By these the Irish, who chose to stay in that kingdom, and take the oaths of allegiance to William, were restored to all their estates which they enjoyed under King Charles. They were freed from the oath of supremacy, as they always look upon the Pope as the head of the church. Not only the French, but as many of the Irish as pleased, had liberty to go over to France, with safe transportation, and free passage. So that about 12,000 choice troops, of the Irish alone, were shipped off, being the original of those Irish forces which have so much strengthened the French Army...... thus ended the second siege of Limerick, having in the two sieges (which may be reckoned as one), held out fifty six days against many furious attacks.

On the 24th of August Graham relates that two men of Lanier's regiment of horse were condemned to death for having robbed Captain Watts, an officer of the same regiment, and a country sutler was likewise condemned to death for having bought the Captain's watch from the troopers.

After the siege of Limerick, WYNNE's Dragoons, and other regiments marched north to winter quarters.

The treaty of Limerick[1] was signed on the 3rd of October 1691, by Sir Charles Porter, Thomas Coningsby Esq., Lords Justices of Ireland, and the Baron de Ginckell, Commander-in-Chief of the English Army in Ireland, on the one part; and the Earl of Lucan, Viscount Galmoy, Colonel Purcel, Colonel Cusack, Sir Toby Butler, Colonel Dillon and Colonel John Brown, on the other part: and with it ended the War in Ireland.

1. By this the Irish were indemnified, with certain qualifications, for their losses, and had restored to them all they enjoyed in king Charles's reign, on condition of taking the oath of allegiance to William and Mary. A free passage to France, with the French troops, was also promised to such as desired to go.

CHAPTER II.

Campaigns in Flanders, 1694-1697.—*Death of James Wynne.—Ross' Dragoons.—Siege of Namur.—Ireland* 1698-1701. —*Some Courts-Martial.*

During 1692 and 1693 WYNNE's Dragoons remained in Ireland, but there is no trace of any particularly interesting events during those two years. On the 15th and 22nd of May, 1694, WYNNE's Dragoons sailed from the Thames for Flanders.[1] Early in June the Regiment joined the Allied armies which were camped near Meldert under the command of King William III.

The following is a list of the British Regiments serving in this campaign :—

HORSE

General. THE EARL OF PORTLAND.
Major-Generals. THE DUKE OF ORMOND.
THE EARL OF COLCHESTER.
THE EARL OF SCARBOROUGH.

Brigadier	Regiment	Squadrons.
L'Etang	Life Guards.	4
	Horse Grenadiers.	1
	Dutch Life Guards.	2
Leveson	Leveson's (2.D.G.)	2
	Wood's (3.D.G.)	2
	Wyndham's (6.D.G.)	2
	Galway's	2
Lumley	Lumley's (1.D.G.)	2
	Langston's (4.D.G.)	2
	Coy's (5.D.G.)	2
	Duke of Leinster's (7.D.G.)	2

[1] The following letters relate to the embarkation of the Regiment :—
To Commissioner of Transport.
"IT IS HIS MAJESTY'S WILL AND PLEASURE THAT YOU PROVIDE SHIPPING FOR COLONEL WYNNE'S REGIMENT FROM THE RIVER THAMES TO WILLENSTADT; WHICH TROOPS ARE TO EMBARK 22ND DAY OF THIS INSTANT MAY."
To Colonel James WYNNE :—
"HIS MAJESTY HAVING THOUGHT FIT THAT THE RECRUIT HORSES FOR FLANDERS SHOULD EMBARK ON FRIDAY 15TH, I DESIRE YOU SHOULD GIVE ME NOTICE AS SOON AS MAY BE WHAT NUMBER OF HORSES WILL BE READY BY THAT TIME."

DRAGOONS

Brigadier	Regiment	Squadrons.
Mathews	Mathews (1.D.)	4
	Scots (2.D.)	4
	Fairfax (3.D. & Hussars)	4
WYNNE	Essex (4.D. & Hussars)	4
	WYNNE'S (5.D. & Lancers)	3
	Cunningham's (7.D. & Hussars)	4
	Eppinger's	5

INFANTRY

Major-Generals. LORD CHURCHILL.
SIR H. BELLASYSE.
RAMSEY.

Brigadier	Regiment	Battalions.
C. Churchill	1st Foot Guards	2
	2 " " (Coldstreams)	1
	3 " " (Scots)	2
	Dutch Guards.	2
Earle	Hamilton's (1st Foot) 1st Battalion.	1
	Selwyn's (Queen's)	1
	Churchill's (Buffs)	1
	Trelawny's (R. Lancs)	1
	Fitzpatrick's (R. Fus.)	1
	Brewer's (12th Suffolk)	1
	Earle's (19th Yorks).	1
Stuart	Granville's (10th Lincoln)	1
	Tidcombe's (14th W. Yorks)	1
	Lesley's (15th E. Yorks)	1
	St. George's (17th Leicester)	1
	Castleton's	1
	Lauder's.	1
	Lloyd's (Northumberland Fus.)	1
	Stanley's (16th Bedford)	1
	Hamilton's (18th R. Irish)	1
	Ingoldsby's (23rd R. W. Fus.)	1
	Tiffan's (27th R. Inniskilling Fus.)	1
	Collingwood's.	1
O'Ffarrel	O'Ffarrel's	1
	Maitland's (25 K.O.S.B.)	1
	Fergusson's (26 Cameronians)	1
	Buchan's	1
	Mackay's	1
	Graham's.	1
On Command	1st Foot. 2nd Battalion	1
	Argyle's	1
	Strathnavor's	1
	G. Hamilton's.	1

The Artillery train consisted of 60 guns and 6 mortars and some 2400 men, which brought the British Contingent of the Allied Army up to some 30,000 men.

A squadron of Horse or Dragoons had a strength of 150 men, and a battalion of Foot 600.

Altogether the Allies had in the field a force of 32,000 cavalry and 57,000 infantry, besides strong garrisons at Ghent, Liége, etc.

An innovation of this year of the wars in Flanders was that the Artillery, some 130 pieces of cannon, was divided amongst the various brigades, instead of marching in one body as they had hitherto done.

During June the King selected Colonel WYNNE to be a Brigadier of Dragoons.

The French army, consisting of 22,500 horse, 6,700 dragoons, 55,300 foot, in all some 84,500 men, under the Dauphin and de Luxemburg, was at this time camped at Gemblours.

The French were sorely handicapped in their preparations for the coming campaign owing to the scarcity of forage and supplies, and to the fact that their treasury had been exhausted by a long series of wars. Also their altered frontier compelled them to find large garrisons for Mons, Charleroi, Namur and Huy. The Allies on the other hand had fewer garrisons to keep up, and had received large reinforcements from England ; hence the strength of the two opposing armies in the field was brought more on an equality than in the Campaigns of the preceding years.

The main hope of the French for the coming campaign was the capture of Liége, and accordingly on the 8th of June they marched to Joudrain with the object of getting between the Allies and Liége.

On the 10th of June the King reviewed the Dragoons of Lord Essex and Brigadier WYNNE ; "the latter wanted two troops that had been left in England, and his horses were very much fatigued in twice crossing the sea the last winter, and their continued motion through Ireland and England, to come to this country."

At the Liége entrenched camps were 24,000 allies, and William had every confidence in their strength. From Meldort he could scarcely prevent the interception of the French, but for better observation he moved to Tirlemont.

On the 10th de Luxemburg marched to St. Tron, which was almost directly between Tirlemont and Liége. At the same time de Boufflers advanced across the Meuse with a corps from Dinant, and encamped near Warem, within a short march of the main army. Another corps under the Marquis d'Harcourt in the Duchy of Luxembourg was ordered to approach the Meuse with instructions to be ready to co-operate.

Both armies remained quiet for the rest of June. Neither side would risk an engagement, the intervening country being so intersected by rivers and streams that the attacker would be placed at a great disadvantage.

During the winter the French had collected a large quantity of siege material at Huy, in readiness for use down the Meuse, but de Luxemburg was of opinion that the siege of Liége was too formidable a task to be undertaken in the face of the Allied Army. Without a second army to cover his operations he could have little chance of success. De Luxemburg apparently had no particular plan of campaign at this time, unless it were that he hoped to prevent the Allies from moving towards the coast.

William on his side was perfectly content to see the French helping him to consume the forage of the district, thereby raising a serious difficulty in their future movements.

In his list of the Allied troops in the Confederate Camp during July, D'Auvergne puts WYNNE's and Mathews' brigades of Dragoons in the Reserve of the line of battle. This reserve consisted of the Dutch and English Dragoons under Major-General Eppinger. The historian writes of them : "you must observe that at first coming to the camp of Mount St. Andrée, all the Dragoons encamped on the Left, where the Left is the post of honour ; 'tis for this reason that the English dragoons in this list fall after the Dutch." He goes on to say : "Of the Dragoons especially I may say that such a body, either for number or the good order they were in, has hardly ever been in the field."

The French on the 1st of July marched to the Jaar, halting between Warem and Tongres, and on the 13th the Allies moved towards Huy, halting at Ramillies. On this, de Luxemburg moved to Vignamont, so covering Huy without losing his hold over Liége. Both sides threw up entrenchments along their positions, and the Allies placed their guns on the Ramillies heights, which enabled them to command the Huy road for a long distance.

D'Auvergne writes that on the 28th of July "we surrounded a party of enemies in a wood. This was a voluntary party of dismounted troopers that had a great mind to ride some of our horses, and to mount themselves at our own cost; they got a partisan with them, who when they came near to our camp inquired whereabouts our horses grazed; a Boor (countrymen and farmers in Flanders were called Boors) told them they were every day by the wood, and that if they would lie there in ambuscade that night they would not fail of horses next morning. They took his advice and the Boor came immediately to advertise my Lord of Athlone of it, who thereupon commanded a detachment of dragoons and some foot to surround the wood and give no quarter." Twenty of the enemy were killed, but at last they gave quarter to seventeen.

Forage was scarce, and de Luxemburg had had to send part of his cavalry beyond the Meuse to find food. The Allies were between the French and their frontier lines on the Scheldt, and as soon as all the forage should be exhausted William could march direct to the Scheldt without further anxiety for Liége.

William not only had a start in the straight race for the Scheldt, but his position prohibited the French from following the same route as himself without affording him an opportunity of fighting; and as soon as Luxemburg should march westwards, the garrison of Liége could join William, and give him an overwhelming preponderance of force. To reach the Scheldt without coming into contact with the Allies, de Luxemburg would have to cross the Sambre, which would increase the start already gained by his opponents.

In anticipation of the race for the Scheldt, both armies had sent their heavy baggage to the rear. WYNNE's Dragoons accompanied that of the Allies, their horses being in such bad condition with the voyage from Ireland and subsequent bad feeding, that they were not fit to take the field. In relation to this D'Auvergne writes:—

"On the 5th August our heavy baggage was commanded away towards Louvain under the escort of Brigadier WYNNE's Regiment of Dragoons, which was sent to garrison in Ghent, because their horses were out of order for reasons we have before alleged."

On the 8th of August the Allies marched to Sombref, where

they halted for a day. The march had been a severe one, and the rest was absolutely necessary.

On the same day de Luxemburg crossed the Mehaigne, and advanced to near Namur. On the tenth he crossed the Sambre near its junction with the Orneau, having bridged it the previous day. After the crossing the army was broken up into nine or ten different corps, and to each was assigned a separate route; all, however, were to halt each day within a certain circumference, so as to facilitate a speedy junction if necessary. Three thousand dragoons had been sent to Charleroi, and these, on hearing of the departure of the Allies from Ramillies, hurried to reinforce the Marquis de la Valette at the Lines.

William marched to Nivelle on the 10th, to Soignies on the 11th, and to near Ath on the 12th.

At Soignies a man who had been detected in attempting to fire the ammunition wagons, suffered the penalty of incendiarism.

Having first been put to the torture, with a view to discover his employers, his right hand was cut off and burnt before his eyes, and he himself was then cast into the flames alive.

De Luxemburg marched on the 12th, his columns being greatly impeded by difficult and woody country, full of streams swollen by rain, and heavy roads. The next day his infantry was so knocked up that it could not go on; but the cavalry under the Dauphin pushed on with great energy, and reached Tournai on the evening of the thirteenth.

On this day William crossed the Dender at Ligne, halted at Frasnes, and sent on General Tettau with 5000 men, to prepare bridges at Hauterive.

De Luxemburg on the thirteenth was marching to Saint Ghislain beyond Mons. His men were so tired that they were ordered to throw away their knapsacks, and, to revive them, a ration of brandy and beer was every now and again served out. At Saint Ghislain de Luxemburg received a message from the Dauphin, to the effect that not only was the Allied army close to the Lines, but Tettau was actually on the Scheldt, and the Elector of Bavaria had crossed the river above Oudenarde. The French General appealed to his men, with the result that the Grenadiers and the major part of the battalion companies at once came forward and expressed their readiness to proceed that night, and the exhausted army struggled on, being plied with brandy at every

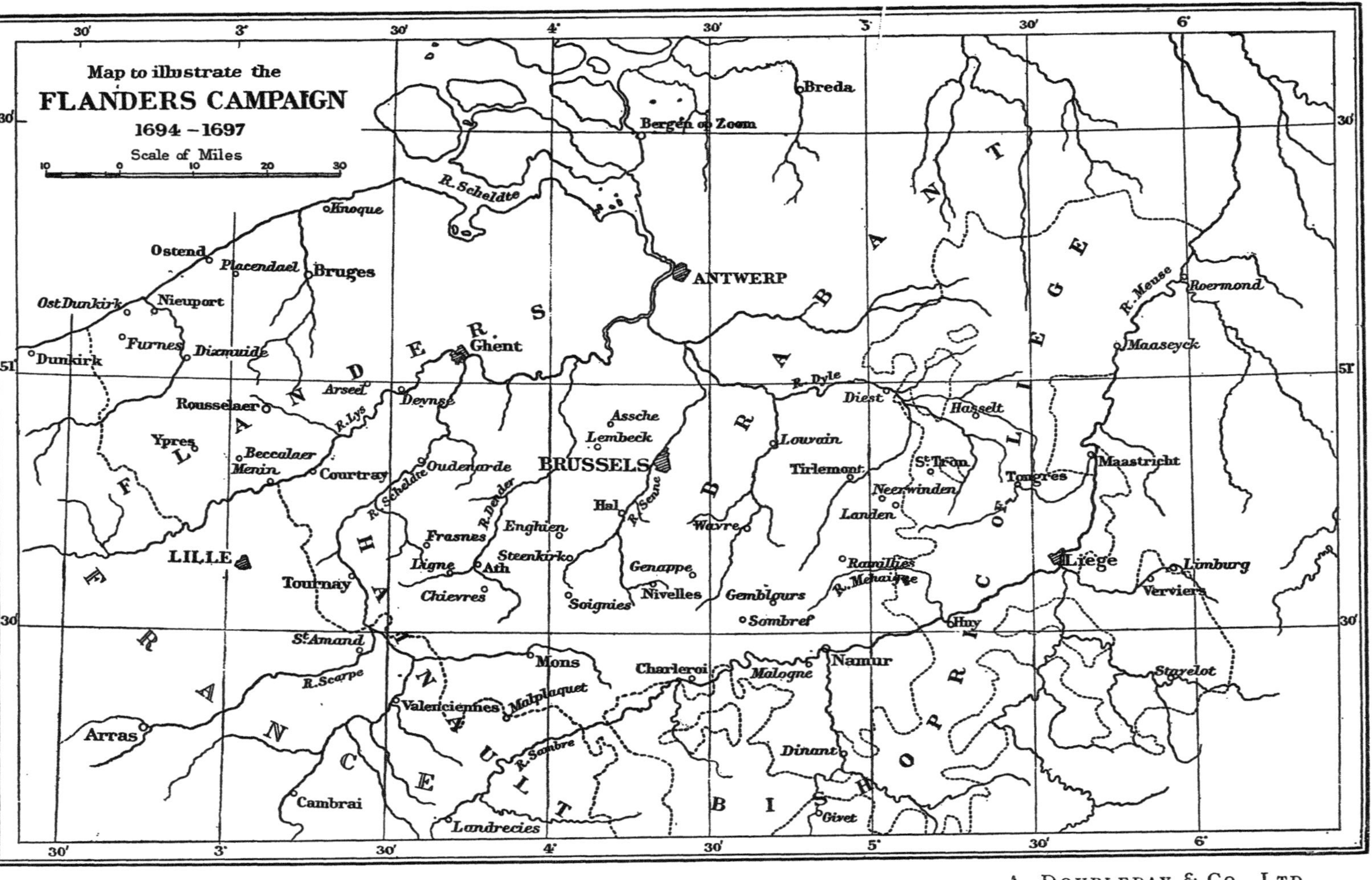

A. Doubleday & Co., Ltd.

halt. The next morning the main body of the French army was in Tournai.

Tettau, on reaching the banks of the Scheldt, was surprised to find de la Valette's force strengthened by the 3000 dragoons from Charleroi; but still more astonished was he to see the brigade of Guards, which had been forced on to Condé, and thence forwarded in boats. Tettau sent for guns and opened fire, but he was effectually hindered from laying his bridges.

On the morning of the 14th William marched to force the river at Hauterive. About midday the Allies reached the river, and to their dismay and disappointment found Villeroi well intrenched on the other side, and the half of the French army winding through the Lines to join him. Before nightfall the whole of de Luxemburg's army was united on the banks of the Scheldt.

The French had left Huy at midday on the 8th, and on the morning of the 14th were ready on the other side of the Scheldt to oppose the passage of the Allies. They had covered, in the five days and a half, a distance of about 120 miles through a difficult country, had crossed the Mehaigne, the Scarpe and the Scheldt, and had twice crossed the Sambre. The distance marched by the Allies in the same time was scarcely 80 miles.

This famous march was a great triumph for de Luxemburg, but the cost was dear. The French cavalry was ruined for any further service, and it is said the French lost some 3000 men : the course of the march could be tracked by the dead bodies of men and horses.

D'Auvergne writes:—"The French Army might have been followed by the scent which they left behind of dead men and horses, which were to be found all along the road it went."

William, foiled in his designs on the Lines, marched to Escanesse and Melden, detaching four brigades to strengthen de Würtemberg's force on the heights of Peteghem, between Oudenarde and the enemy on the other side of the river.

The next day William crossed at Oudenarde, and halted half way between that place and the Lys.

De Luxemburg moved to the Lys, halting at Courtrai; and on the 17th he crossed that river, and encamped with his right on Courtrai and along the Lys to the Heulle river, while his left stretched at right angles with the Lys as far as Moorseele.

De la Valette occupied the Lines from the Scheldt to the

Lys with some 5000 men; and de Villeroi those from the Lys to the town of Ypres with some 12,000 more; while 15,000 men were placed in garrison at Furnes, the place being made ready for defence.

William crossed the Lys and encamped at Rousselaer, his object being to detain the French Army North of the Scheldt by threatening Furnes and Fort Knoque.

"Our heavy baggage," writes the historian, "came up to the camp on the 19th August; it had been sent away on the 5th under the convoy of Brigadier WYNNE's Dragoons. They marched by Louvain, Vilvarde and Ghent, and so joined us in this place."

On the 24th the heavy baggage was sent back to Ghent.

On the 7th of September all the English Horse and Dragoons came up to Rousselaer from their camp at Wonterghem. "WYNNE's Dragoons, that had been sent from the camp by Mont St Andrée to Ghent, to get their horses in better case, being now refreshed from the fatigues they had endured, came up to the camp along with them, and on the 8th they were all sent to canton upon the villages between our right and Dixmude." The English Horse and Dragoons were reviewed by the King on the 10th of September near Dixmude, and the troops "appeared in very good order" after their trying marches.

Huy, in the meantime, was besieged by a detached portion of the Allied Army under the Duke of Holstein, to whom the garrison capitulated after a short and feeble resistance.

During the summer of this year an expedition was fitted out from England for the purpose of harrying the French Coast, but without any goodre sult. On the 6th of June an attempt on Brest resulted in complete failure, the troops being re-embarked with great difficulty.

From the middle of September the Allied Army began to move into their winter quarters. WYNNE's Dragoons, together with the regiments of Eppinger, Mathews, Livingstone, Essex and Cunningham, were quartered in the villages between Ghent and the Sas Van Ghent.

The main feature of the campaign of 1694 was the famous march by de Luxemburg from the Meuse to the Scheldt, which William had regarded as a physical impossibility.

This great general was an extremely ugly man with a deformed figure. Upon its being reported to him that William had once exclaimed "What! am I never to beat this hump-backed fellow?" de Luxemburg observed, "How should he know the shape of my back; for it is certain that I never turned it to him." He died before the Campaign of 1695 opened.

With regard to the bloodless nature of the campaign, D'Auvergne, writing of an attack by the French on some bread wagons, says the officer commanding had the misfortune to be mortally wounded; "which is the only Officer of our Army I have heard of, that has been killed this campaign by the Enemy."

Each year of the war had been hitherto one continued success to the French, and a series of discouragements and disasters to the Allies, but the turning point had at length been reached. The operations of the French had been offensive; but several reasons combined to deprive them now of this advantage, and to reduce them to the defensive. Not only was the great General de Luxemburg dead, and a comparatively untried general, the Maréchal de Villeroi, invested with the command, but the drain on the French resources was becoming more and more felt each year. Added to this, the Allies had at length learned the importance of beginning the campaign in good time, and with an adequate force. They had determined to concentrate their whole efforts upon Flanders for this campaign, and no endeavours were spared to assemble as large an army as possible, and to get into the field as early as the season would permit.

The French Lines now comprehended all the country within one long stretch of defence from Namur to Furnes. From Namur they ran to Charleroi: from the Sambre to the Haisne at Mons and Condé, thence to the Espierre, whence they were carried to the Lys and on to Ypres: and on from Ypres to Dunkirk viâ the Knoque Fort and Furnes. During the winter 20,000 navvies, covered by a Corps d'armée, had been employed in renewing and strengthening the Lines, the whole of which was connected along its length by forts and redoubts at intervals, while, where a natural frontier such as the Haisne and the Dunkirk canal was wanting, entrenchments had been run up. But the maintenance of this great extent of frontier within an enemy's territory compelled the French General to keep to the defensive.

During April a letter was sent to Brigadier WYNNE to supply three men to guard the King's yacht at the Hague.

In April both armies began to assemble. The main army of

the Allies under King William, with the Prince de Vaudemont and Duc de Würtemberg as second in command, assembled at Deinse on the left bank of the Lys : and a force under the Elector of Bavaria encamped at Assche with its left towards Brussels. The King's Army consisted of about 11,000 cavalry and 42,000 infantry, and included all the English regiments except about a dozen that were with the Elector, whose force numbered some 15,000 cavalry and 20,000 infantry.

The following is a list of the British Contingent :—

CAVALRY

Lieut.-Generals.	M. D'Auverquerque.
	The Duke of Ormond.
Major-Generals.	Earl Rivers.
	Eppinger.

Brigade	Regiments	Squadrons
L'Etang.	Life Guards. 1st Troop	1
	" 2nd "	1
	" 3rd "	1
	Horse Grenadiers.	1
Lumley	Lumley (1 D.G.)	3
	Langston (4 D.G.)	2
	Wyndham (6 D.G.)	2
	Schomberg (7 D.G.)	2
Coy	Wood (3 D.G.)	2
	Coy (5 D.G.)	2
	Leveson (2 D.G.)	2
	Galway's Horse.	3
Mathews	Livingstone's (Scots Greys)	4
WYNNE	Essex (4 D. and Hussars)	4
	WYNNE (5 D. and Lancers)	4
	Cunningham (7 D. and Hussars)	4

D'Auvergne in his line of battle at the Arseel Camp shows WYNNE's brigade, consisting of his own and Essex's Dragoons, as being in the first line of the Left Wing. A squadron of dragoons at this time consisted of one hundred men.

FOOT.

General.	Duke of Würtemberg.
Lieut.-Generals.	Count Nassau.
	Count Noyelles.
	Sir Henry Bellasyse.
Major-Generals.	Churchill.
	La Melonière.
	Ramsay.

Brigade.	Regiments.	Battalions.
Guards	1st Foot Guards	2
	2nd " (Coldstream)	1
	3rd " (Scots)	2
Erle	Hamilton (1st Royals)	1
	Selwyn's (Queens)	1
	Churchill (Buffs)	1
	Trelawny (4th)	1
	Stanley (16th)	1
	Erle (19th)	1
Fitzpatrick	Fairfax (5th)	1
	Fitzpatrick (7th Fus.)	1
	Fred Hamilton (18th)	1
	Ingoldsby (23rd)	1
	Collingwood's	1
	La Melonière	1
Collier	Hamilton (1st Royals)	1
	Columbine, Rada (6th)	1
	Granville (10th)	1
	Seymour	1
	Saunderson	1
	Collier	1
O'Ffarrel	O'Ffarrel (21st)	1
	Lauder	1
	Mackay	1
	Morton	1
	Strathnaver	1
	Geo. Hamilton	1

WITH THE ELECTOR'S FORCE.

General. The Earl of Athlone.

CAVALRY.

Brigade.	Regiment.	Battalions.
Lloyd	Mathews' (1 D.)	4
	Lloyds' (4 D. and Hussars)	4

FOOT.

Brigade.	Regiments.	Squadrons.
	Brewer (12th)	1
	Tidcombe (14th)	1
	Leslie (15th)	1
	St. George (17th)	1
	Maitland (25th)	1
	Ferguson (26th)	1
	Tiffeney (27th)	1
	Graham	1
	Lorne	1
	Buchan	1

The total of the British Contingent was 29,100 men, of whom 5700 were cavalry. Besides the two armies under William and the Elector, the Allies had the following :—

Ellenberg's corps of 13,000 men at Dixmude.

Brandenburg Contingent of 16,000 men at Liége.

Count de Tilly with about 4000 men also near Liége. These brought the total Allied Army in the field up to about 124,000 men.

On the morning of the 31st May the King reviewed the army at the camp of Arseele. "Our English Cavalry...... which made a very gallant show, the horses being in very good order, and the men very well clothed and armed."

The French assembled in three bodies: the first under de Villeroi near Menin; the second under de Boufflers, between the Lys and the Scheldt: the third under M. de Montal, in the neighbourhood of Furnes.

On the 2nd of June William marched from Arseele to Rousselaer, and on the next day to Becelaer. On this march D'Auvergne writes, "We made detachments of 4000 dragoons towards Menin, which was not far from our left, to cover our march. They met with a party of the enemy's dragoons, which they pursued to the very palisades of Menin, and brought back 23 prisoners."

At the camp the dragoons of the Left faced to Moorsleede, covering the King's horse.

The whole of the French forces, with the exception of a small flying column under the Marquis d'Harcourt, were between the Scheldt and the sea. D'Harcourt was on the Meuse in observation of the Brandenburg and Liége contingents, which had now been joined at Liége by a large body of cavalry from the main armies.

During the winter the French had much facilitated the defence of their Lines by cutting what were termed "routes royales" for purposes of speedy communication: these "routes royales" were perfectly direct, nothing being allowed to interfere with their course; trees, houses, walls and villages were ruthlessly demolished if they stood in the way.

William now detached the Duc de Würtemberg with eight battalions under Churchill, together with some artillery and a pontoon train, to join Ellenberg and threaten Knoque Fort. From the 8th to the 16th June de Würtemberg operated against Knoque, finally withdrawing in silence to Dixmude. At the same time the Elector made feints of attacks on the Lines of the Lys and Scheldt.

When de Würtemberg marched to Knoque, William returned to Rousselaer; and then pushed speedily on towards the Meuse, escorted only by the Life Guards and some Dutch cavalry, leaving de Vaudemont with the army of observation, and instructing de Würtemberg and the Elector to march on Namur.

The design was, to draw all the French forces westwards, and then suddenly to invest Namur. To this end an immense quantity of siege material had been collected at Maestricht, and all the boats on the Meuse had been pressed. De Tilly, with the Liége and Brandenburg contingents, had gone to Falaise on the Mehaigne; and on the 18th the Earl of Athlone, with the cavalry from the Elector's Army, went to Tirlemont en route to join them. On the 20th de Vaudemont moved to Grammen on the Lys, and there remained. The Elector's corps left Tieghem on the 18th, and marched to St. Livens, and thence to Ninove, to Halle, to Genappe, and to Le Masy, each stage being a day's march.

In the meantime the Earl of Athlone, leaving some cavalry at La Falaise, (which lies a little north of Namur), marched towards Charleroi, crossed the Meuse at Castelet, and then doubled back and marched straight towards Namur, sweeping the river of boats as he went: the effect of this feint on Charleroi was to deceive d'Harcourt, and to induce him to weaken his forces by throwing a body of dragoons into Charleroi. Athlone, on arrival in front of Namur, occupied posts on the North stretching from Floreff on the Sambre across to the Meuse. The Elector meanwhile crossed the river to Malogne, where he arrived on the 23rd. The same day the Brandenburg contingent closed in on the South. This last move was, however, a day too late; for de Boufflers, who had made forced marches by Solre and Phillipsville, was able to reach the town a few hours later with seven regiments of dragoons, some artillery and sappers. The King arrived in camp the same day, and on the 23rd the investment of Namur was complete.

Namur was one of the most formidable strongholds in the world: and this strength it owed partly to nature, and partly to its strong defences. Its garrison consisted, (after de Bouffler's reinforcement), of nineteen battalions of foot, eight regiments of dragoons, a company of miners, and a company of gunners, in all about twelve or fourteen thousand men: and of artillery there were 120 guns and 8 mortars: and ample stores of munitions, money and provisions. The Comte de Guiscard was the Governor, and de Boufflers took the general command of the troops.

Immediately after the investment, Lord Athlone was detached with the cavalry to consume the forage between Mons and the Sambre, while the investing army awaited the arrival of reinforcements and of the guns and siege material. By the 28th of June these had arrived and the trenches were opened.

With the siege of Namur we have little to do, as the cavalry were not employed in the investment. We note that on the 6th of July there was an assault in which the Brigade of Guards bore the brunt of the attack, and greatly distinguished themselves. The result of the affair was that the Allies were left masters of the whole of the heights of Bouge, with a loss of not fewer than 4000 men killed and wounded in the evening's work. Another assault took place on the afternoon of the 17th, and from that date to the surrender of the town on August the 27th there was continuous fighting, in which the British regiments bore the lion's share.

To return to the operations which had in the meantime taken place in the neighbourhood of the Scheldt and Lys.

We left de Vaudemont at Grammen on the Lys, whither he had marched on June the 20th, and where he was joined by the Duc de Würtemberg.

On June the 18th, when William marched from Becelaer to Rousselaer, de Villeroi crossed the Lys and encamped at Harlebeck, and shortly afterwards marched to Pottes beyond the Scheldt, at the same time detaching bodies of cavalry towards the Sambre to harass the convoys of the Allies.

At Rousselaer on the 14th June "the King had notice the Enemy were sending a party of Dragoons to attack our Bread Wagons commanded from Ghent. Immediately a detachment of Dragoons was ordered to wait for them at Moorsleede by which they were to march under the command of [Briga]deer Wynne, they mett the party, which had ... ... our hands, had we not attacked them too. [In the] action Brigadeer Wynne was wounded in the Legg, Lieut. Webb and 10 private men killed. Of the enemy one Hussar Captain and eight men made prisoners, what was killed of the enemy is uncertain by reason this action happened in the dark." [1]

The following account of this affair is taken from D'Auvergne:—

1. Add. MS. 18,776, fol. 4.

The Dragoons under WYNNE were " advised they (the enemy) were to pass that evening (on the 14th) by Moorsleede. " The Dragoons overtook them and had orders to dismount and attack them, " which they did with a great deal of courage. " The enemy were commanded by a Lieutenant-Colonel and were about four hundred strong ; and " they made some barricades of wagons, which they defended no longer than to gain time to make their escape ; but notwithstanding, a captain and thirty men remained prisoners, and they had several killed and wounded. The fire was very hot for half-an-hour. Brigadier WYNNE was wounded in the knee, which though it was not esteemed very dangerous at first, yet he died afterwards of this wound at Ghent, being generally regretted. The King has since given his Regiment of Dragoons to Colonel Ross, Lieutenant-Colonel of the Regiment, who has been one of the King's Aides-de-camp the three last campaigns. Captains Collins and Holdgate likewise wounded, both officers of dragoons. Lieutenant Webb was killed. " The next day the convoy came safe to camp.

On the 3rd of July, de Villeroi marched to attack de Vaudemont, whose presence on the Lys prevented him leaving the Lines for the relief of Namur. Leaving Pottes at ten o'clock at night, he arrived at Rosebeck at nine the next morning, having during the night marched some four and twenty miles, and crossed the Scheldt and the Lys, besides two smaller streams.

The Allies, with their usual neglect of scouting precautions, were only aware of de Villeroi's approach when some of their advanced posts had been made prisoners by the French.

De Vaudemont's numbers were not much more than half of his opponent's. He despatched his baggage to Ghent and awaited the French General's attack. Three English regiments were ordered up from Deinse, the right was thrown back so as to occupy the higher ground about Arseele, and the front was intrenched. The British position extended from the right at Arseele to the left at Gothem, where the Mandel river joins the Lys. The Mandel protected the left flank, while in the front were numerous ravines, streams and woods.

De Villeroi halted at the village of Denterghem on the morning of the 4th, while his troops effected some sort of clearance of the ground between themselves and the Allies : and the design was formed of surrounding de Vaudemont under cover of the very obstacles on which he was relying for protection from attack. A large body of French cavalry under de Montal and Berwick, got on to the high road at Thielt and so round the right flank of the Allies: the infantry meanwhile extended along the front, and at intervals little explosions were made along the line to maintain communication and regulate the advance.

De Vaudemont, however, perceived the intentions of his adversary. His situation was critical; to retire might prove even more disastrous than the attempt to hold on to his position, but he finally saved the situation by an extremely clever retirement. Giving orders for the troops in front to continue at work on the entrenchments, and directing the Artillery on his left to keep up a perpetual cannonade, he caused a number of houses along his front to be set on fire, and then with the greatest possible caution and silence withdrew his Artillery from the front and from the right, and despatched it to Deinse. At the same time he moved the cavalry of the right wing, alternated with the infantry regiments of Collier's brigade, to his right rear between Arseele and Vinckt, as if to check de Montal's approaching attack on that flank; but suddenly M. d'Auverquerque marched off with the cavalry straight upon Ghent by a narrow road through the woods, while the infantry, with their pikes and colours trailed, took the road along the rear to Deinse. During this movement the Prince de Vaudemont, the Duc de Würtemberg and a number of English Staff Officers remained formed in line, in order to keep up an appearance of the presence of cavalry. Lastly, the cavalry of the left, together with the Dutch infantry, drew off towards Deinse.

So silently was the withdrawal carried out that two companies of Tiffeney's (27th) foot who had been left at their post in the front near Arseele to keep up the deception to the last, had not a notion of their abandoned situation, until an aide-de-camp brought them the order to withdraw.

The French Generals nearest the Allies, on discovering what was being done, sent off word to de Villeroi, but not one of them would take the responsibility of acting without orders. By the time de Villeroi received the news, only the rearguard of the Allies was to be seen retiring in the distance. The French pursued, but the rearguard contested every point of vantage and so gained time for the retreat of their main body, while cavalry trumpeters were sent to the rear of the main body and sounded repeatedly, as if the force were drawing up for battle, thereby causing the pursuers to halt and form for attack. These constant delays, together with the advent of nightfall, put an end to the pursuit.

Ross' Dragoons were apparently engaged in this action. D'Auvergne writes that " this great and renowned retreat is as fine a piece of the Art of War as can be read of in history, and which can hardly be paralleled in it ; which has showed more the Art Conduct, and Prudence of a General, than if the Prince had gained a considerable victory. "

The next day de Vaudemont re-united his forces at Ghent, and de Villeroi returned to his camp between Rosebeck and Rousselaer.

On July the 6th de Vaudemont, with the smallest portion of his force, camped at Oostaker near Ghent, while Bellasyse and de Wirtemberg marched through Bruges to Placendael, where the Nieuport and Ostend canals join.

With Sir Henry Bellasyse's force of twelve battalions marched Ross' and Cunningham's dragoons. On the 7th he halted his infantry to refresh near Bruges, but sent his dragoons on to Placendael, whence, on being joined by the infantry, the whole force marched viâ Newendam to Nieuport.

De Vaudemont's aim was now the defence of the line of the Nieuport, Ostend, Bruges and Ghent canals, and the guarding of the country to Brussels.

Bellasyse reached Nieuport on the 8th, and de Würtemberg following him, the defence of the canal from that place to Placendael was provided for ; and de Villeroi, who had advanced a body of troops beyond Dixmude with intentions against Nieuport, was forestalled. De Würtemberg still further confined the movements of the French by opening the sluices about Nieuport, and putting the whole country under water.

Baulked at Nieuport, de Villeroi detached a force under de Montal to invest Dixmude. Major-General Ellenberg (a Danish officer) was in command of that place, and had a sufficient force of Danish and English troops for its defence, and plentiful supplies and munitions. Not twenty-four hours had elapsed, however, after the trenches were opened, before Ellenberg beat a parley, and, in spite of the opposition of some of the officers, and to the great disgust of the soldiers, next day surrendered the place. Numbers of the men broke their weapons, and Lord

Lorne's Scotsmen tore their colours to pieces rather than give them up.

The French met with equally good fortune at Deinse, where O'Ffarrel surrendered without a shot being fired.

Ellenberg was subsequently beheaded, and O'Ffarrel broke with ignominy and imprisoned, and many of the other officers cashiered by sentence of a court martial.

Finding it hopeless to compel William to raise the siege of Namur by attempting the capture of the large fortresses between the Scheldt and the coast, de Villeroi marched eastwards. Leaving the Comte de Montal with 6000 men to protect the lines, he moved on the 25th July to Avelghem on the Scheldt; and the next day with a large portion of his army to Enghien viâ Renaix; whereupon, de Vaudemont marched by way of Dendermonde to Dieghem, just N. of Brussels, "having twenty-six battalions with him and all the English horse and dragoons, except the dragoons of Ross and Cunningham, which remained with Lieut.-General Bellasyse," whose force now marched to Bruges, and on the 30th to Ghent, and subsequently joined the army at Brussels.

De Würtemberg joined de Vaudemont at Dieghem on the 30th with twelve battalions, having first provided for the safety of the Western fortresses by inundations and reinforcements.

On the 1st of August de Villeroi marched to Halle, and next day camped between Gaesbeck and Anderlecht; but he found his intentions on Brussels anticipated, for de Vaudemont had inundated the front of the town from the Senne to Monterey, and had entrenched along the river and canals from Monterey to Vilvorde; the Earl of Athlone was also approaching with a large body of troops from Namur, while William himself had left Namur with a further reinforcement, leaving the Elector to carry on the siege. De Villeroi was forced to content himself with the barbarous satisfaction of bombarding the town of Brussels.

The shelling began on the 3rd August, and about it Parker writes: "Five days the bombardment continued, and with such fury that the centre of that noble city was quite laid in rubbish. Most of the time of bombarding I was on the counterscarp, where I could best see and distinguish; and I have often counted in the air at one time, more than twenty bombs; for they shot whole bombs out of their mortars all together. This, as it must needs be terrible, threw the inhabitants into the utmost confusion. Cartloads of nuns, that for many years before had never been out of the cloisters, were now hurried about from place to place to find retreats of some security. In short the groves and parts remote were all crowded : and the most spacious streets had hardly a spectator left to view the ruins."

All this time the siege of Namur was being carried on.

On August the 7th de Villeroi desisted from the wanton bombardment of Brussels and marched to Enghien, and on the 9th to Soignies, where he halted some days to receive orders from Paris, detaching d'Harcourt to Solre on the Sambre to get together any troops available on that frontier.

The moment de Villeroi retired from Brussels, de Vaudemont effected a junction between Waterloo and Genappe with Athlone, and on the 10th he joined the army before Namur.

De Villeroi, on receiving his orders from the Court, marched to Nivelles; and on the 16th proceeding towards Ornlau, he encamped between St. Amand and Sombref. On the 18th August he continued his march to Gemblours, where he took up a position with the Bois de Grand-Lez on his left, the river along his front, and his right near Tougrines.

King William had returned to Namur, and on hearing of de Villeroi's movements took command of the covering army, secured de Vaudemont's junction with him, and camped behind the village of St. Denis, with his left intrenched as far as Isne-les-dames.

De Villeroi arrived in front of this camp on the 18th, and made every sign of intending to remain for at least a day; but at 11 o'clock the same night he struck camp in the profoundest silence, crossed the Orneau, and arrived at dawn in the thick woods which alone now separated him from the Allied Camp. Here he found every precaution for defence had been made, and he was reluctantly forced to return to the other side of the Orneau.

On the 20th de Villeroi moved to the Mehaigne, and halted with the village of Grand Rosiere in his rear. He had found the Allied position too formidable for attack, and was obliged to remain an inactive witness of the calamity he was powerless to prevent.

This catastrophe was not long in arriving.

At noon on August the 20th an exploded barrel of powder flashed from the British quarters at Salsines, the signal for a general assault on the fortress of Namur.

From the first line of trenches towards the Terra Nova marched four English sergeants, each with fifteen men. Immediately behind them came the grenadiers of the Guards under Colonel Evans, and closely following them were the grenadiers of the other regiments. Lord Cutts commanded and personally led the attack. St. George's (17th) and Mackay's foot supported the attack, and Fred Hamilton's (18th) and Buchan's regiments were in reserve.

At the same time 3000 Bavarians marched out of the second parallel towards the breach of the Coehorne. Two thousand Brandenburgers assailed the point of the Coehorne, and were in touch with the attack of two thousand Dutch on the Cassotte. Six hundred men were also detailed to assault the basse-ville which lay below the castle.

The breach assigned to the English was more exposed to fire than the others, and the access to it was more difficult; while between it and the besiegers' trenches was an open space of more than half a mile. Across this valley of death the British grenadiers marched exposed to a terrible fire from front and flank. They mounted the breach supported only by St. George's foot; the other three regiments having been delayed. Nearly every Officer of the Grenadiers was killed or wounded, and Lord Cutts himself was severely wounded in the head.

When the three supporting regiments at length came up, the men were dispirited, and all the *élan* of the first attack had evaporated ; but the troops again advanced, and again made their way on to the breach. Fred Hamilton's regiment got quite within the breach, and planted their colours on the ramparts. But now for the first time it was discovered that the enemy had erected within the breach an interior intrenchment which was untouched and unassailable. The English troops retired ; and, as they went, a large body of French foot and dragoons came down between the Coehorne and Terra Nova and fell upon them, while a cross-fire was poured upon them from both fortresses. Lord Cutts, returning so soon as his wound was dressed, saw it was suicidal to remain, and gave the order to retreat. The British attack had failed.

Meanwhile, the Bavarians got into difficulties in their attack on the breach in the Coehorne, and after two hours' fighting found themselves on the point of being beaten back, although they still held their ground upon the glacis. It was at this critical moment that Lord Cutts, as he retreated from the Terra Nova, perceived the state of affairs : calling for two hundred volunteers to form a forlorn hope, he was promptly answered, and Lieutenant Cockle of Mackay's was selected to lead them. This Officer's instructions were to attack the face of the salient angle next the breach, without firing a shot; and, if he could master the palisades, to lodge himself in the covered way. Mackay's regiment was next to the front, with the Ensigns bearing their colours at their head.

Liberal promises of promotion and rewards were made, and the soldiers were encouraged by handfuls of gold being given to those who particularly distinguished themselves.

Cockle succeeded in surmounting the palisade; and, beating the enemy back on the covered way, he turned their own guns on to them. He was ably supported by the four English regiments, and the Ensigns of Mackay's marched boldly up and planted their colours on the ramparts, which so fired the men that not only was the covered way gained, but held.

The Brandenburgers and Dutch were also successful in their efforts, and by 5 o'clock in the afternoon the fighting was over, and the Allies remained effectually lodged within the enemy's works.

King William had watched the whole affair with the greatest anxiety, and having assigned to the British troops the post of most danger, he especially observed their conduct. He marked his approbation of the bravery of Fred Hamilton's (18th) regiment in the second, and almost hopeless, attack on the Terra Nova, by conferring upon it the title of "The Royal Regiment of Ireland," together with its present badge and motto. Lieutenant Cockle was shortly rewarded with money and promotion.

The loss on this day amongst the British force alone, consisting of four regiments and seven hundred grenadiers, was :—

	Killed	Wounded
Officers	29	55
Men	275	790

On the 26th of August the garrison surrendered and filed out between two long lines of the Allied troops with all the honours of war, colours flying, arms carried, bullet in mouth, and matches lighted, and six guns following with the baggage. The garrison began the siege with 13,000 troops, and marched out under 5,000 strong, and when the victors entered the place they found it absolutely reeking with the putrid stench of dead men and horses.

The Allied Army before Namur next marched towards Brussels, whereupon, de Villeroi, detaching a force to take care of Dinant, marched by way of Charleroi and Mons towards the Lines.

Bellasyse's force marched to Ghent upon the capitulation of Namur.

Nothing, however, came of these movements, and the Allies commenced to move into winter quarters, the dragoons being quartered in the villages between Ghent and Sans van Ghent, except the dragoons of Ross, who were quartered in the villages between Bruges and Damme.

During the winter the French King expended his resources in preparation for an invasion of England. This abortive design had no effect on the Allied Army in Flanders beyond causing the recall home of some twenty battalions, ten of which, however, returned without even landing in England. Two regiments, Erle's (19th) and Bellasyse's (22nd), were captured by the French during the passage home and carried into Dunkirk. While the Allies were thus compelled to weaken their army, the French on the other hand were reinforced by their forces from the Duchy of Savoy, peace having been concluded with that State.

On the 4th March, in a raid made on Givet by a body of Cavalry under the Earl of Athlone, joined by a detachment from the garrison of Namur, an immense quantity of forage accumulated by the French was destroyed.

During the first week in May the opposing armies began to assemble.

Ross' dragoons, with those of Livingstone, were the dragoons of the 2nd Line of the Right Wing, the brigadier being Mathews.

De Villeroi with his main army was between Menin and the Scheldt, while de Bouffler's corps was collected about the Orneau. Also, two flying columns were assembled, under de la Mothe and de Montal, towards the sea; another in Luxembourg under d'Harcourt, and a fourth about Dinant under the Comte de Guiscard. In all, the French forces reckoned 173 battalions and 223 squadrons, or about 120,000 men. These numbers were so overwhelming that the Allies contented themselves with forming two corps of observation, one at Tirlemont under Prince Nassau-Saarbruck, and the other at Affleghem near Alost, under the Prince de Vaudemont.

On May the 9th de Villeroi, marching along the Lys from Courtrai reached Deinse, and at the same time de Boufflers encamped at Fleurus; whereupon Prince Nassau-Saarbruck retreated from Tirlemont to Parck Camp.

De Villeroi contented himself at Deinse with foraging and consuming the country, but none the less every measure was taken by the Allies to secure the line of canals against him.

On May the 27th William arrived in camp and took command of the Army of the West. By this time Nassau-Saarbruck had been joined by the Brandenburg, Cologne, and Liége contingents, and had advanced as far as Wavre, while de Boufflers had retired to Charleroi.

On June the 1st William marched with a strong force from Ghent to Wavre.

On June the 9th the King marched from Wavre to Conroy (halfway between Wavre and Gemblours), but de Boufflers' attitude all along the Sambre was such, that William hesitated before making any fresh move. On the 27th, however, he advanced to Gemblours, and there awaited the arrival of the contingent under the Landgrave of Hesse. On the 15th July the Landgrave with 15,000 men arrived at Namur, and William advanced to Sombref to meet him.

De Boufflers' fears at this time were that the Allies would either force the Sambre, and thus make their way on to French territory towards Dinant and in rear by Charleroi, or else, repeating their tactics of 1694, make a sudden rush against the lines of Espierre or for the siege of Mons.

The following moves of the two armies need not be gone into at length. The attitude of both sides was simply that of waiting for something to turn up. The French Army was the stronger. The aggressive therefore lay with them, while the Allies could not initiate any action of decision. On August the 16th William left the Army. De Vaudemont and de Villeroi had meanwhile remained observing each other on the canals of Ostend and Bruges, contenting themselves with foraging and reconnaissance.

On the 20th of August "the enemies beat the 'General' very early in their camp, which we could hear very easily from ours; the Prince [Vaudemont] thought that it was in order to march, and accordingly ordered the 'General' to beat on our side, and the army marched forthwith towards Bruges..... The same day that we marched up so near to Bruges, the Prince ordered a detachment of horse and dragoons to march on the other side of the canal, to observe the enemies, whom we expected to be on the march too. Captain Cornwallis of Ross' Regiment commanded the dragoons, who fell in at Oostcamp with a party of the enemies posted in a defile ; and though this was a considerable disadvantage, yet Captain Cornwallis charged them with his dragoons sword in hand so vigorously that he killed several of the enemies, and brought off all his detachment safe out of the defile (except for two killed) and a prisoner that was wounded in the action : the Dutch horse being informed that several of the enemy's squadrons lay between Oostcamp and Lophem to cover the foragers, would not venture in to the dragoons' assistance, else we might have taken the whole party."

Another extract from D'Auvergne, under date 31st August:—

"The Prince, having advice that all the French troops under Villeroi were now upon their march towards Torhout, ordered the infantry of his army, with the dragoons of Eppinger, Ross, and Mirmont, to pass the canal. The infantry, having passed the canal, encamped in two lines, two and twenty battalions in the first line and sixteen in the second, with the Right near the canal of Ostend, and the Left at St. Michael ; the dragoons of Eppinger and Mirmont encamped upon the right of the first line, while those of Ross upon the left at St. Michael near the canal of Bruges."

About the 29th September the firing of the enemy's guns created a stir in the Prince's camp, upon which "Eppinger and Mirmont's dragoons on the right and Ross' on the left were ordered to mount and pass the canal of Ostend, to march towards Placendael." It was found, however, to be a foraging party, and that there was no danger to Placendael, and the Prince "countermanded the dragoons."

Meanwhile the army was beginning to go into winter quarters, the campaign having been got through without loss, though at the end of July Huy had an exceedingly narrow escape of capture.

The dragoons of Ross and Livingstone were quartered for the winter in the Palis de Nort, beyond Bruges.

The campaign of 1697 was almost as devoid of incident as that of the previous year.

During April the Allies occupied themselves in making a line of defence from Ostend by Bruges and Brussels to Namur; for the protection of this work the Elector of Bavaria assembled a force at Deinse, while the rendezvous of the main army was at Bois-Seigneur-Isaac.

On the French side de Villeroi took command of an army of observation, while Maréchal de Catinat undertook the siege of Ath; and de Boufflers assembled a corps on the Meuse.

The strength of the French force was as follows:—

De Villeroi	60,000
De Boufflers	56,000
Catinat	40,000

On May the 16th both the Allied armies united at St. Quintin-Lenneck, under the command of the King, who had arrived on the 14th.

De Boufflers had approached de Villeroi, who now had under his hand in case of a general action some one hundred and twenty thousand men besides those engaged in the siege, while William had some fifteen or twenty thousand fewer.

With this preponderance of force, and with dispositions so much in favour of de Villeroi, William regarded it impracticable to relieve Ath, and on the 22nd marched to Genappe viâ Oudenarde, while the Elector returned to Deinse.

On the 28th of May Ath surrendered with all the honours of war.

On June the 12th de Villeroi marched to Gammerage and de Boufflers to Enghien, and both made ready to march conjointly on Brussels. If the French could start from Enghien before William became aware of it, they were sure to reach Brussels before him; in which case they would be able to proceed to cut off all communication between the Eastern and Western portions of the Allied line of defence.

But fortunately for the Allies, they for once had scouts on the look-out, and these movements were known to William the same afternoon. The King, grasping the situation, marched by the one road open to him. Starting three brigades of infantry between four and five o'clock, he despatched the artillery two hours later, the baggage at ten, and the rest of the infantry at eleven. He personally superintended the despatch of the troops, and at midnight he rode with four regiments of the dragoons to the front to reconnoitre, and to be ready for the army on its arrival. The main body of the cavalry left Genappe at daylight, and so covered the rear.

Before ten next morning the Allies had occupied the camp of Anderlecht; and when de Villeroi and de Boufflers appeared presently on the heights of Anderlecht and on the Assche road, they saw the Dutch and English flags waving over the coveted ground, and knew that they were foiled. Besides providing against an attack on the camp, the Allies placed Brussels in readiness for a siege. The failure of the design upon the Capital was the concluding act of the War in Flanders.

For some months negotiations had been proceeding, and on the 11th of September peace was signed at Ryswick.

After the Peace of Ryswick steps were taken by Parliament to reduce the strength of the Standing Army, and to make matters worse the Government failed to find the money necessary to wipe off the arrears of pay due to the troops. Discontent and insubordination became rife, and the effect of disbanding such a large body of men was disastrous. The country was flooded with gangs of unemployed men rendered desperate by the refusal of Government to settle their just dues.

A new Parliament met in December, and William urged the necessity of maintaining an army sufficient for the needs of England, and the imperative duty of settling the arrears of pay.

The House thereupon passed an Act fixing the Army to be maintained in England at 7,000 men, and that in Ireland at 12,000, Ireland bearing the cost of her own troops. The Army was to consist entirely of British subjects, thereby excluding William's Dutch soldiers.

With the exception of the regiments specially named, all corps were to be disbanded by the next 25th of March.

To complicate matters, the Mutiny Act, which had expired in the previous April, was not re-enacted, and the military authorities were powerless to enforce discipline.

The following is a list of the regiments of the British Army continued in force by the proclamation.

IN ENGLAND.

HORSE.

3 troops of Guards.
1 troop Horse Grenadiers.
Oxford's Regt. (Blues).
Lumley's (1 D.G.)
Wood's (3 D.G.)
Arran's (5 D.G.)
Wyndham's (6 D.G.)
Schomberg's (7 D.G.)
Marchfield's.

DRAGOONS.

Raby's (1 D.)
Lloyd's (3 Hrs.)
Essex's (4 Hrs.)

FOOT.

1st Regt. Foot Guards.
2nd " " "
Selwyn's (Queen's).
Churchill's (Buffs).
Trelawny's (King's Own).

IN IRELAND.

HORSE.

Harvey's (2 D.G.)
Langston (4 D.G.)

DRAGOONS.

Ross (5 Lancers.)
Echerin (6 D.)
Conyngham (8 H.)

FOOT.

Orkney (R. Scots).
Columbine (6th).
Fairfax (5th).
Webb (8th).
Stewart (9th).
Granville (10th).
Hanmer (11th).
Brewer (12th).
Jacob (13th).
Tidcombe (14th).
Howe (15th).

Stanley (16th).
Bridges (17th).
Fred Hamilton (18th).
Erle (19th).
Geo. Hamilton (20th).
Bellasyse (22nd).
Ingoldsby (23rd).
De Tinzar (24th).
Tiffan (27th).

IN SCOTLAND.

HORSE.

1 troop guards (afterwards disbanded).

DRAGOONS.

Royal Regiment (Scots Greys).
Jedburgh's (7 H.).

FOOT.

Scots Guards.
Colyear's.
Scots Fusiliers (21st).
Maitland's (25th).
Geo. Hamilton's.
Strathnavar's.

IN CHANNEL ISLANDS.

O'Hara's (7th Fusiliers).

Ross' Dragoons to consist of 8 troops, making a total of 362 men.

The Dragoons disbanded out of the Regiment received a total sum of £272.19.2¾ amongst them.

Soon after the Peace of Ryswick Ross' Dragoons had returned to Ireland where they enjoyed a term of home service until March, 1702.

Quarters of the Army, 27th June, 1698. Colonel Ross' Dragoons are quartered as follows :—

1 Troop at Mullingar and Ballynalackbridge.
1 „ Longford and Castleforbes.
1 „ Castlebar and Foxford.
1 „ Birr, Ballyboy and Ballyloghuane and the adjacent villages.
1 „ Roscommon, Athleage and Castlereagh.
1 „ Loughrea.
1 „ Boyle and Elphin.
1 „ Sligo.

Warrant of six months' allowance of powder to the Army.

Dragoons. { Colonel Ross'.
" Echlin's.
" Conyngham's.
Marquis de Mirmont's.

The allowance for each troop being half a barrel of powder.
Establishment of the Regiment.

34 Officers.
8 Sergants.
16 Corporals.
8 Drums.
8 Trumpeters and Hautbois.
288 Private men.

362 Total.

The following confirmation of a Court Martial dated in Dublin on the 2nd of December 1698 is interesting as shewing the cruel methods of punishment in vogue in the Army at the time. Six of Ross' Dragoons were convicted of mutiny and were sentenced "to run and be whipped three several times by an entire regiment of Foot drawn out for that purpose on three several days on St. Stephen's Green."

This punishment was termed the gatloup. The Troops ordered to carry it out were paraded with open ranks, each man being furnished with a stout switch; the ranks were faced inwards, and the prisoner, stripped to the waist, was marched up and down the lanes of men, each man striking him on the "naked back, breast, arms, or where his cudgel should light" as he passed; and in order to drown the cries of the "patient," drums were beaten during the punishment.

At the same Court Martial sixteen more men of Ross' Dragoons suffered the gatloup, while another ten were to be present "stripped of their clothes." This latter was a common punishment of the time, the lesser offenders having to be present to witness the punishment of the more guilty, undergoing all the disgrace short of the actual infliction.

While on the subject of Military Law, we notice that on the 15th of March 1698 a petition was presented to the House of Commons by soldiers of Ross' Dragoons against the Major, among other things, for beating a soldier for remonstrating about his accounts, with a thick cane, to such a degree as to damage him for life. Evidently the Officer could not be amenable to ordinary martial law for this. The allegation was considered proven by the House.

"Drumming out," too, was a punishment of the time, for in the same year sixteen of Ross' Dragoons were to lose their horses, clothes, etc. and to "be declared broken and disbanded and for ever incapable to serve His Majesty, and then to be trooped out of the Garrison with drums." [1]

Quarters of the Army, 4th June, 1700.
Colonel Ross' Dragoons quartered at :—

1 troop at Mallow.
2 " Clonmel.
1 " Rosse and Rossebarcan.
1 " Cashell.
1 " Carrickneshure.
1 " Cappoquin and Lismore.
1 " Thurles.

Quarters of the Army, 1701.
Colonel Ross' Dragoons quartered at :—

1 troop at Colooney.
1 " Boyle.
1 " Athlone.
1 " Longford.
1 " Longhrea.
1 " Castlebar.
1 " Roscommon.
1 " Castlereagh.
1 " Headford.

1. Dublin State Papers.

CHAPTER III.

Ross' Dragoons embark for the Netherlands.—
Campaigns of 1702-1703.—Marlborough and the Dutch Deputies.—
Narrow escape of Marlborough.

On the last day of the year 1701 the King, in a masterly speech to Parliament, pointed out the position in which the nation found itself by the action of the French King in violating the treaties and accepting the Crown of Spain on behalf of his grandson ; and Parliament was at last aroused when Louis XIV proclaimed James's son as King of England, thus ignoring the Treaty of Ryswick, by which William's claim to the Crown had been recognised.

The regiments of the Army were brought up to war strength, orders were given for a force to be sent to the Low Countries, and for nine additional regiments of foot to be raised.

War became inevitable, and an address was presented by the Commons to the King " that no peace shall be made with France until His Majesty and the nation have reparation for the great indignity offered by the French King in owning and declaring the pretended Prince of Wales King of England, Scotland and Ireland. "

William at this time was in an enfeebled state of health, and falling from his horse at Hampton Court on February the 20th, broke his collarbone. The shock was fatal, and he gradually sank and expired on the 8th of March. He was succeeded by the Princess Anne, daughter of James II.

Marlborough, who had been appointed Commander-in-Chief of the Allied Forces, arrived at the Hague in the middle of May.

The general plan of operations was as follows :—

" A German army on the upper Rhine was to threaten Eastern France, the Prussians and Dutch were to besiege Kaiserwerth. The main army, 35,000 strong, under the Earl of Athlone, was to hold the frontier of Holland from the Rhine to the Meuse, at the same time covering the siege of Kaiserwerth. A fourth army, collected near the mouth of the Scheldt, threatened the country round Bruges." [1]

The main French Army of 60,000 men was in the neighbourhood of Liége. Tallard with 13,000 men was detached on the Upper Rhine to raise the siege of Kaiserwerth, while another French force was to oppose the fourth allied army in the neighbourhood of Bruges.

Kaiserwerth capitulated on the 15th of June.

Boufflers, on the 10th of June, made a sudden dash to cut off Athlone, who was encamped twenty miles away near Cleve with 25,000 men, from Nimeguen. By a forced march, accompanied by a running fight, the Allies were able to forestall the French design. Athlone next withdrew across the Waal.

On July the 2nd Marlborough joined the Army, accompanied by two Dutch Deputies. These, throughout the long war now commencing, except during 1704, frequently used their influence to obstruct the operations of the British Commander who was also much troubled by the jealousies of the various Allied commanders.

On arrival at Nimeguen Marlborough concentrated 60,000 men, 12,000 of whom were British troops, the British contingent consisting of seven regiments of horse and dragoons, fourteen battalions of foot, and fifty six guns.

Marlborough then crossed the Waal and encamped at Ober-Hasselt, about six miles from the French camp, where he was delayed until the 26th of July.

Meanwhile Brigadier Ross had received an order that six troops of the Regiment under his command were to be employed beyond the seas, the remaining two to remain in Ireland. The Regiment of eight troops at this time consisted of 362 Officers and men.

Ross' Dragoons embarked at Dublin on March the 16th, and the ships being dispersed at sea, presumably by a storm, they were landed at different ports in England. The Regiment assembled at Northampton at the end of April, and soon after left England for Holland.

The pay of the two troops left behind in Ireland was fixed at the following rates for each troop :—

[1] Knight.

	s	d	
Captain	10.	0	per diem
Lieutenant	5.	0	„
Cornet	4.	0	„
Quarter Master	3.	0	„
1 Sergeant	2.	6	„
2 Corporals, each	1.	6	„
1 Drummer	1.	6	„
1 Hautboye	1.	6	„
36 Dragoons, each	1.	2	„

The following entry is found in the " Marlborough Despatches" dated 24th July :—" The English train, with the three regiments of Dragoons commanded by Colonel Ross, will arrive from Breda to-morrow night at Bois-le-Duc ; an express was sent them yesterday to halt at that place until further orders. "

On the 27th :— " The ways were so bad on Tuesday by the continued rains, that the generals thought fit to defer the march of the army from Ober Hasselt for that day ; however, most part of the baggage went over the Meuse in the afternoon upon three bridges just below the Grave, and yesterday the whole army followed. "

" Some squadrons of the enemy appeared yesterday morning at a good distance while the army was going over the Meuse, but did not advance so as to interrupt our passage. " [1]

In five marches Marlborough arrived at Hamont, threatening Brabant. This move caused Boufflers to fall back hurriedly towards the Demer, calling Tallard from the Rhine to join him.

" On the 2nd August the French army, exhausted and worn out by forced marches, was encamped in a very unfavourable position at Lonovur, and lay practically at Marlborough's mercy, but the Dutch Deputies forbade an attack, and the French were enabled to cross the Demer at Diest unmolested. " [2]

On the 10th August:— " This night a thousand men were commanded to make the roads, and all the horse and dragoons are commanded to make each a facine to fill up a bog which lies between us and the French; and to-morrow as soon as the roads are made, the Army will march towards the enemy to oblige them to decamp. " [3]

Marlborough was now making preparations for the siege of Venloo.

[1] Marlborough's Despatches.
[2] Knight.
[3] Marlborough's Despatches.

The French army again escaped the possibilities of a defeat on two occasions. On the first, the Dutch General Opdam deliberately refused to carry out his orders to attack, and then when Marlborough proposed attacking Boufflers, the Deputies insisted on further delay, and the French Army got away.

On the 24th both armies were in camp near Helchteren, and during the afternoon their artillery were engaged. All night long both camps remained under arms, and on the night of the 26th the Allies prepared to attack the French. During the night, however, the enemy retired in "great disorder, whereupon at daylight Marlborough followed them with twenty squadrons of the Right, as the Earl of Athlone did with the same number from the Left; but only Brigadier Wood with some squadrons of the Right managed to catch up the French rear guard of three or four squadrons which he entirely broke." [1]

Meanwhile the siege of Venloo had been going on, and on the 18th of September the castle was carried by storm; Lord Cutts and the English Grenadiers, behaving with the greatest gallantry, were the first to enter.

Stevenswaert was not invested, and on the night of the 1st of October, "the besieged having abandoned the Counterscarp, and our men being ready to mount the breach, they beat the chamade, and desired to capitulate." [2]

Maeseyk and Ruremond next fell into the hands of the Allies. Boufflers, anxious for the safety of Cologne and Bonn, despatched Tallard back to the Rhine, remaining himself in the vicinity of Tongres. Becoming anxious about Liége, the French General marched for that place about the 9th of October. He, however, found himself anticipated by Marlborough, for on arrival in front of the town he found the Allied Army drawn up on the very ground he had intended occupying. Again the Dutch Deputies interfered, and prevented Marlborough from throwing himself on the enemy, who during the night made good their retreat to within the lines of Landon.

The town of Liége at once surrendered to the Allies, and on October the 23rd the citadel was stormed and taken, the British troops again figuring conspicuously in the assault.

The capture of Liége brought the campaign of 1702 to an

[1] Marlborough's Despatches.
[2] *Ibid.*

end, and Marlborough set out for the Hague. At Maestricht he embarked on a boat with an escort of twenty five men, while a party of fifty horse rode along the banks of the river. These latter contrived to lose themselves, and at midnight an ambuscade of the enemy surrounded and seized the Duke's boat. Marlborough luckily was not recognised, and after the baggage had been looted the party was allowed to proceed. He was in a great degree indebted for his escape to the presence of mind of one of his attendants named Gill, who, happening to have in his possession a passport which had been granted by the French General for the use of General Churchill, who had not employed it, put it into Lord Marlborough's hands, who thus passed for his brother. Marlborough afterwards rewarded Gill with an annual pension of fifty pounds.

On his return to England the Queen conferred a Dukedom on Marlborough.

In the early spring of 1703 the British contingent in Flanders was reinforced by four regiments of foot.

Knight in his " *History of the Buffs,* " obtaining his appreciation from Allison's " *Life of Marlborough,* " finds the situation in Flanders at the end of April to be as follows:—

" The French King instead of confining the war to one of posts and sieges in Flanders and Italy...... resolved to throw the bulk of his forces into Bavaria and operate against Austria from the heart of Germany, by pouring down the valley of the Danube. The advanced post held there by the Elector of Bavaria in front, forming a salient angle, penetrating as it were into the Imperial dominions, and the menacing aspect of the Hungarian insurrection in rear, promised the most successful issue to this decisive operation. For this purpose Marshall Tallard, with the French Army on the Upper Rhine, received orders to cross the Black Forest and advance into Swabia and unite with the Elector of Bavaria. " " Marshal Villeroi, with 40 battalions and 29 squadrons, was to break off from the army in Flanders and support its advance by a movement on the Moselle, so as to be in a condition to join the main army on the Danube, of which it would form, as it were, the left wing ; while Vendomme, with the army of Italy, was to penetrate into the Tyrol and advance by Innspruck or Salzburg. The united armies, which, it was calculated, after deducting all the losses of the campaign, would muster 80,000 combatants, were then to move direct by Lintz and the valley of the Danube on Vienna, while a large detachment penetrated into Hungary to support the already formidable insurrection in that Kingdom. The plan was grandly conceived ; it extended from Verona to Brussels, and brought the forces over that vast extent, to converge to the decisive points in the valley of the Danube. " . . . .

" But if the plan was ably conceived on the part of the French Cabinet, it presented from the multiplicity of its combinations serious difficulties in execution, and it required to insure success a larger force than was at their disposal...... Marlborough, by means of secret information which he obtained from the French headquarters, had got full intelligence of it, and its danger to the Allies if it succeeded struck him, as much as the chances of great advantage to them, if it could be baffled. Louis had contemplated offensive operations in the Low Countries as well as in other quarters ; and Marshal Villeroi...... even flattered himself he would be able to regain possession of the fortresses on the Meuse before the Allies were in a condition to take the field." Marlborough's force was stronger than that of the enemy.

" With these, however, he meditated offensive operations of the most important kind. His design was to make a grand attack on Antwerp, and after taking it to reduce Ostend, which would have opened up a ready communication with England...... But he could not prevail on the States to adopt so vigorous a plan, and by them he was compelled, much against his will, to begin his operations with the siege of Bonn, a considerable fortified town on the Lower Rhine. "

" Having been obliged to adopt this secondary plan of operations, Marlborough set about its execution with his usual vigour and activity. He landed at the Hague on the 17th March ; and having completed his arrangements there, he set out for Bonn at the head of forty battalions and sixty squadrons, with one hundred guns, leaving Auverquerque with the remainder of the army to form a corps of observation between Liége and Bonn. "

" Marlborough arrived in the vicinity of Bonn on the 20th April, but thanks to the dilatoriness of the Dutch, it was not until the 3rd May that the trenches could be opened. "

Meanwhile the following British troops, under Lieutenant-Generals Lumley and Charles Churchill, assembled at Aerschot, whence they marched to Ruremond on the Meuse, arriving at Maesyck by the 7th of May :—

Lumley's Horse	(1 D.G.)
Wood's ,,	(3 D.G.)
Cadogan's ,,	(5 D.G.)
Wyndham's ,,	(6 D.G.)
Schomberg's ,,	(7 D.G.)
Raby's Dragoons	(1st D)
Teviot's ,,	(2 D)
Ross' ,,	(5th Lancers)

1st Battn. (1st Guards)
Orkney's (R. Scots)
*Portmore's (2nd Queen's)
Churchill's (Buffs)
Webb's (8th King's)
*Stewart's (9th)
North and Grey's (10th)
*Stanhope's (11th)
*Barrimore's (13th)
Howe's (15th)
Stanley's (16th)
*Bridge's (17th)
Fred Hamilton's (18th)
Rowe's (21st)
Ingoldsby's (23rd)
Marlborough's (24th)
Ferguson's (26th)
Huntingdon's (33rd)
Meredyth's (37th)

"In order to raise the siege of Bonn, Villeroi determined to suddenly threaten Maestricht and to then turn on Liége and to endeavour to carry that place before Marlborough could arrive to its assistance."

The French army on the 8th of May invested Tongres with 3000 men. The garrison consisted of only two regiments of foot, a British and a Dutch, and after a gallant resistance for 28 hours they were compelled to surrender.

Meanwhile Lumley and Churchill made a forced march of 35 miles and effected a junction with Auverquerque at Maestricht on the 9th.

On the 15th of May Villeroi appeared before Maestricht with the intention of attacking the Allies, but not liking the look of things, he retired to Tongres.

Meanwhile Bonn had surrendered to Marlborough, and the Duke hastened to join Auverquerque's army at Maestricht, six

* Sent to Portugal at the end of this year.

battalions of foot and three regiments of English dragoons following him from Bonn.

"Upon his Grace's coming out of Maestricht (to the camp), the cannon was discharged round the town and in the evening the army and artillery was drawn out, and a triple discharge made of all the cannon and small arms for the taking of Bonn."

The following letter, dated the 27th May, 1703, from Marlborough to the Duke of Queensberry, shewing the arrears of pay etc. due to a regiment of this time, is interesting:—

"My Lord,

Having read the enclosed memorial from Colonel Rowe, who produces a certificate from the treasury that there is upwards of £5500 due to his regiment......, I could not refuse recommending his request to your Grace, since it must needs be a very great hardship to the regiment to have so great an arrear, and that it would much contribute to the service if some parts of it were paid, to enable the Colonel to better clothe his regiment, and the officers to support themselves in the army.

I am, My Lord,
Yours Grace's etc.
M."

On the 25th of May the army under Marlborough marched to Hannut, and the enemy, remaining on the defensive, and conforming their movements to Marlborough's, gradually fell back on Huy. Marlborough now contemplated the capture of Antwerp as well as Huy, but the noncompliance with his orders on the part of the Dutch General Cohorn, caused delay, and forced him to change his plans and to attack Antwerp forthwith.

The Allies struck their camp at Hannut on the 20th June, and, repassing the Jecker, crossed the Demer at Hasselt. The French, on discovering what was happening, marched by Diest on Antwerp. The success of Marlborough's plans unfortunately depended on the co-operation of several distinct corps advancing from opposite directions. The Dutch Generals Cohorn, Spaar and Opdam again failed to carry out their orders, the latter allowing himself to be surprised and surrounded, with a loss of 4000 killed and wounded, 600 prisoners and eight guns; which disaster completely upset Marlborough's plans for the capture of Antwerp. But in spite of this he conceived a fresh attack on the place.

Marlborough's main army was at Bavin on the 6th of July, and on the 20th advanced to Hoogstraeten. The French, however, declined the battle. Next morning the Allies again advanced in order of battle, but once more the French declined to fight, and fell back into the lines covering the town, Marlborough with 400 horse following them right up to the entrenchments. He was eager to attack, but his propositions were, as usual, vetoed by the Dutch Deputies.

In a letter to Mr Stepney, dated 11th of August, Marlborough writes about the "disappointment we have met with in our designs on Flanders and against Antwerp; this is partly owing to M. de Cohorn's stubbornness and the dissensions among those generals, which has obliged us to return towards the Meuse, where we are now making all necessary preparations for attacking Huy."

On the 15th of August the Duke's Army was encamped at Val-Notre-Dame, within half a league of Huy, and the next day the siege was commenced by the Prince of Anhalt and Brigadier Hamilton.

Ross' Dragoons in the brigade commanded by Brigadier Ross, and fifteen battalions under Lieutenant-General Somerfeldt, on the 16th crossed the Mehaigne, and encamped near the Meuse to secure the bridge, and to preserve the communication between the main army and the troops carrying on the siege on the other side.

Huy surrendered after a fourteen days' siege, and on the following day, at a Council of War, Marlborough strongly urged the attacking of the French lines. The Dutch, however, would not hear of this course, but insisted on the siege of Limburg.

Marlborough having to give way, the 5th of September found the Allied Army encamped at Hannut, less than two leagues from the French grand camp.

The Allied cavalry were set to collect and make fascines, as if an assault were about to be made on the lines, while Marlborough reconnoitred the position. The French, expecting an immediate attack, manned the defence, and Marlborough was able to perceive that the position was too strong. The following day the Allies fell back to St. Trond, to cover the attack on Limburg, which fell into their hands on the 27th.

The Duke of Marlborough shortly after left the Army for the Hague and England. The troops went into winter quarters at the end of October.

Marlborough was so disgusted at the perversity and obstruction of the Dutch authorities, that on his arrival in England he was determined to resign the appointment of Commander-in-Chief, and it was only the personal intervention of Queen Anne that induced him to reconsider his decision.

In the Order of Battle for the campaign of 1703, Ross' brigade of Dragoons were in the Right Wing of the 1st Line and consisted of the following regiments.

Raby's Dragoons	(1 R. D.)
Tuviot's "	(Scots Greys)
Ross' "	(5th Lancers)
A Foreign regiment.	

CHAPTER IV.

*Campaign of 1704.—Change of title.—March into Germany.—
Battle of the Schellenberg.—Blenheim Campaign of 1705.—
The forcing of the French Lines.*

It will be remembered that Ross' Dragoons landed in the Low Countries in 1702 with six troops only, two being left behind in Ireland. In letters to the Duke of Ormond and to Lieutenant-General Earle, dated St. James', 6th of January, 1704, these two troops are ordered by Marlborough to " be forthwith put into the best condition that may be, and sent to the sea coast in order to be embarked. " Every effort was made in England during the winter to bring the regiments in the Low Countries up to strength.

We now come to the granting of a special title to the Regiment by Queen Anne. Since its formation it had been known by the names of its commanding officers—the Dragoons of the gallant Wynne and of Ross. The latter officer petitioned that his Regiment should be known as The Royal Dragoons of Ireland, and in a letter dated Whitehall, 22nd February, 1704, Marlborough writes to the Duke of Ormond :— " Brigadier Ross acquaints me that your Grace has given your consent to his regiment's having the title of Royal Dragoons of Ireland, wherein, however, I shall not venture to do anything until I have it from you ; " and in a subsequent letter dated the 9th March, Marlborough writes:— " Brigadier Ross is sensibly obliged to your Grace for your readiness in gratifying him in his request. "

For the coming campaign the French plans were to follow up the successes of the previous year in Germany. Villeroi was to remain in the Low Countries while Tallard with 45,000 men was to join the Elector of Bavaria, who had some 45,000 Bavarians and French, at Ulm. In addition the French had 10,000 men on the Moselle, available to re-inforce either Villeroi or Tallard.

Marlborough, foreseeing the French strategy, determined to

carry the war into the heart of Germany, leaving the defence of the Low Countries to the Dutch.

These plans he confided only to those whom he could implicitly trust.

In April Marlborough, who had already visited the Hague during this year, again set out for the Continent, accompanied by his brother Churchill, who had been promoted General to command the British contingent.

It was only with the greatest difficulty that he persuaded the Dutch to consent to his marching on the Moselle. "Indeed, it was only on his threatening to proceed with the British troops alone that they gave way."

The following is a list of the British contingent of the 1704 campaign :—

Lumley's Horse	(1 D. G.)
Wood's do.	(3 D. G.)
Cadogan's do.	(5 D. G.)
Wyndham's do.	(6 D. G.)
Schomberg's do.	(7 D. G.)
Hay's Dragoons	(Scots' Grays).
Royal Dragoons of Ireland	(5 Lancers).
1st. Battn.	(1st. Foot Guards).
Lt. Gen. Hamilton's Foot.	
Churchill's	(Buffs).
Webb's	(8 th.)
Lord North & Grey's	(10 th.)
Howe's	(15 th.)
Stanley's	(16 th.)
Hamilton's	(18 th.)
Rowe's	(21 st.)
Ingoldsby's	(23 rd.)
Marlborough's	(24 th.)
Ferguson's	(26 th.)
Meredith's	(37 th.)

Also 34 guns, 4 howitzers and 21 pontoons.

Having no Dutch troops with him, Marlborough was not troubled with the Dutch Deputies, who had impeded him so much in 1703; and by appointing his brother to command the contingent, he silenced the pretensions of Foreign commanders.

During the middle of March the Dutch army of 60 battalions and 100 squadrons, made Maestricht their rendezvous, and remained in that neighbourhood under the command of Auverquerque throughout the campaign.

On the 8th of May the British troops commenced their famous march into Germany.

From Maestricht Marlborough writes on the 11th of May to Mr. St. John: "I may venture to tell you (though I would not have it public as yet) I design to march a great deal higher into Germany."

On the 17th May the British contingent crossed the Maes at Ruremond. The infantry and artillery under Churchill continued their march by Sinzig and Andernach, arriving at Neuendorf close to Coblentz on the 26th; while, on the same day Marlborough, having gone with the mounted troops to inspect Bonn, crossed the Moselle and Rhine at Coblentz.

On the 21st, Marlborough, writing to Sir Charles Hedges says:— "We have certain advice to-day that the Maréchal de Villeroi passed the Meuse on Monday last with thirty battalions and forty five squadrons of the best troops in the Low Countries, with orders, as I am informed, to observe me wheresoever I march." In addition to these troops Tallard had succeeded in sending a re-inforcement of 10,000 men to the Elector of Bavaria.

Churchill's force of infantry and artillery marched from Neuendorf to Branbach; the Duke with the Cavalry marched to Schwalbach and thence to Castel, opposite Maintz.

Churchill resumed his march on the 30th, and on the 3rd of June encamped at Castel, where they were inspected "by the Elector and the various allied generals present, and their appearance, discipline, and excellent equipment, and the remarkably orderly manner in which the march had been conducted, excited universal surprise and admiration."

Marlborough with his horse and dragoons had meanwhile crossed the Maine on May the 30th, and marched via Zwingenberg and Weinheim, to Ladenburg, where they crossed the Neckar and encamped.

The infantry and artillery, resuming their march, encamped a little below Heidelberg on the 8th of June, having marched by Zwingenberg, Weinheim and the Neckar.

Marlborough and the cavalry left Ladenburg on June the 6th, and marched via Wisloch, Eppingen and Gross Gartach, crossing the Neckar on the 9th, and arriving at Mondesheim the following day. On June the 11th the cavalry marched to Gross Heppach, where they were reviewed by the Duke in the presence of the

Prince Eugene, "who was loud in his praise of their remarkable efficiency." "Their good order and fine appearance after so long and rapid a march, and the excellent condition of their clothing, accoutrements, and horses" especially struck him with admiration.

The Prince Eugene is reported to have said: "My Lord, I never saw better horses, better cloaths, finer belts, and accoutrements; yet all these may be had for money, but there is a spirit in the looks of your men, which I never yet saw in any in my life." Marlborough was not to be outdone in politeness and is said to have replied: "Sir, if it be as you say, that spirit is inspired in them by your presence."[1]

The Cavalry with Marlborough next marched to Ebersbach, and on the 16th of June reached Gross Saxenheim, and thence on the 21st "to a point between Lannsheim and Urspring." "The next day he formed a junction with the Army of Prince Louis of Baden near Westerstetten," and on the 24th the combined force encamped with their right at Elchingen and their left at Langenau, where they halted to allow Churchill's columns to join them, which they did on the 27th of June.

The enemy, under the Elector of Bavaria, continued in camp at Dillingen and Lavingen, two leagues away.

The relative strength of the two forces was as follows:—

Allies, under Marlborough and Prince Louis of Baden:—

96 battalions.
202 squadrons.
44 field guns.
4 howitzers.
24 pontoons.

Elector of Bavaria (including the garrison of Donawert):—

88 battalions.
160 squadrons.
90 cannons.
40 howitzers and mortars.
30 pontoons

The object of Marlborough's recent movements had been to obtain the fortified town of Donawert, at the junction of the river Wernitz and the Danube, the possession of which place would give the allies a bridge over the Danube and afford them a place

[1] *The Field of Mars.*

of arms for the invasion of Bavaria, the road into which country it covered."

The recent northerly movements of the Allies, however, betrayed the Duke's intentions to the Elector of Bavaria, who, to secure the passage leading through Donawert, detached a force of 10,000 infantry and 2,500 cavalry, under General Count d'Arco, to occupy the hill of Schellenberg, which commanded the town from the north bank of the Danube.

Marlborough at once urged Prince Louis to consent to an advance upon Donawert before reinforcements from France could reach the Elector.

On the 30th of June the Allied camp at Giengen was struck, and the army marched to Lauthausen and Balmershofen, and on the following day to Amerdingen and Onderingen. These movements left no doubt in the Elector's mind as to the object of the Allies, and he sent a strong force to the aid of d'Arco on the Schellenberg, which position was now strongly intrenched.

Marlborough and Prince Louis on alternate days assumed supreme command of the Army, and on July the 2nd it was Marlborough's turn. He saw the necessity of attacking the position on this day at all risks, and not leaving it for his colleague to attempt the following day. For Prince Louis would probably waste the day, and allow the enemy to receive reinforcements and complete their defences.

Having established a field hospital at Nordlingen for the wounded, Marlborough detailed a force of 35 squadrons and 130 men out of each battalion of the left wing to form the van of the attack.

Preceded by a mounted force of several squadrons under Quarter-Master-General Cadogan to mark out a camp, and by the pioneers and a pontoon train, the detachment marched off at 3 a.m. on the 2nd of July. The foot, 5850 strong, was commanded by Lieut.-General Goor. The 35 squadrons were commanded by Lieut.-Generals Lumley and Hompesch, and under them Major Generals Schuylemberg, Wood, Count Erbach, Vittinghoof, and Brigadiers the Prince of Saxe Heibourg and Bothmar.

The remainder of the Army followed at 5 a.m. under the command of Prince Louis in two columns, the objective being a height that was between Ebermergen and Wernizstein, while the artillery, also in two columns, marched to Harburg with instructions to wait there, for further orders, without unharnessing their teams. All the baggage followed the artillery, and strict orders were given that no baggage whatever should interfere with the march of the columns.

About 8 a.m. Cadogan with the advanced party arrived within two miles of Donawert, driving in the enemy's "grand guard" divided into three squadrons, while the Quartermasters of the regiments marked out a camp by Ebermergen.

At 9 a.m. Marlborough came up, and reconnoitring with the advanced party towards Donawert, discovered that the enemy had made two fronts, as if they expected to be attacked in two different places. Prince Louis, Lumley and Goor joined the Duke in this reconnaissance, "and they were now so near the enemy that they were exposed to their great shot, which began to pour upon them very plentifully from several of their great works." However, this did not prevent them from making a careful reconnaissance of the ground and entrenchments, and they also discovered a camp in preparation on the other side of Donawert.

At 12 noon Goor's detachment arrived at the Wernitz, and soon after the whole army came up, when Goor's force was ordered to cross the river by a stone bridge and to form up on the other side, and the pontoons were immediately placed for the whole army to follow.

The Duke having judged it absolutely necessary to attack the enemy at once, before reinforcements could arrive, ordered the 35 squadrons to go into the adjoining woods to get fascines for the infantry.

From a hill called Boschberg, Goor's detachment, supported by eight entire battalions under Major-Generals Withers and Beinheim, and with eight more battalions under Count Horn, as a reserve, were ordered to the attack.

About 4 p.m. these troops reached the bottom of the Schellenberg, where they made a short halt to receive the fascines from the horse.

To describe the situation :—The Schellenberg hill was about two English miles in circumference at the base, and had generally a gradual ascent. At the top was a plateau, half a mile across, where the enemy were encamped in several lines. The intrenchments began at Donawert on the south, and ran round the top of the hill to a point where it joined the Danube on the other side. The hill was more accessible on the northern side and there were

The hill was more accessible on the northern side, and there were also the Nieuburg woods; the entrenchments on this side were consequently stronger.

At 6 p.m. Marlborough ordered Goor to commence the attack. The detachment then moved up the rising ground in six lines, four being of foot and two of horse, "the English being on the left and close to the wood."

The fascines were then distributed to officers and soldiers alike, who were ordered to carry them until they could throw them down in the enemy's entrenchments, and to move on steadily and reserve their fire. In spite of the enemy's heavy artillery fire, the foot advanced resolutely and calmly to within eighty paces of the intrenchments, and were gallantly sustained by the horse and dragoons; Lumley keeping close with eighteen squadrons in the first line, and Hompesch bringing up the other seventeen in the second. The enemy now used case shot, doing great execution. Goor was killed, and a great many other officers. The troops, however, kept gallantly on, and the enemy essaying a bayonet charge, were driven back by Her Majesty's Guards. The enemy, finding they were not attacked elsewhere, now threw all their strength against the English attack. Some of the attacking infantry shewing signs of retiring, "Lieutenant-General Lumley and Major-General Wood with the first line of horse and dragoons moved up, and stood so close, and animated the foot so much by their brave example, that they rallied and went on again."

"The horse were now, however, so near a mark for the enemy's shot, that a great many fell or were disabled." "Major-General Wood received a wound in his breast from a musket shot, as did Colonel Palmer in his body. Count Erbach and Colonel Cadogan had their horses shot under them. The Hereditary Prince of Hesse was shot in the breast, and so was the Prince of Saxe Heidelberg. Brigadier Bothmar and a great many other horse officers were likewise wounded. But for all this, both the lines of horse continued firm, and encouraged the infantry to keep their ground and to press the attack with renewed vigour."

And so the battle raged. Lumley eventually ordered Lord John Hay's dragoons to dismount and charge on foot, which they were preparing to do when the enemy gave way and the infantry entered the trenches. The Cavalry now pursued, and putting the enemy to flight, killed a great many and captured thirteen colours.

In writing his despatch of the battle of Schellenberg to Mr Secretary Harley under date 3rd of July, Marlborough says:—

" All our troops in general behaved themselves with great gallantry, and the English in particular have gained a great deal of honour in this action, which I believe was the warmest that has been known for many years, the horse and dragoons appointed to sustain the foot standing within musket shot of the enemy's trenches most of the time. "

Milner puts the British losses during this action at 32 officers and 420 men killed, and 83 officers and 1001 men wounded, which was about one-third of the losses of the Allies. Amongst the casualties of the Allies were 8 general officers killed and 9 wounded.

In this action Major Caldwell and Cornet Hamilton of the Royal Irish Dragoons were wounded, and four men of the Regiment were killed and nineteen wounded.

The French Armies in the Low Countries meanwhile had crossed the Rhine, and hearing of the Elector's defeat at Schellenberg, Tallard hurried on and joined hands with the Bavarians.

On August the 9th Prince Louis of Baden, much to the relief of Marlborough, left the Allied camp with 15,000 men for the siege of Ingolstadt.

Marlborough now decided to give Tallard battle. The dangers of attacking a superior force were pointed out to him, and he replied that he knew the difficulties, but a battle was absolutely necessary: " and, " he said, " I rely on the discipline of my troops. "

Alison, in his *Life of John, Duke of Marlborough* gives the strength of the Allies before the battle of Blenheim as 66 battalions, 164 squadrons and 66 guns, in all 56,000 men. The French and Bavarians he gives as 84 battalions, 147 squadrons, a total of 60,000 men.

The two hostile camps lay some five miles apart. The intervening ground was a plain of varying breadth bordered by a line of woods and the Danube. "This plain is cut by a succession of streams running down at right angles to the Danube, no fewer than three crossing the line of march between the Kessel and the French position. The first of these, the Reichen, cuts a ravine through which the road passed close to the village of Dapfheim ; and Marlborough, seeing that at this point the enemy could greatly embarrass his advance, sent forward pioneers to level the ravine, and occupied the village with two brigades of British and Hessian infantry. "

At two o'clock on the morning of August the 13th the Allied Army passed the Kessel in a dense white mist. They marched in eight columns, the two outer ones on each flank consisting of cavalry, and the inner ones of infantry. At Dapfheim the army halted, and the two outlying brigades, reinforced by eleven more British battalions, formed a ninth column under Lord Cutts, whose orders were to cover the march of the artillery along the great road on the extreme left, and in due time to attack Blenheim. With this column went Major-General Wood and Brigadier-General Ross "with fifteen squadrons of Dragoons to sustain them," the first line of cavalry being formed by Ross' Dragoons, and the second by part of Wood's brigade.

Marlborough, on the left, occupied the ground from the Danube to Oberglau, while Eugene prolonged the line to Lutzingen. At 6 a.m. the French advanced posts were driven back, and at 7 o'clock the Allies were on the Nebel, and in full view of the enemy's camp.

Marlborough in writing home his despatch says that the enemy "we found did not expect so early a visit." In truth they did not. Their cavalry had dispersed to gather forage, and no precautions had been taken against attack. The French outposts came hurrying back and all was confusion. Tallard with his defective eyesight was able to make out the red coats of Cutts' British soldiers, and he knew that there on his right the heaviest fighting was to be expected. He therefore lost no time in occupying Blenheim.

The story of Blenheim has been told so often that but a slight account will suffice here.

Amongst his dispositions, Tallard had barricaded with wagons that side of Blenheim next the Danube, as being the most open against the British horse. From the village of Blenheim to that of Oberglau were posted 80 squadrons of French cavalry and two brigades of infantry. The last named village was also occupied by fourteen of the enemy's battalions, among which were three regiments of Irishmen commanded by the Marquis de Blainville.

On the side of the Allies, Lord Cutts with his twenty battalions still continued on the left of all towards Blenheim, drawn up in four lines; and Wood and Ross, with their fifteen squadrons, were in two lines behind that body of foot. As before mentioned, Ross' brigade of Dragoons were in the first line of cavalry while Wood's formed the second line.

About eight o'clock, Tallard's guns opened fire, whereupon Eugene hurried off to his command on the right, and Marlborough personally superintended the posting of his batteries. "The chaplains came forward to the heads of regiments and read prayers; and then the Duke mounted and rode down the whole length of his line. As he passed, a round shot struck the ground under his horse and covered him with dust. For a moment every man held his breath, but in a few seconds the calm figure with the red coat and the broad blue ribbon reappeared, the horse moving slowly and quietly as before, and the handsome face unchangeably serene."

At last, at about half past twelve, an aide-de-camp galloped up from Eugene to say that all was ready. Marlborough at once directed Cutts to commence his attack on Blenheim, and ordered all the lines to cross the Nebel.

It was nearly one o'clock when Cutts' leading brigade, with Brigadier Rowe at the head, advanced across the stream to Blenheim. At thirty paces' distance they were received by a murderous fire, but Rowe had given orders that no shot was to be fired until he struck the palisades, and that the village was to be carried by cold steel. The British, without firing, reached the palisades which Rowe struck with his sword, and the troops, pouring in a volley, rushed forward and endeavoured to drag down the timbers and force an entrance. In a few minutes the gallant Rowe fell mortally wounded; his Lieutenant-Colonel and Major were killed in the attempt to bring him off, and the brigade, shattered to pieces, fell back in disorder. As they retired they were charged in flank by several squadrons of gendarmes, who captured the colours of Rowe's regiment, but, pursuing too far, were stopped by a terrific fire poured into them by the Hessian Foot, and driven back pell mell by a charge of five British squadrons (apparently of Ross' brigade). These were, however, attacked in turn by a superior force of fresh cavalry and forced across the rivulet. Again the Hessian infantry poured a tremendous fire into the pursuing cavalry, and utterly routing them, retook the lost colours of Rowe's regiment.

Marlborough, to avoid a useless sacrifice of life, ordered the infantry regiments to remain under cover and to keep the defenders of Blenheim occupied, while Wood's and Ross' brigades were directed to join the centre, where he intended to make his main effort.

As soon as the first line of infantry had formed on the far side of the Nebel, Marlborough ordered the cavalry over. They reached the stream in good order, but got into difficulties in the morass between the two branches of the stream. The British squadrons had the most difficult place on the left, and in addition were heavily shelled by the enemy's guns near Blenheim. However, the horse struggled on, and with the aid of fascines, got over, and formed up in front of the infantry. The cavalry now met with an extremely warm reception, the enemy's infantry and artillery firing heavily into their left flank from Blenheim, while the French cavalry thundering down the slope of the hill, charged them in front.

The English squadrons were overwhelmed, and driven back nearly to the banks of the stream, where the infantry fire checked the victorious French. Then the Prussian general Bothmar fell upon the disordered French with the second line cavalry, and drove them in confusion behind the Maulweyer, where for some time they held their own, but were penned in behind the stream, the head of which they dared not pass for fear of being charged in flank.

The Duke of Marlborough sent for five more squadrons from Major-General Wood to strengthen the British cavalry, and these on coming up with Brigadier Ross, passed the Nebel. The rest of the cavalry was meanwhile crossing the rivulet with some difficulty; for on its far side was an enemy already formed and supported by several guns. Yet, by the brave example and the diligence of the officers, and the eagerness of the men, all passed over by degrees and held their ground. The English cavalry got over on the left under Lumley; that of the Dutch was in the centre, and the Danes on the right. Bulow with the second line of cavalry followed, and stretched away towards Oberglau.

On the right, the Danish and Hanoverian cavalry were now engaged in a severe struggle near Oberglau, and the Prince of Holstein-Beck commenced his attack on that village. He was met by a fierce counter-attack from the famous Irish Brigade, and things would have gone ill with him had not Marlborough hastened up with fresh infantry and artillery, and forced the enemy back into Oberglau, thus securing the passage for the centre of the Allied Cavalry.

It was now three o'clock; and the Duke sent Lord Tunbridge

to Prince Eugene's wing for news. From his A.D.C. he learnt that the Prince was holding his own and no more.

By four o'clock Marlborough had got the whole of his left wing of the Allied Army across the Nebel, his cavalry being drawn up in two lines in front of the infantry, the latter being ranged at intervals to allow of the cavalry passing through, in case of repulse.

On the opposing side the French had intermingled nine battalions of foot from the second line amongst their horse, and against these Marlborough sent three Hanoverian battalions and a battery of artillery.

For a long time the young French infantry stood the storm of shot. To relieve them, Tallard ordered the squadrons on their left to charge, but they refused, and fled before a charge of the Allied Cavalry, who rode into the hapless battalions of infantry and swept them out of existence. At the same time, a wide gap was left in the French line by the cavalry on Marsin's right, who, having seen their flank exposed, fell back on his centre. About five o'clock Marlborough ordered the "charge," and placing himself at the head of the Allied Cavalry, the two lines of cavalry pressed up the slope, sword in hand, to the attack. The French, firing a feeble volley from the saddle, broke, and falling back on their supports, carried all away with them in confusion, and fled wildly in the direction of Hochstadt and the Danube, pursued by Hompesch's division of horse, who drove them into the river, where hundreds were drowned, hundreds cut down, and a vast number taken prisoners.

Meanwhile, Marsin and the Elector on the French left, seeing the collapse of Tallard's army, set fire to Oberglau and retreated, followed by Eugene.

All this while the village of Blenheim still held out and gave employment to the English infantry. Directly the Allied Cavalry had beaten and cleared the field of that of the enemy, General Churchill had marched the foot upon the village and surrounded it. The finest troops of France were locked up in Blenheim without orders of any kind. At last they tried to break out to the rear of the village, but were headed back by the Royal Scots Dragoons. They made a final attempt to cut their way out towards Oberglau, but here they were checked by the Royal Irish Dragoons—who had been sent from the pursuit to Blenheim—and were compelled to seek cover behind the houses and enclosures.

Churchill now ordered Cutts to make another attack on the side of the Nebel, while Orkney assailed the church-yard on the west with eight battalions, and Ingoldsby with four more, and the Royal Irish Dragoons endeavoured to force their way into the village by the opening opposite Oberglau. Supported by artillery fire they succeeded in entering; but the French made such a determined resistance that they had to retire. The attack was about to be renewed when the French called a parley, and after some demur, surrendered at about eight o'clock in the evening, twelve squadrons of dragoons and twenty four battalions laying down their arms.

The field now being entirely cleared of the enemy, the Allied army was drawn up with the left to Sondersheim and the right towards Morselingen; "and the soldiers were to lie all night upon their arms on the field of battle," though as a matter of fact, several regiments quickly possessed themselves of the enemy's tents and the food they contained, while a hundred fat oxen, killed and skinned (apparently the day's rations for the French Army) proved a boon to the famished soldiers.

The total loss of the Allies amounted to four thousand five hundred killed, and seven thousand five hundred wounded, of which number the British lost six hundred and seventy killed, and some fifteen hundred wounded.

The day was closing on the field of Blenheim "when Marlborough borrowed a leaf from a commissary's pocket-book and wrote a note in pencil to his wife, the message and the handwriting both those of a man who is quite tired out."

"13 August 1704.

"I have not time to say more, but to beg you will give my duty to the Queen, and let her know her army has had a glorious victory. Monsr. Tallard and two other generals are in my coach, and I am following the rest. The bearer, my aide-de-camp, Colonel Parke, will give her an account of what has pass'd. I shall doe it in a day or two by another more at large.

Marlborough."

The day after the battle the whole of the victorious army marched by Hochstadt and camped between Steinheim and Wittislingen.

No regimental lists of the casualties amongst the rank and file appear to exist.

In the Blenheim Roll Call, the following officers of the ROYAL IRISH DRAGOONS, are shewn as having been wounded :—

Captain Hugh Caldwell
Cornet Jno. Hunter
Cornet Edw. Hamilton
Quarter-Master Jno. Skelston
Adjutant David Ross.

In the "*Annals of Queen Anne*," the following evidently inaccurate statement appears : "the Dragoons suffered so little, that only the Adjutant of Ross' Regiment was much wounded."

In his letter to Mr. Secretary Harley, dated at Hochstadt 14th August 1704, giving a report of the battle, Marlborough writes : "and in the village of Blenheim, which the enemy had intrenched and fortified, and where they made the greatest opposition, we obliged twenty-six battalions and twelve squadrons of dragoons to surrender themselves prisoners at discretion. We took likewise all their tents standing, with their cannon and ammunition, as also a great number of standards, kettle-drums, and colours in action, so that I reckon the greatest part of M. Tallard's army is taken or destroyed. The bravery of all our troops on this occasion cannot be expressed ; the generals as well as the Officers and soldiers behaving themselves with the greatest courage and resolution, the horse and dragoons having been obliged to charge four or five several times."

Grose, in his "*Military Antiquities*," writing of the ROYAL IRISH DRAGOONS, says : "In consequence of the marked good behaviour of this Regiment at the battle of Hochstet (Blenheim), in August 1704, three additional troops were put upon the establishment, making its strength to consist of nine troops. The three kettle-drums which were captured from the French by the Regiment at this memorable engagement, were directed by the Duke of Marlborough to be carried at the head of the ROYAL DRAGOONS of IRELAND."

The following list of Officers of the ROYAL DRAGOONS of IRELAND, and the bounties they received after the campaign, is from the "Blenheim Bounty Roll." It is to be noted that Officers and men who were wounded received double the amount allowed to their rank.

BATTLE OF
BLENHEIM
August 13th. 1704

Statute Miles
½ ¼ 0 ½ 1

British and Allies.
French and Bavarians.

Morselingen
Lutzingen
Schwenenbach
Berghausen
Ober Glanheim
Weilheim
Wolpertstetten
Unter Glanheim
Deisenhofen
Schweningen
To Donauwörth
Sonderheim
Blindheim or Blenheim
Greinheim
HÖCHSTÄDT
From Dillingen
DANUBE
DANUBE
PRINCE EUGENE OF SAVOY
EL. OF BAVARIA
Ml. MARSIN
MARSHAL TALLARD
DUKE OF MARLBOROUGH

HISTORY OF THE FIFTH LANCERS.
A. DOUBLEDAY & CO., LTD.

COLONEL

Charles Ross	£105.	0.	0.	Bounty

LIEUT: COL: COMMANDING

Owen Wynne	£ 78.	10.	0.	„

MAJOR

Robert Hunter	£ 61.	10.	0.	„

CAPTAINS

Jno. Hill.	£ 46.	10.	0.	„
Ric. Gore.	£ 46.	10.	0.	„
Hugh Caldwell (Wounded)	£ 93.	0.	0.	„

CAPTAIN LIEUTENANT

Robt. Drury	£ 27.	0.	0.	„

LIEUTENANTS

Chas. Beatty	£ 27.	0.	0.	„
Jno. Johnston	£ 27.	0.	0.	„
Mat. Watts	£ 27.	0.	0.	„
Danl. Boisragon	£ 27.	0.	0.	„

CORNETS

Jno. Dunbar	£ 24.	0.	0.	„
(Alex) Abercromby	£ 24.	0.	0.	„
Jas. Hamilton	£ 24.	0.	0.	„
Jas. Poé	£ 24.	0.	0.	„
Jno. Hunter (Wounted)	£ 48.	0.	0.	„
Edwd. Hamilton (Wounded)	£ 48.	0.	0.	„

QUARTER-MASTERS

...... Brown	£ 16.	10.	0.	„
(Geo.) Mackean.	£ 16.	10.	0.	„
(Ric.) Johnston.	£ 16.	10.	0.	„
Ric. Dunbar	£ 16.	10.	0.	„
Jno. Skelston (Wounded)	£ 33.	0.	0.	„
David Ross (Wounded)	£ 33.	0.	0.	„

CHAPLAIN

Simon Babe	£ 20.	0.	0.	„

ADJUTANT

David Ross (Wounded)	£ 30.	0.	0.	„

SURGEON

Wm. Cocksedge.	£ 18.	0.	0.	„

NON-COMMISSIONED OFFICERS AND MEN

12 Sergeants each	£ 2.	10.	0.	„
18 Corporals „	£ 2.	0.	0.	„
268 Dragoons „	£ 1.	10.	0.	„

(Including drummers and hautboys).

In connection with the battle of Blenheim it is to be noted that the French call it Hochstet, after the village (Hochstadt) near Blenheim. There is an old print now in the possession of the officers of the Regiment, published in 1800, and dedicated to General Lord Rossmore, who was at the head of the Regiment from 1787 to 1799, in which is depicted a 5th ROYAL IRISH DRAGOON bearing in a corner a green standard with the Harp and Crown and the word Hochstet.

It is unnecessary to follow the events of this year any further, suffice it to say that in Novemher the English troops were sent into Winter Quarters for the rest they had earned so well.

The result of the famous campaign of Blenheim, rightly grouped with Creçy, Poitiers, Agincourt and Waterloo, was a crushing blow to the French. For well nigh forty years had their armies triumphed in every quarter of Europe. But now an English general had administered a decisive and humiliating defeat. Marlborough had outwitted the Marshals of France by his march to the Danube, had twice attacked a vastly superior force, and had utterly destroyed one army, and driven the other in a headlong flight to the Rhine.

" Blenheim " is born on the Standards, colours and appointments of the following regiments which fought in that battle and are still existent in the British Army :—

1st Dragoon Guards	The Buffs.
3rd " "	Liverpool Regiment.
5th " "	Lincolnshire "
6th " "	E. Yorkshire Regiment.
7th " "	Bedfordshire Regiment.
R. Scots Greys	R. Irish Regiment.
5th Lancers	R. Scots Fusiliers.
Royal Artillery	R. Welsh Fusiliers.
Royal Engineers	South Wales Borderers.
Grenadier Guards	1st Scottish Rifles.
Royal Scots	1st Hampshire Regiment.

The order of Battle for the Campaign of 1704 places the British Cavalry in the 1st Line of the Left Wing, the IRISH DRAGOONS being brigaded with the Scots Dragoons.

The following is an extract from a letter to Mr. St. John from Marlborough, dated 22nd October, 1704, at Weissemburg:—" But I must observe to you, in reference to our English Horse and Dragoons, that having clothed entirely new this year, and lost a great many men with all their accoutrements in the two actions, the Officers of horse humbly hope Her Majesty will be pleased to allow them twenty pounds a horse, and the dragoons fifteen, towards enabling them to repair this great loss."

The Campaign of 1705 does not call for much note.

In the order of Battle of the Allied Armies for this campaign, the British contingent under Marlborough were in the first line of the Right Wing. The ROYAL IRISH DRAGOONS with the Scots Dragoons, each regiment consisting of three squadrons, were brigaded together on its extreme right. Ross was now a Major-General.

The regiment was engaged in the forcing of the French Lines on the 18th of July.

For this expedition Marlborough made up a force of some eight thousand men, divided into two columns. One column was placed under the command of the Count of Noyelles, and consisted of twelve battalions (four of which were English), and the first line Cavalry of the Right Wing, the whole of which were British squadrons belonging to the :—

Royal Scots Dragoons.
ROYAL IRISH DRAGOONS.
The present 1st Dragoon Guards.
3rd " "
5th " "
6th " "
7th " "

and six fieldpieces and workmen, and materials for making bridges.

Lieutenant-General Scholten commanded the other column, and had with him the 2nd Line Cavalry of the Right Wing, and also workmen and materials for bridging.

The utmost secrecy was observed in the preparations, even the corps composing the force knowing nothing of each other or of the work before them.

At 9 o'clock on the night of July the 17th, the two columns moved off in the strictest silence. Noyelles marched towards the castle of Wanghe before Elixheim, while Scholten marched on the village of Neerhespen. The instructions given to both commanders were to seize the barriers in the openings in the French Lines opposite the points they were marching on, and with as little noise as possible to overpower the guards defending them. If the columns should not succeed in surprising the enemy, and should find the French prepared for them, they were to advance in full force without waiting for orders, and attack with all the vigour they could. Marlborough with the main army marched an hour after the two leading detachments in support of them. The distance to be covered was about ten miles.

It was an extremely dark night, and Noyelles' guides missed their way, which resulted in a delay of two hours or more, so that his column did not arrive near Wanghe until between three and four o'clock in the morning. He, however, immediately ordered a captain with sixty grenadiers, supported by a Colonel with all the other grenadiers of the twelve battalions, to cross the river Geete and capture the barrier. As the Grenadiers approached the castle, the garrison of thirty men abandoned it and retreated within the Lines. The grenadiers gained the castle and at once pushed on over the river to the barrier at the opening in the Lines ; but this the enemy also abandoned on the approach of the grenadiers. The infantry, following, forded the river and scrambled over the ramparts into the French Lines with such determination and celerity, that three regiments of French Dragoons, camped near by, had not time to oppose them, but hastily retired to Leau. Meanwhile, the opening in the Lines having been enlarged, bridges were quickly thrown over the Geete, and the Cavalry, who had apparently found the banks of the river too steep for fording, crossed with all haste.

Marlborough arrived on the scene during the passage of the Cavalry. But meanwhile, the alarm had been given, and the enemy got together a force of forty or fifty squadrons and eight guns, which they drew up in two lines; and some twenty battalions of infantry were coming up quickly to support their horsemen.

Marlborough took in the whole situation at a glance, and having got over nearly the whole of the cavalry of both columns, he formed them in two lines, and with the British cavalry leading in the first line, amongst them being the IRISH DRAGOONS, he led them personally against the enemy, sword in hand. The French fired a feeble volley from their saddles and broke in confusion. But soon after they managed to rally, and rode a counter attack against the flank of the British squadrons and broke them in their turn. Marlborough, who was riding on a flank, was cut off with his trumpeter. A Frenchman, apparently famed for neither good horsemanship nor swordmanship, galloped furiously at the Duke, and, striking a blow at his head, missed his mark, lost his balance, and was captured by the Duke's trumpeter. The fighting for a little while was fast and furious, but the allied squadrons rallied, and once more charging the French, rode them down and broke them past all reforming, also overthrowing part of the infantry who were coming up in support, and capturing the guns.

Marlborough at once sent on a detachment of dragoons, the IRISH DRAGOONS being amongst them, to pursue the enemy : and they had the good fortune to overtake and capture a good part of their baggage.

Marlborough with the allied army now followed the enemy in his retreat.

The Lines were of a most formidable description, and Marlborough's capture of them was due firstly to his drawing the main body of the French towards the Mehaigne, whither for that purpose he had sent d'Auverquerque with his army, while with the remainder of the Allies he suddenly, and at night, fell upon the two most unguarded posts ; and secondly to the bravery of his cavalry.

The French losses were some two thousand men, amongst the prisoners being two lieutenant-generals, two major-generals, and the entire regiment of Monlve. The Allies also captured eighteen pieces of cannon,—eight of which were triple-barrelled, and were sent to England to be copied.

CHAPTER V.

*Campaign of 1706.—Ramillies.—Marlborough again escapes.—
Capture of the Regiment du Roi.—Campaign of 1707.*

Marlborough, who had been home to England during the winter, returned to the Hague in April, and found a most discouraging state of affairs. The Dutch were backward in their preparations; the contingents of Prussia and Hanover were apparently not forthcoming ; Prince Lewis of Baden was sulking ; everybody was ready with a separate plan of campaign ; and there had been great difficulty in providing the British force with horses to replace the large numbers which had died from sickness.

Villeroi lay safely entrenched with a French Army behind the river Dyle. He knew that the Prussian and Hanoverian contingents had not yet joined Marlborough, and that the Danish cavalry had refused to march to him until their wages were paid ; and he resolved to risk a general action. On May the 19th he left his lines for Tirlemont, on the Great Geete.

Marlborough hastily made arrangements for the payment of the Danish troops, and, concentrating the British and Dutch at Bilsen on the Upper Demer, moved south to Borchloen, at which place the Danes joined him, swelling his force to sixty thousand men, an army but little inferior to that of the enemy. On the same day came the news that Villeroy had crossed the Great Geete and was moving on Judoigne. Marlborough decided to attack him.

At 1 a.m. on Whitsunday the 23rd of May, Quarter-Master-General Cadogan, with six hundred cavalry and all the quarter-masters and camp colours of the army, was dispatched to the Great Geete to mark out a camp for the army by the village of Ramillies; and some two hours later Marlborough followed with the army divided into eight columns. About 8 a.m. on a foggy morning Cadogan's force rode up the heights of Merdop, whence they dimly descried troops moving in front of them. A message was sent to Marlborough, who hastened up and ordered the cavalry on through the mist.

Shortly after 10 o'clock the fog rolled away, and the whole of the French army was seen marching towards them.

Marlborough had appeared to the French a day before he was expected. Villeroy, however, promptly drew up his army on the line of the villages of Taviers, Ramillies, Offus and Autréglise, in two lines facing due East, the line being somewhat over four miles in length.

Villeroy's left, extending from Autréglise (Anderkirch) to Offus, consisted of infantry backed by cavalry. His centre, composed of infantry, stretched from Offus to Ramillies: while on his right, on the plain between the Geete and the Mehaigne, were massed more than one hundred and twenty squadrons of cavalry, with some battalions of infantry. The village of Ramillies covered the left of this expanse, and was defended by twenty battalions and twenty guns, the village being surrounded by a ditch. On the right, the villages were defended by detachments of infantry, and in Taviers were guns. It appeared to Marlborough that Taviers was too far from Ramillies for the maintenance of a cross-fire of artillery, and the cavalry of the French, left secure against attack behind the marshes of the Geete, was for that same reason incapable of aggressive action. Marlborough determined to turn the French right, and, to prevent its being reinforced, opened the action by a demonstration against the French left.

The British Force was once more in the Right Wing of the confederate Army. The following regiments being in the FIRST LINE:—

	Scots Greys Dragoons.		
	The ROYAL IRISH DRAGOONS.		
The present	1st	Dragoon	Guards.
"	5th	"	"
"	7th	"	"
"	6th	"	"
"	3rd	"	"
	Eighteen Dutch Squadrons.		
The present	1st	Battalion.	1st Guards.
"	1st	"	Royal Scots.
"	16th Foot.		
"	26th Cameronians.		
"	28th Foot.		

The present 23rd Royal Welsh.
" 8th Foot.
" 3rd Buffs.
" 21st Royal Scots Fusrs.
" Evans' Foot.
" Macartney's Foot.
" Stringer's "
" 15th Foot.

SECOND LINE

The present 2nd Battalion Royal Scots.
" 18th Royal Irish.
" 29th Foot.
" 37th "
" 24th "
" 10th "
and Foreign Cavalry and Infantry.

The infantry of the Allied Right now moved in two lines towards Offus and Autréglise, with all the pomp of war. At the river they halted, and appeared to be busy with pontoons.

The mass of British scarlet appeared to indicate to Villeroy where Marlborough was about to launch his attack. He accordingly strengthened his left with several battalions from his right and centre. The Duke, seeing the white coats of these battalions streaming away to the French left, ordered the infantry of his right to fall back to some heights in their rear. The second line halted on the crest, and facing about, covered the ground with the well known scarlet, while the first line marched on until out of sight, and then, covered by the hill from the view of the French, hurried with all possible speed to the opposite flank. "Many British battalions stood on that height all day without moving a step or firing a shot, but none the less paralysing the French left wing."

About 1.30 p.m. the guns of both armies opened fire, and soon after four Dutch battalions carried Taviers and Franquinay. Twelve more battalions were ordered to attack Ramillies, while the cavalry of the left slowly advanced. Franquinay, a village on the Mehaigne, had been galling the left wing cavalry with a flanking fire, and this annoyance having been disposed of, Auverquerque led on his horse and charged.

The Dutch after routing the first French line, were driven back by the second; but some fresh squadrons under Marlborough

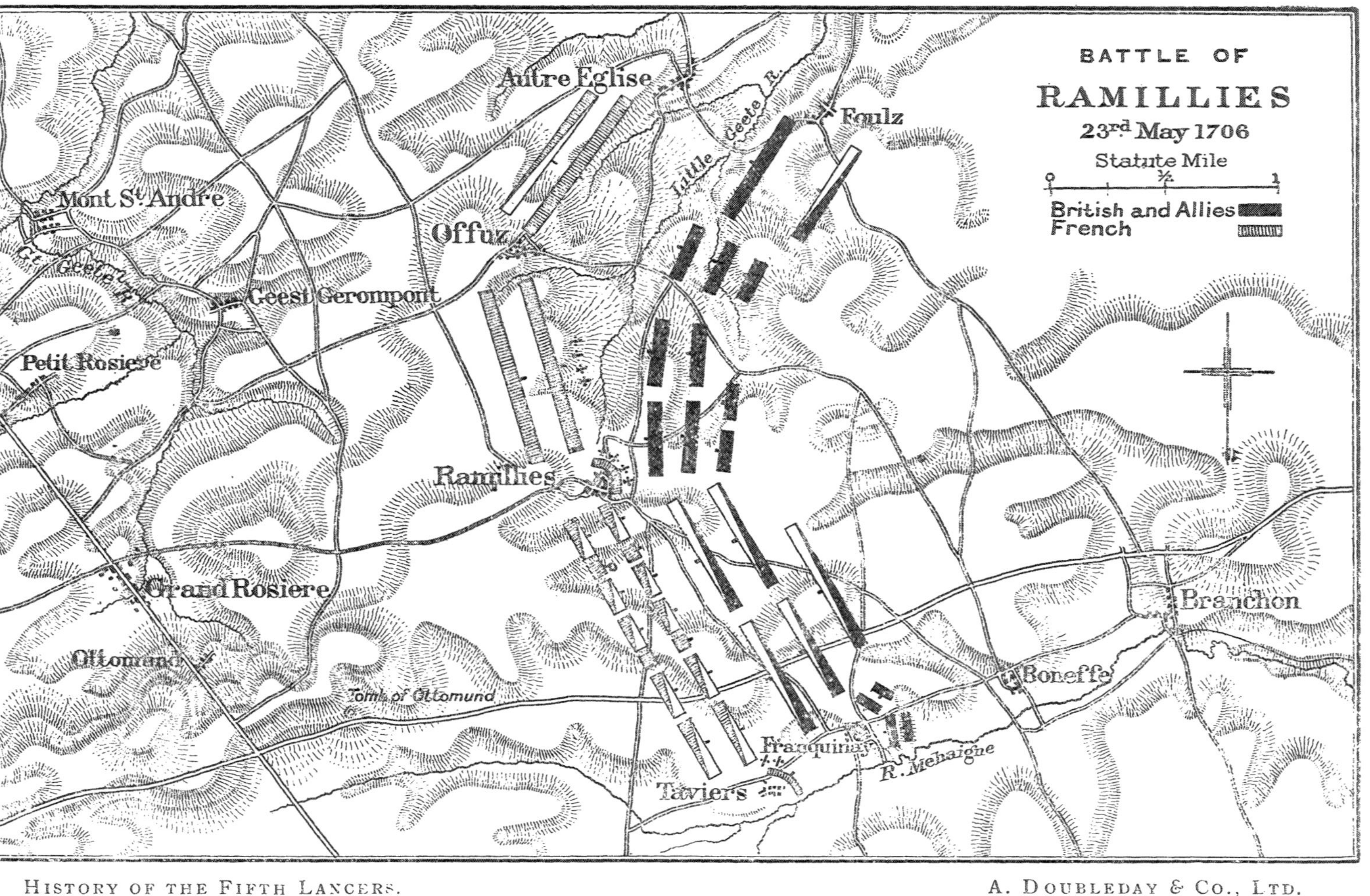

HISTORY OF THE FIFTH LANCERS. A. DOUBLEDAY & CO., LTD.

himself checked the advance of the French. Marlborough then ordered up every squadron of the Right wing, except those of the British cavalry. The Duke was now in the thick of the fight, and being thrown from his horse, which escaped, was furiously attacked by some French dragoons, and in imminent danger of capture. His aide-de-camp, Captain Molesworth,[1] dismounted at once and gave his horse to the Duke, enabled him to escape, and remained to face the enemy alone. The French, however, were so intent upon pursuing the Duke that Molesworth escaped with a few sabre-cuts. He then recovered the Duke's horse and rejoined him. While Marlborough mounted, the equerry who was holding his stirrup had his head carried away by a round shot.

Meanwhile the infantry attack on Ramillies had fully developed; and the fresh horse ordered up from the right by Marlborough came along at top speed and formed up in rear of the re-forming lines. Before they could get into action, however, the Danish horse and Dutch guards were getting round the French right flank. Then the rest of the Allied horse rode against the French front and a fierce fight ensued. The famous French Household Cavalry (the Maison du Roi) were cut to pieces, and in spite of Villeroy's efforts, the whole of the French horse were driven in headlong flight off the field, leaving the infantry to their fate. Villeroy now endeavoured to use his cavalry of the left to cover the retreat of the infantry, in which he only partially succeeded, as the horsemen were much obstructed by the baggage which encumbered the ground. The French troops in Ramillies now gave way, and the battle was won. The British dragoons, amongst them being the ROYAL IRISH DRAGOONS pushed their way into the village of Autréglise, and made "a terrible slaughter of them," while the Dutch and the Danes pursued those that fled to the left "and made an abundance of prisoners," and "those that fled to the right were chased by the regiments of Lumley, Hay, and Ross." The last two regiments, the Scots Dragoons and the IRISH DRAGOONS fell in with, and captured the entire King's Regiment (Regiment du Roi): "of whom having killed many, the rest threw down their arms and begged quarter, which was generously granted."

The French army now broke up in panic and fled in all directions. The British cavalry, practically fresh, took up the

[1] Molesworth was in 1737 the Colonel of The ROYAL IRISH DRAGOONS.

pursuit, and the red-coated troopers pressed on, playing havoc with the beaten enemy. Not till two o'clock in the morning did they pause, having by that time reached Meldert, fifteen miles from the battle field. The loss of the French during the battle and pursuit of Ramillies was some fifteen thousand men in killed, wounded and prisoners; eighty standards and colours, fifty guns, and an enormous quantity of baggage. The loss of the Allies was between four and five thousand killed and wounded. This loss was chiefly amongst the Dutch and the Danes, for excepting in the pursuit by their cavalry, the British contingent was but little engaged.

The following were the casualties amongst the Allied Cavalry at Ramillies :—

	Killed.	Wounded.
Colonels.	2	3
Lt.-Colonels.	—	3
Majors.	4	3
Captains.	10	24
Lieutenants.	6	27
Cornets.	4	28
Subalterns.	8	18
Troopers and Dragoons.	343	695
Horses.	990	351

Writing of the ROYAL IRISH DRAGOONS at Ramillies, in his Military Antiquities, Grose says :

" In consequence of this Regiment (5th DRAGOONS), assisted by the Scots Greys, making two battalions of the regiment of Picardie prisoners of war, and cutting a third to pieces, before it could secure a retreat behind a line of horse that were galloping to bring it off, both corps were distinguished from other cavalry regiments by being permitted to wear Grenadier caps."

In this action, Mr. Ellis of the Regiment, captured Count Horn, a French Lieutenant-General.

The main army, having slept for two or three hours, started after the pursuing cavalry at 3 a.m., and steadily followed the demoralised French troops until the 27th May, when Marlborough gave his weary troops a halt at Grimbergen; the French retiring into Ghent.

" Ramillies " is borne on the standards, colours and appointments of the following regiments of the British Army :—

1st Dragoon Guards.	Liverpool Regiment.
3rd "	Lincolnshire Regiment.
5th "	E. Yorkshire Regiment.
6th "	Bedfordshire Regiment.
7th "	R. Irish Regiment.
R. Scots Greys.	R. Scots Fusiliers.
5th Lancers.	R. Welsh Fusiliers.
Royal Artillery.	S. Wales Borderers.
Royal Engineers.	1st Scottish Rifles.
Grenadier Guards.	1st Gloucester Regiment.
Royal Scots.	1st Worcester Regiment.
The Buffs.	1st Hampshire Regiment.

On the 1st June, from the camp at Meerlebeck, Marlborough despatched Major-General Ross with 600 of his Dragoons to Bruges, with letters inviting the magistrates of that city and of the Franc to submit themselves to the obedience of their lawful sovereign, King Charles III of Spain. On his approaching Bruges, the French battalion there retired, and the Magistrates expressed their desire to submit themselves to Charles III.

The enemy also having quitted Damme, fifty of Ross' Dragoons took possession of that place. The Dragoons remained in both these places until relieved by Dutch regiments.

Marlborough sent the following letter to Major-General Ross, dated 3rd June, 1706, from Meerlebeck :

" I received your letter early this morning: the magistrates both of the Franc and the city of Bruges have been with me since, and made their submission to King Charles........ Upon the encouragement you give me, we have ventured to write to the Governor of Ostend. You will find enclosed two letters for him, both under flying seals ; that signed by myself alone is only by way of summons. You will march that way with your whole detachment, or such part as you think fit, and send him in that letter. If you find he is inclined to surrender, you may then send him in the other letter as a security likewise from the States. Since the writing of these letters, Oudenarde has likewise surrendered. A French battalion that was in garrison has leave to retire, but the Governor, with two Spanish battalions, have declared for King Charles. This you may make use of as an argument to induce the Governor of Ostend to follow so good an example.

" I find the people of Bruges are apprehensive those of Ostend should let the sea in upon them ; this you should endeavour to prevent either by fair means or by threats.

I am, etc. M.

P. S. We shall march to-morrow, and pass both the Scheldt and the Lys to take the camp of Deynse and Nivelle, and be so much nearer. You need not hasten your return to the army as long as you can be victualled where you are."

On June the 4th Marlborough wrote from Nivelle to Major-General Ross :—

"I have received your letter of this day with the account of the stores and ammunition which the enemy left at Damme, which are sufficient marks of a precipitate retreat. I am glad you give me hopes of Ostend, and shall expect the success of your summons with great impatience. We shall be at Aerzeele to-morrow, which is a good deal nearer to you; it may be, our approach may influence those parts.

I am, etc. M.

The Governor of Ostend, however, refused to surrender, and Marlborough decided to besiege the place. He was delayed in this by his artillery being five or six days behind him. The troops destined for the siege, under Auverquerque, arrived before Ostend on June the 17th, and a British squadron, under Sir Stafford Fairborn, which had been sent by Her Majesty to lie off Ostend and Nieuport to help in reducing the garrisons of those places, was already on the coasts ; but it was not until the last day of June that Auverquerque was able to commence his attack. After a week's hard fighting Ostend surrendered, the garrison marching out on the 8th of July.

After this followed the siege and capture of Menin by British battalions, and the capture of Ath, and with the latter ended the campaign of Ramillies, "one of the most brilliant in the annals of war, wherein Marlborough in a single month carried his arms triumphant from the Meuse to the sea."

The year 1707 was barren of any great military operations in the Netherlands, and proved a great disappointment after the glories of Ramillies. Owing to numerous complications on the part of the Allies, Marlborough spent the greater portion of the season in camp before a superior French force at Meldert.

For this campaign the 5th Dragoons were again brigaded with the Scots Dragoons ; the Brigadier being Stairs.

Towards the end of the summer Marlborough was twice within an ace of surprising the French who had withdrawn some of their forces towards Provence. On August the 12th the Confederate Army marched to Nivelle to attack the French, who, at midnight, had moved from Gosseliers to Seneff. It was ordered that the attack should be made in the early morning. Lest the enemy should endeavour to get away during the night to their camp at Cambron, Count de Tilly, with forty squadrons of horse and dragoons, commanded by Albemarle, d'Erbach and Ross, with a strong detachment of grenadiers, was ordered to watch the French, and in the event of their decamping, to fall upon their rear and endeavour to keep them employed until the main army could come up. The enemy moved during the night, and in spite of all Tilly could do, even with strong reinforcements from Marlborough, it was found impossible to bring them to an engagement.

The following extract of a letter written by Marlborough to the Earl of Cardigan, on December the 19th, in reply to his Lordship's request that the son of the late Major-General Brudenell should be given a company in a regiment on service in Spain, is interesting as indicating the manner in which commissions were given in the army at this time. The Duke writes: "I have so just a sense of the father's good services that I shall be always glad to embrace any opportunity of showing it to his family; but your Lordship tells me he is not above five years old. This giving him a command in the army would be directly contrary to the rules the Queen has prescribed to herself in that matter; besides that the enquiry the Parliament is making of the Officers absent from their commands in Spain makes it yet the more difficult."

CHAPTER VI.

Campaign of 1708.—Oudenarde.—Malplaquet.

For the campaign of 1708 Parliament voted money for an additional ten thousand men, and of the new battalions raised three were sent to Flanders.

It is interesting to note that all regiments took the field this year with their new colours, bearing the cross of St. Andrew blended with that of St. George; the Union of England and Scotland having taken place the previous year.

Early in the year CHARLES Ross was promoted to be a Lieutenant-General, and another troop was added to his Regiment under the following order:—

> " Anne R. January 12th, Whitehall.
>
> It is our Will and Pleasure that a Troop shall be raised and added to our Dragoons whereof Our Well Beloved Lieutenant-General Charles Ross is Commander. To consist of two Drums and 60 privates and 3 Corporalls.
>
> St. JOHN "

This order apparently brought the Regiment up to the nine troops as mentioned by Grose in his " Memoirs, " for which it was so long conspicuous. Ross was also commanded to fill up the Regiment from 54 to 60 men in each troop. For this purpose £1500 was paid under the head of " Some extraordinary charges of War incurred and not provided for by Parliament. Buying one hundred horses and accoutrements to augment the Regiment of Dragoons commanded by Lieutenant-General Ross."

An over abundance of unnecessary transport was evidently as much of a nightmare to General officers of this time as it is to-day, for we find Marlborough writing from St. James' in January to

Mr. (Lieutenant-General) Lumley, in command of troops in Flanders, on the subject of the approaching campaign, " as to what you mention in your last relating to the wheel carriages, I am very sensible of the great embarras and trouble they are to us upon all occasions, and therefore desire you would earnestly recommend it from me to all the officers to have as few as possible. "

The French, who in the early spring had made a futile endeavour to land a force in Scotland, marched at the end of May from their rendezvous south of the Haine, North, to the Forest of Soignies. Marlborough had landed in April, and promptly concentrated his army at Hal and summoned Eugene to him.

Marlborough's army numbered but eighty thousand men, while that of the French main army in Flanders under the Duke of Burgundy and Vendôme was little under a hundred thousand. Ever since the Ramillies campaign the French had been drawing troops from all quarters to Flanders.

From Soignies the French manœuvred towards Waterloo, as if to threaten Louvain. Marlborough met this move by a forced march to the river Dyle, where he remained inactive for a month, waiting for Eugene. On the night of the 4th of July the French marched Westward and crossed the Senne at Hal, whence they detached small forces to Bruges and Ghent, and within twenty four hours they were in possession of both towns ; the result of bribing the authorities in these places.

At 2 a.m. during the same night, Marlborough moved his army and crossed the Senne at Anderlecht, and after a most trying march was almost within reach of the French. During the night of July the 6th a report reached the Duke that the enemy were approaching to attack him. This report was false, and before the truth could be ascertained, the French had crossed the Dender, in great haste to get away from the Confederate army. Marlborough's cavalry was soon in pursuit, but failed to do more than capture some prisoners and baggage. The French army now encamped at Alost, where it threatened Brussels. Marlborough thereupon moved to Assche, midway between Alost and Brussels, in order to allay the panic in the capital. Here Eugene's force joined the army.

The French meditated a descent upon Oudenarde, for the recovery of the line of the Scheldt, and were already across the Dender and had destroyed the bridges, having had a start of Marlborough. The Duke, however, being unable to leave Brussels exposed, remained perforce where he was, bnt ordered the Governor of Ath to collect what troops he could and garrison Oudenarde, which was promptly done.

On the 9th of July Vendôme sent a force to invest Oudenarde, and with the main army moved to Lessines, to cover the siege. Meanwhile, however, Marlborough had started at 2 a.m. on the 9th, and, by a splendid march via Herfelingen, was in occupation of Lessines on the arrival of the French. Baulked in their plans, the French wheeled about, and moved North West towards Gavre on the Scheldt.

At dawn on the 11th, Marlborough sent Cadogan with 30 squadroons and 16 battalions and 24 guns, to prepare and cover the passage of the Scheldt below Oudenarde for the Army. By 10.30 a.m. Cadogan had reached the river. Presently parties of French horse appeared to the North. Their advanced guard had leisurely crossed the river at Gavre, six miles further down the river, and were marching in happy ignorance of the presence of an enemy, which however, was quickly dispelled by a dash of Cadogan's squadrons. Vendôme, on receiving the report, and seeing that the mass of the Allied Army was on the wrong side of the Scheldt, gave orders to take up a position parallel to the river. The Duke of Burgundy objected to the line ordered by Vendôme, and ordered a position to be taken up on the heights of Huysse, in rear of the river Norken, from Asper to Wannegem, and some two miles from the Scheldt. Unfortunately, seven battalions ordered by Vendôme to occupy the village of Heurne, in ignorance of the change of orders, marched down to the village of Eyne instead of Heurne, in the centre of Vendôme's proposed line, where they remained backed by a few squadrons.

Marlborough, who had early arrived on the scene, meanwhile sent orders hurrying up the main body, and at 2 p.m. the head of the infantry had reached the Scheldt. The march of the seven French battalions into Eyne had been marked by Cadogan, who got the whole of his advanced guard across to the left bank. His British infantry brigade, supported by the other two, moved against Eyne, while the Hanoverian cavalry rode to the rear of the village and cut off all hope of retreat.

The British were soon hotly engaged, and the French, making a poor resistance, had three of their battalions captured entire, and the remaining four killed or taken piecemeal in their flight. The cavalry under Prince George of Hanover, afterwards King George II of England, now charged the few French squadrons in rear of the village, routed them, and drove them across the Norken. The Prince gallantly led the charge and had his horse shot under him, and ever after maintained that Oudenarde was the hottest thing of his life.

The Duke of Burgundy now made every preparation for the defence of his position behind the Norken, but when four o'clock had arrived and the Allied army was not yet in order of battle, he changed his mind; he advanced the whole of his centre and right, pushing his right cavalry across the stream. Marlborough judged the attack would be against his left, and against Cadogan's isolated battalions about Eyne. Two Prussian regiments of this force had been pushed forward half a mile beyond Eyne to Groenewald, and were promptly re-inforced by twelve more from the advanced guard. The British Cavalry were formed up on the left, on the heights of Bevere, and the Prussian horse on the right near Heurne; and the rest of the Army were gradually crossing the Scheldt.

At about 5 p.m. thirty French battalions advanced on Groenewald, which was only held by Cadogan's two advanced regiments of Prussians. They, however, gallantly held their own among the hedges until re-inforced by twenty battalions under the Duke of Argyle. The fresh battalions formed in succession on the left of the Prussians, and the fighting became most severe. The ground was so enclosed that the struggle became a duel between battalions. At one time the French, outflanking the left of the Allies, drove them back, but Marlborough's fresh infantry kept arriving on the scene, and, by prolonging the line on the left to the South, defeated the movement.

Eugene was now given eighteen battalions, and entrusted with the command of the right. With this accession of strength, he was able to relieve Cadogan's corps, and even to pierce the first line of the enemy's infantry; whereupon the Prussian cavalry charged the second line, only to be driven back by the French Household Cavalry with heavy loss.

Marlborough, meanwhile, was slowly pressing on with the Hanoverian and Dutch infantry on the left. The French, contesting every inch of ground, were gradually forced back to Diepenbeck, where they stood fast, and all the efforts of the Dutch and

Hanoverians could not shift them. Marlborough now directed the gallant old Dutch cavalry leader, Auverquerque, to take the cavalry of the left under cover round the French right, and take them in their rear. Auverquerque carried out his orders, and fell upon the rear of the unsuspecting French, and although some of their Household Cavalry and dragoons made a short stand, he rode them down, and pressing rapidly, the French right was fairly surrounded.

During all this time the French left, for some unaccountable reason, had remained motionless and inactive on the far side of the Norken. The Duke of Burgundy now tried to bring them forward, but it was too late. Vendôme himself, at the head of the infantry, failed to make the slightest impression, and the cavalry dared not advance, for the ground in front of them was bad, and the whole of the British Cavalry, who had been withdrawn from their first position, stood watching, ready to swoop down on them should they begin to move.

Darkness came on, and the Allies wormed themselves closer and closer round the French right. At nine o'clock, fearful lest his own troops should engage each other in the darkness, Marlborough ordered a halt, and to cease firing. Many of the French seized the moment to escape, but presently all the drums of the Allies began to beat the French retreat, and the Huguenot officers, shouting "A moi, Picardie ! A moi, Roussillon ! " gathered the relics of the scattered French regiments around them, and so captured several thousands of prisoners. Vendôme endeavoured to keep the army together, but, Burgundy having ordered a retreat, the French ran off in confusion and disorder towards Ghent. Another hour of daylight, Marlborough always declared, would have enabled him to finish the war.

The total loss of the Allies was about three thousand killed and wounded. The British contingent suffered little, losing only 4 officers and 49 men killed, and 17 officers and 160 men wounded; for the infantry, though early engaged, suffered slightly, while the cavalry, being employed to watch the inactive French left, scarcely suffered at all.

The French lost six thousand in killed and wounded, and nine thousand prisoners, and their army was completely shaken and demoralised for the remainder of the campaign.

The Allies lay on their arms on the field for the night, and at dawn forty squadrons, mostly British, started in pursuit.

Of the battle of Oudenarde, Fortescue says it was undoubtedly the most hazardous action that Marlborough over fought. His troops had started at two o'clock on Monday morning, and had covered fifty miles, including the passage of two rivers, when they came into action at two o'clock on Wednesday afternoon. This march of eighty thousand men over fifty miles of the bad roads of those days, and with heavy packs, in sixty hours, was no mean performance. Finally, the army had to pass the Scheldt in the face of the enemy, and ran no small risk of being destroyed in detail.

"Oudenarde" is borne on the standards, colours and appointments of the following regiments of the British Army:—

1st Dragoon Guards
3rd " "
5th " "
6th " "
7th " "
Royal Scots Greys
5th Lancers
Royal Artillery
Royal Engineers
Grenadier Guards
Coldstream "
Royal Scots
The Buffs
Liverpool Rgt.
Lincolnshire Rgt.
E. Yorkshire Regt.
Bedfordshire Rgt.
R. Irish Rgt.
R. Scots Fusiliers
R. Welsh Fusiliers
S. Wales Borderers
1st Scottish Rifles
1st Hampshire Rgt.

The army halted to rest for a couple of days where it lay, during which time news was received that the French army on the Moselle was marching, with all haste, to occupy the lines constructed by the French to cover their frontier from Ypres to the Lys. At midnight of the 13th of July, Marlborough despatched Count Lottum, a distinguished Prussian Officer, with fifty squadrons and thirty battalions, to capture these lines. The news that Lottum had carried out his orders reached the Duke when following with the main army. The next day, the 14th of July, Marlborough's army was encamped along the Lys between Menin and Commines, and actually on French territory. The Allies next undertook the siege of Lille. Their preparations at last being complete, a huge convoy left Brussels on its seventy-five mile march to Lille, arriving there on the 12th of August without the loss of a single wagon.

The garrison of Lille amounted to fifteen thousand men, commanded by the brave old Marshal Boufflers. Vendôme at Ghent in the North, and Berwick at Douay in the South, with their enormous army of one hundred and ten thousand men, failed to prevent Marlborough, with his inferior strength of eighty four thousand men, from prosecuting a successful investment. This long and famous siege lasted to the 9th of December, when the gallant commander and his heroic garrison marched out with all the honours of war, having lost eight thousand men, or more than half its number, and having cost the Allies no fewer than fourteen thousand.

There were but five British battalions regularly employed in the trenches, and it is worthy of note that the first advance to attack was opened by a single English soldier, Sergeant Littler, of the first Guards, who swam across the Marquette to a French post which commanded the passage of the stream, and let down the drawbridge.

During the siege Marlborough's covering army was engaged in watching the armies of Vendôme and Berwick, who employed themselves in harassing the Allies' convoys proceeding to the siege. On the evening of August the 29th, we find that Ross and his Dragoons were ordered to cover the march of a convoy of seven hundred and fifty wagons of provisions and ammunition from Ath to the Army and the siege, which duty they successfully performed, the convoy arriving before Lille on the morning of the 30th. The covering army on the 18th of September defeated the French at the action of Wynendale.

After the fall of Lille followed the capitulation of Ghent and Bruges, and therewith the end of the campaign, and Marlborough and Eugene at last sent their troops into winter quarters.

For the campaign of 1709 the French had greatly strengthened their army by withdrawing troops from all quarters to Flanders, and had placed in command "their only fortunate general, that very able soldier and incomparable liar, Marshal Villars." To cover Arras, "the northwestern gate of France," Villars had thrown up the strong lines of La Bassée, extending from Douay to the Lys, behind which he lay "entrenched to the teeth." By June, Marlborough and Eugene, with an army of one hundred and ten thousand men, lay encamped near Lille. From there, the Allies moving South, made Villars anticipate a direct attack upon his Lines for the purpose of a march into France. On the evening of the 26th of June the Allies struck their camp, and marched towards the French Lines, before which the army expected a bloody action at dawn. After advancing for some time, to the general surprise, the columns were ordered to change their direction to the left, and, marching eastward, the soldiers at dawn saw "the grey walls and the four spires of Tournay before them," and found they were about to invest one of the strongest fortresses of France. The garrison of Tournay had, unfortunately for the French, been weakened by Villars, for that unsuspecting general had, upon Marlborough's apparent move against his Lines from Lille, summoned the greater part of it to his assistance at the Lines. On the 7th of July the trenches were opened, and by September the 3rd, in spite of Villars' demonstrations, the town and citadel were in the hands of the Allies. Only seven battalions of the British contingent were employed in the investment.

Before the close of the siege, Marlborough and Eugene, leaving the besieging force before Tournay, had marched the main army towards the Lines at Douay. The French lines being too formidable to be forced, Marlborough sent Lord Orkney, with twenty squadrons and all the grenadiers of the army, silently eastward towards St. Ghislain. Three days later a force under the Prince of Hessen-Cassel followed; a few hours later Cadogan, with forty squadrons; and at midnight the remainder of the army. Twenty-six battalions were left before Tournay with orders to watch Villars, and not to move until he did.

Too late, Villars discovered that he had again been duped, and that Marlborough, instead of ramming his head against the formidable French Lines, proposed an easier entry into France round their eastern end, across the Trouille. He at once sent a detachment to Jemappes, the nearer entrance across a natural barrier of forest which cut him off from the weakly garrisoned fortress of Mons; but the detachment came too late, for the Allies were before them. The Allied army had invested Mons, and on September the 7th Villars and his whole army had arrived on the scene, and encamped a couple of miles to the west of the forest barrier, between Montreuil and Athis. Here he was joined by the veteran Marshal Boufflers, whose arrival caused such a tumult of rejoicing in the French Camp that the Allies, not knowing what the clamour might portend, advanced westward into the plain of Mons, and bivouacked in order of battle between Ciply and Quévy, leaving only a small investing force before Mons.

The French, however, did not move, but remained threatening both the passages of the forest. We have seen that the Northern crossing was at Jemappes, while the southern one was at Malplaquet. That night Villars sent detachments to occupy the southern crossing, and by midday of September the 9th the whole of his army was taking up its position across this opening. Marlborough immediately moved forward, but at a Council of War, for some extraordinary reason, it was decided not to attack. Villars at once prepared his position for defence, and next morning Marlborough proposed to attack him, but was again obstructed by the Dutch Deputies. The attack was fixed for the morrow, which delay Villars did not fail to turn to account in the way of entrenching.

Villars' position was on the high ground about a mile in advance of the villages of Campe du Hamlet and Malplaquet. His extreme right occupied the forest of Laignières, the natural obstacles of a thick and tangled cover being strengthened by abattis. From the edge of the wood he constructed a triple line of entrenchments, which ran a third of its way across the passage, where the defence was continued by a line of nine redans. From these redans stretched a swamp, backed by more entrenchments, to the wood of Taisnières. Several cannon were mounted on the entrenchments, and twenty guns were before the redans. The woods of Taisnières and Sart projected beyond the French front, and formed a salient and re-entering angle on Villars' left, and here entrenchments and abattis were constructed, and guns placed at various points to enfilade an advancing enemy. In rear of all, the French Cavalry was drawn up in several lines. Ninety-five thousand men was the strength of the French, and the Allies numbered about the same.

The dawn of the 11th of September broke in a dense heavy mist. At three o'clock prayers were said in the Allied camp, and shortly after the artillery moved into position.

Marlborough and Eugene had decided to feint against the French right, and to force home a true attack against their left front and flank, for the reason that the wood of Sart ran out some distance beyond the fortified angle, and would serve to conceal the movements of troops against the extreme left flank.

The French Cavalry, being massed behind the entrenchments, could not take any part in the action until the defences were forced.

The Allies massed forty guns against the French left, and covered them from enfilade fire by an epaulment ; twenty-eight pieces were placed against the French right, and the remainder distributed amongst the brigades. Count Lottum was directed against the eastern face of the salient angle of the Taisnières wood with twenty-eight battalions supported by the British and Hanoverian cavalry. Forty battalions under General Schulemberg moved against the northern face of the wood, and were backed by Eugene's cavalry, while a little to Schulemberg's right, General Gauvain with two thousand men was to press on the French left flank in rear of their entrenchments. In rear of Schulemberg, Lord Orkney had fifteen British battalions drawn up in a single line, ready to advance against the centre as soon as Schulemberg and Lottum had done their work, and thirty squadrons under Auvergne.

Far away to the right was General Withers, with five British and fourteen foreign battalions, and six squadrons, ready to turn the extreme left of the French at the village of La Folie. The Prince of Orange, with thirty-one battalions, chiefly Dutch, and twenty-one Dutch squadrons under the Prince of Hesse, stood ready behind the small wood of Tiry to make his feint against the French right.

The cavalry was ordered to sustain the foot, but to keep out of the range of grape shot, and, on the central entrenchments being forced, they were to press forward, form on the other side of the entrenchments, and "drive the French army from the field." Eugene was to command the right, and Marlborough the left of the army.

At half past seven the fog lifted, and the guns of both armies opened fire. The Allied advance began by the divisions of Orange and Lottum moving in two dense columns up the glade. The Prince of Orange's men halted just out of range of grape shot, while Lottum's column pushed on under heavy fire to the rear of the forty gun battery, and deployed to the right in three lines. About nine o'clock a salvo of the forty guns gave the signal for the attack. Lottum's and Schulemberg's divisions, each in three lines, advanced perpendicularly to each other, and Gauvain's force moved into the wood unperceived, while Orkney extended his scarlet battalions across the glade. Both Schulemberg's and Lottum's attacks were repulsed with very heavy losses. The former then resumed the attack with his second line, but unsuccessfully, for the French regiment of Picardie stubbornly held its own. Orkney sent three British battalions to reinforce Lottum, who attacked again. The Englishmen moved on Lottum's left, and scrambled on through a swamp in spite of being threatened by twelve French battalions on their left, who, however, retired on catching sight of Auvergne's supporting squadrons led by Marlborough himself. Lottum's men attacked the entrenchments in front, while the British brigade succeeded in turning the flank, and after desperate fighting and heavy losses, the French were forced back into the wood. Then the Picardie regiment, exposed to the double attack of Lottum and Schulemberg, fell back on to the regiment of Champagne, and these two gallant corps made yet another stand behind an abbatis. Eventually they had to give way in disorder, and, owing to the denseness of the wood, the struggle resolved itself into a succession of small parties fighting desperately from tree to tree. The entrenchments on the French left had been won.

The Prince of Orange, who had been waiting for orders to commence his feint, lost patience, and without orders, opened, not a false, but a real attack against the French right. Orange himself led the attack, and at the head of the famous Blue guards, and the Highlanders in the Dutch service, advanced under a most appalling fire of grape and musketry. His horse being shot dead, he continued on foot, and the gallant Highlanders and Dutchmen pressed on, and, in spite of whole ranks being swept away by the fire of a French battery on their left flank, they carried the first entrenchment with a rush. While they halted to deploy, the French rallied, and charging down on the Prince, drove him and his force headlong back. This impatience of the Prince of Orange had cost the Dutch a loss of some six thousand men killed and wounded, and the Blue guards had been annihilated, while the supporting battalions had suffered little less severely.

Meanwhile Schulemberg and Lottum continued to push their attack, and now, on the extreme Allied right, Withers was advancing. Villars, seeing this danger on his left flank, called up his Irish brigade, and other regiments from the centre, and launched

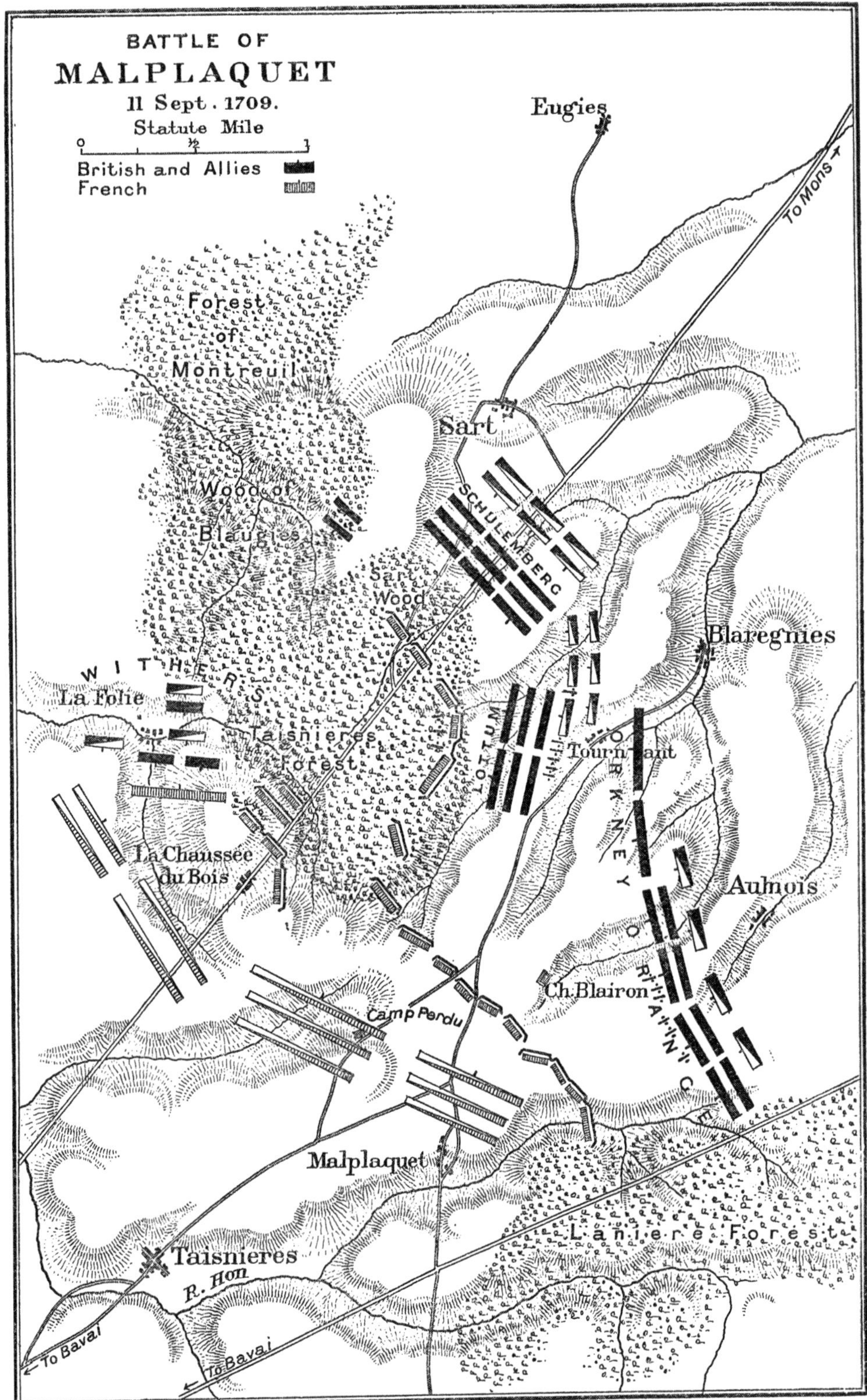
BATTLE OF
MALPLAQUET
11 Sept. 1709.
Statute Mile
0
½
1
British and Allies
French
Eugies
To Mons
Forest of Montreuil
Sart
Wood of Blaugies
SCHULEMBERG
Sart Wood
Blaregnies
WITHERS
La Folie
Taisnieres Forest
LOTTUM
Tournant
ORKNEY
La Chaussée du Bois
Aulnois
Ch. Blairon
Camp Perdu
ORANGE
Malplaquet
Laniere Forest
Taisnieres
R. Hon
To Bavai
To Bavai

them upon the British and Prussians. Wither's battalions were forced back some way by the impetuous Irish, whose formation was eventually broken by the density of the forest. "Then up came Withers, just when he was wanted. The Eighteenth Royal Irish of the British army met the French Royal Regiment of Ireland, crushed it with two volleys by sheer superiority of fire, drove it back in disorder, and pressed on." Villars was badly wounded and carried insensible from the field, but notwithstanding his fall, and the fact that they had been driven from their entrenchments and from the wood on their left, the French still barred the advance of the Allies; but Marlborough's time had now come.

The forty-gun battery was moved forward, and Lord Orkney led his British battalions against the redans, which, after a severe fight, he captured; and on the left of the British, the Prince of Orange again advanced, and cleared the whole of the defences in the glade. Now came the turn of the Allied cavalry. The Dutch squadrons were first past the entrenchments, but were driven back to their edge by the veteran Boufflers at the head of the French Gendarmerie. Here, however, Orkney's British infantry lining the parapet, three times forced the Gendarmerie to retire. Meanwhile, the central battery of guns advanced and supported the infantry by a cross fire, and Marlborough at the head of the British and Prussian cavalry, fell upon the Gendarmerie. Boufflers now came along with fresh squadrons, and leading the French Household Cavalry, crashed into Marlborough, and threw his horsemen into disorder. Then Eugene, coming up with fresh squadrons, threw the Imperial Horse into the melée, and drove the French back. Simultaneously the Prince of Hesse charged the infantry of the French right, and with the help of the Dutch foot, kept it isolated from the bloody fight raging in the centre. Then Boufflers saw that the day was lost and sounded the retreat, which was carried out in admirable order, for the Allies were too exhausted to pursue.

"Thus ended the battle of Malplaquet, one of the bloodiest ever fought by mortal men."

The loss of the French was some twelve thousand men, and five hundred prisoners, fifty standards and colours, and sixteen guns. The Allies lost not less than twenty thousand in killed and wounded, due chiefly to the mad attack of the Prince of Orange.

The British contingent out of their twenty battalions lost nineteen hundred men. The Royal Irish Dragoons were again brigaded with the Scots Greys during this campaign.

The Regiments of the British Army who now bear "Malplaquet" on their colours and appointments are:—

1st Dragoon Guards	The Buffs
3rd " "	Liverpool Regiment
5th " "	Lincolnshire "
6th " "	E. Yorkshire "
7th " "	Bedfordshire "
2nd Dragoons	R. Irish "
5th Lancers	Yorkshire "
Royal Artillery	R. Scots Fusiliers
Royal Engineers	R. Welsh "
Grenadier Guards	S. Wales Borderers
Coldstream "	Scottish Rifles
Royal Scots	Hampshire Regiment

Malplaquet has been described as a grand action. The French were equal in numbers to the Allies, and occupied a position which was described at the time as a fortified citadel. They were commanded by an able general, and yet they were driven back, and forced to leave Mons to its fate.

On the third day after the fight the Allies returned to the investment of Mons, which capitulated on the 9th of October, and the campaign came to an end.

The following is a list of the officers of Her Majesty's Royal Regiment of Dragoons of Ireland present of Malplaquet:—

CAPTAINS.

Lieutenant-General Chas. Ross. Colonel-Staff.
Brevet-Col. Jno. Hill. Lt.-Col. Commanding.
Brevet-Major Ric. Gore.
Robert Drury.
Bt.-Major Jno. Warre.
Jno. Usher.
Jno. Johnston.

LIEUTENANTS.

Chas. Beatty — Captain-Lieutenant.
Jno. Mann.

Jas. Poé.
Ed. Hamilton — Staff. A.D.C.
Beverley Newcomen.
Jas. Hamilton.
Michael Parker
Alex. Abercromby.
Jno Hunter.

CORNETS.

Wm. Cocksedge
Robt. Barker.
Jno. Skelston.
Geo. McKean.
Fras. Bogges.
Edw. Hill.
Wm. Ross.
Jno. Gough.

QUARTER-MASTERS.

Jas. Knox.
Jer. Balfour.
Rich. Dunbar.
Mor. White.
Jas. Welsh.
Jno. Evans.
Ric. Johnston.

ADJUTANT.

Wm. Ross.

CHAPLAIN.

Simon Babe.

SURGEON.

Jas. Scott.

CHAPTER VII.

Campaigns of 1710-1711.— Marlborough's Stratagem.— His disgrace in England.—Ormonde's campaign of 1712.—Peace of Utrecht.

Negotiations for peace coming to naught, Marlborough opened the campaign of 1710 by a rapid movement over the lines of La Bassée and the capture of Douay, the siege of which place lasted from the 22nd of April to the 26th of June.

He next turned to the siege of Bethune, and the Generals of that siege being under some apprehensions from the detachments the enemy had sent towards St. Omer, Lieutenant-General Ross with twenty one squadrons (amongst them being his own regiment), and nine battalions, was ordered on the 21st of August to march and encamp between the Allied main army at Viar Brudin and the siege ; where he would be at hand to prevent any design the enemy might have " to give us a disturbance. "

The next day Marlborough wrote to Ross that he was very well satisfied with his having thrown bridges over the stream in his front ; and that he had taken every care that Ross' force should be supplied regularly with bread.

General Schulemberg being somewhat retarded in his attack during the siege by want of fascines, Marlborough writes to Ross to help that General to make fascines, adding that " the service will be no great fatigue to your men, I need not tell you they will be paid for it. "

Bethune surrendered on the 28th of August, and after that Ross' force joined the army on its march towards Aire, camping between Teroneune and Lillers on the 4th of September.

The capture of Aire and St. Venant on the Upper Lys, after a desperate resistance, closed the campaign for the year.

On the 20th of December Marlborough wrote to Mr Sweet:— " The village of Steesch in the mairie of Bois-le-duc complains of exactions committed by Ross' regiment of dragoons in their march to winter quarters in November 1709, amounting to 1122 odd gilders ; whereof 236 gilders they allege to have been obliged to pay in money, to be exempted from further quarters. The Regiment having neglected the repeated directions to satisfy the same," Mr. Sweet is authorized to pay to the village the sum of 720 gilders and "to place it to the account of the Regiment, they having neglected the repeated directions."

With regard to the opening of the campaign of 1711, Fortescue says :— "The French, fully aware of the political changes in England, had during the winter made extraordinary exertions to prolong the war for yet one more campaign, and to that end had covered the northern frontier with a fortified barrier on a gigantic scale. Starting from the coast of Picardy the lines followed the course of the river Canche almost to its source. From thence across to the Gy or southern fork of the Upper Scarpe ran a line of earthworks, extending from Oppy to Montenancourt. From the latter point the Gy and the Scarpe were dammed so as to form inundations as far as Biache, at which place a canal led the line of defence from the Scarpe to the Sensée. Here more inundations between the two rivers carried the barrier to Bouchain, whence it followed the Scheldt to Valenciennes. From thence more earthworks prolonged the lines to the Sambre, which carried them at last to their end at Namur."

Beyond a couple of moves, the rival armies were inactive during May and June. The French, safe behind their formidable lines, would not come out and fight ; and the Allied army, which had been greatly reduced from one cause and another, was not strong enough to attempt an attack. To pass the lines Marlborough therefore had to resort to stratagem. "The inundation on the Sensée between Arras and Bouchain could be traversed only by two causeways, the larger of which was defended by a strong fort at Arleux, the other being defended by a redoubt at Aubigny half a mile below it." The Arleux fort Marlborough knew he could take and demolish, but he also knew that so soon as he had left it Villars would certainly retake it and rebuild it. He therefore schemed to induce Villars to demolish it himself. With this view a strong force under General Rantzau was detached to capture the fort, and the General successfully carrying out his orders on the 6th of July, the Duke ordered that the captured works should be greatly strengthened, while a large force under the Prussian General

Hompesch was posted three miles away on the glacis of Douay, as a further protection.

Two days later Villars surprised and made a determined attack on Hompesch's force, and was only repulsed with considerable difficulty, whereupon the Duke re-inforced Hompesch as if to shew his regard for the safety of Arleux, and pushed forward the new works there with renewed vigour. When these new works were completed, Marlborough left but a weak garrison in the fort, and on the 21st of July led the rest of the Army away two marches to the west, encamping opposite the lines between the Canche and the Scarpe. Villars moved west parallel with the Allies, but before starting, detached a force to attack Arleux. The Commander of the fort sent an urgent message to the Duke for help, whereupon Cadogan was sent with a strong force to its relief; before, however, he had gone half way, he returned with the news that Arleux had surrendered.

Villars was elated beyond measure — and Marlborough correspondingly cast down. He declared in public, with much passion, that he would be even with Villars. The Duke's ill temper was not appeased by the news that Villars had razed the whole of the works of Arleux to the ground. All the time and trouble Marlborough had spent over the new works had been wasted, and he angrily vowed he would avenge the insult to his army; and declared his intention of a direct attack on the French entrenchments. Villars, on the 26th of July, detached a force to make a diversion in Brabant. This step seemed to drive Marlborough distracted. He sent a force of ten thousand men under Lord Albermarle to Bethune to check its march, and the whole of his heavy artillery and baggage to Douay; and so weakened an army already greatly inferior to the French. With his weakened forces he repaired the roads that led to the enemy's entrenchments, and with much display of sulkiness and temper advanced a day's march nearer to the lines. His own troops could not understand such proceedings from Corporal John; and while they watched him with amazement, Villars was in a transport of delight. "He drew every man not only from all parts of the lines but also from the neighbouring garrisons towards the threatened point," and asked nothing better than that Marlborough should be mad enough to attack him.

On the 2nd of August the Duke was within a league of the

Lines, and during that day and the next set the whole of the cavalry to work to collect fascines. The quote from Fortescue :— " At nightfall of the 3rd he sent away all his light artillery, together with every wheeled vehicle, under escort of a strong detachment, and next morning rode forward with most of his generals to reconnoitre the lines...... He had now thrown off all his ill-temper and was calm and cool as usual, indicating this point and that to his officers. 'Your brigade, General, will attack here, such and such brigades will be on your right and left, such another in support, and you will be careful of this, that, and other.' The Generals listened and stared; they understood the instructions clearly enough, but they could not help regarding them as madness." But none of them noticed General Cadogan slip away from the crowd and gallop off to camp at top speed. The orders for the morrow's attack were issued, and not a man in the Army could fail to see how hopeless the enterprise was. A direct attack on the lines itself was an over bold venture, and to attempt it with an army half depleted, and in the absence of all the artillery, appeared insane. "Again Marlborough's violent and unprecedented outburst of surliness and ill temper was difficult to explain ; and the only possible explanation was that the Duke, rendered desperate by failure and misfortunes, had thrown prudence to the winds and did not care what he did. " A few there were, however, who clung to the hope that the chief who had so often led them to victory, might still have some surprise in store.

But meanwhile, Cadogan, with forty cavalry soldiers, had left the camp for Douay, five leagues away. "There he found Hompesch ready with his garrison, now strengthened by detachments from Bethune and elsewhere to twelve thousand foot and two thousand horse, and told him the time was come. Hompesch thereupon issued his orders for the troops to be ready to march that night. Still the main army under Marlborough knew nothing of this, and passed the day in dismal apprehension till the sun went down, and the drummers came forward to beat the tattoo. Then a column of cavalry trotted out towards the Allied right, attracting every French eye and stirring every French brain with curiosity as to the purport of the movement. Then the drums began to roll ; and the order ran quietly down the line to strike tents and make ready to march immediately. " The Cavalry having distracted the French vigilance to the wrong quarter, returned unseen by the enemy. "At nine o'clock the whole army faced to its left and marched off eastwards in utter silence, with Marlborough himself at the head of the vanguard." Hour after hour did the army march on this memorable night of the 4th of August, and by 5 a.m. had reached the Scarpe. There they found pontoon bridges already laid for them, and on the further bank were waiting the field artillery ; and soon after a despatch arrived for Marlborough. A message was passed down the columns of weary soldiers : "Generals Cadogan and Hompesch crossed the causeway at Arleux unopposed at three o'clock this morning, and are in possession of the enemy's lines. The Duke desires that the infantry will step out." The infantry responded to a man ; the right wing of horse halted to form the rearguard, while the Duke at the head of fifty squadrons pushed on at a trot.

Villars heard of Marlborough's march only two hours after he had started, but was so bewildered by the Duke's intricate manœuvres that he did not awaken to the true position for another three hours. He then at the head of the Household Cavalry rode away so hard to the east, that he finished near Oisy with but one hundred of his troopers, who were captured by Marlborough's outposts, Villars himself only escaping the same fate by a miracle. Meanwhile, Marlborough's infantry were pressing forward desperately ; hundreds dropped under the severity of the march, but the survivors were still ahead of the French when they turned off the Arleux causeway, and by five o'clock in the afternoon the whole force was drawn up between Oisy and the Scheldt, and within striking distance of Arras, Cambrai and Bouchain. They had covered some forty miles in eighteen hours, and Villars was defeated ; a warning to Generals who put their faith in fortified lines. So ended what is perhaps "the most remarkable, and certainly the most entertaining feat of the Duke during the whole war."

Marlborough's succeeding manœuvre was the capture of Bouchain, which place surrendered under the very eyes of Villars on the 23rd of September, and "the last and not the least of Marlborough's campaigns came, always victoriously, to an end."

The Duke at this time was out of favour at home. To quote from Fortescue : "The most brilliant manifestation of military skill was, however, powerless to help him against the virulence of faction in England. The passage of the lines was described as the crossing of the kennel, and the siege of Bouchain as a waste of lives. In May the House of Commons had addressed the Queen for inquiry into abuses in the public expenditure, and when the Duke arrived at the Hague in November, he found himself charged with fraud, extortion, and embezzlement. The ground of the accusation was that he had received in regular payment from the bread contractors during his command sums amounting to £63,000. Marlborough proved conclusively that this was a perquisite regulary allowed to the Commander-in-Chief in Flanders as a fund for secret service, and he added of his own accord that he had also received a deduction of two and a half per cent from the pay of the foreign troops, which had been applied to the same object. But this defence, though absolutely valid and sound, could avail him little. His reasons were disregarded, and on the 31st of December he was dismissed from all public employment."

The Duke of Ormonde was now appointed Commander-in-Chief in Marlborough's place, and to him the Ministry in England confirmed the very perquisites which the House had just declared to be unwarrantable and illegal.

Fortescue writes: "Effrontery and folly such as this are nothing new in representative assemblies, but it is significant of the general attitude of English civilians towards English soldiers, that not one of Harley's gang seems to have realised that this vindictive persecution of Marlborough was an insult to a brave army as well as a shameful injustice to a great man."

To continue quoting from Fortescue: "It is not necessary to dwell on the operations, if such they may be called, of the Duke of Ormonde. He did indeed take the field with Eugene, but under instructions to engage neither in a battle nor a siege, but virtually to open communications with Villars.... Ormonde was directed to suspend hostilities for two months, and to withdraw his forces from Eugene. Then the troubles began. The auxiliary troops in the pay of England flatly refused to obey the order to leave Eugene, and Ormonde was compelled to march away with the British troops only.... The British and the Auxiliaries were not permitted to speak to each other.... The parting was one of the most remarkable scenes ever witnessed. The British fell in, silent, shamefaced, and miserable; the auxiliaries gathered in knots opposite to them, and both parties gazed at each other mournfully without saying a word. Then the drums beat the march and regiment after regiment "tramped" away with full

9 Lieutenants.
9 Cornets.
9 Quarter Masters.
1 Chaplain.
1 Adjutant.
1 Surgeon.
9 Sergeants.
18 Corporals.
9 Trumpets.
9 Hautbois.
180 Privates.

1764 Sec. of State's Letters for Ireland.

August. Alteration in the clothing of the regiments of cavalry in Ireland;

The Officers and men of the 5th Royal Irish Dragoons are to have epaulettes on the left shoulder instead of shoulder knots and are to wear light boots.

Height of men from 5 feet 8½ inches to 5 feet 10½ inches.

Height of horses 15 hands to 15 hands 1 inch.

1770 and 1771 Muster Rolls

The 5th Dragoons were stationed at Athlone (head quarters), Navan, Belturbert, Roscommon.

1772 Muster Rolls

The 5th Dragoons were stationed at Athlone, Dublin, Donaghmore, Roscommon, Belturbet, Cappoquin.

1775 and 1776 Muster Rolls

The 5th Dragoons were stationed at Clonmel, Clogheen, Kilkenny, Carrick, Casheel, Tallow.

In January, 1776, a whole troop was drafted to the Infantry, and in the December of the same year a whole troop deserted.

1777-8-9 Muster Rolls

The Stations of the Regiment were Roscommon, Athlone, Ballyraget, Kilkenny, Longford, Fermoy, Carlow, Carrickfergus and Belfast.

Three troops were stationed at Phillipstown.
Three " " " " Tullamore.
Three " " " " Kilkenny.

1780 Muster Rolls

Sir Joseph Yorke's troop at Dublin.
Lt.-Col. James Stewart's troop at Dublin.
Major King's " " Dublin.
Cap. Hunt's " " Ballyraget.
Capt. Hugh Massey's " " Dublin.

1782 Muster Rolls

At the " Man of War " Oct. 22nd, 1782.

" Sir,

I have the honour to acquaint you, that agreeable to my instructions, I was proceeding with the three troops to Drogheda, and was within half a mile of the town stopped by the Mayor and Recorder of that city, informing me that everything was perfectly quiet and that if the troops were to go into the town they had reason to fear very disagreeable consequences might arise. They therefore directed me to return with the troops to this place and I arrived here at 5 o'clock this morning, where I shall remain for further orders. Enclosed I have the honor to send you the Mayor and Recorder's order for my returning here. The accommodation here is extremely bad and insufficient for both men and horses.

(Sgd) James Allen,
Capt. 5th Dragoons.

To
The Right Honble,
The Commander-in-chief.

Soldiering in Ireland about the year 1784 was of a pleasing nature to the individual soldier, and is thus described by John Francis Smet, Surgeon of the 8th Light Dragoons :—

" Cavalry Corps in Ireland were extremely select, as from the very low establishment, it was in the power of the Colonels of choosing among a number of young gentlemen of distinction who might wish to get a commission, and who all could easily afford to add a hundred pounds a year to their pay. The warrants were also purchased at a high price, often by the sons of gentlemen for as much as five hundred guineas. The privates were always young men well recommended and whose connections were known. Indeed, the dragoon service was at that time extremely easy and pleasant, so much so, that when a vacancy happened, several desirable recruits always offered, and the man selected in general, got no more than one shilling bounty. Two thirds of the officers had in general leave of absence for the greater part of the year. Many of the dragoons were often on furlough, who were sometimes allowed to take their horses with them to their parents' houses, and generally wore their own clothes while with their friends. The horses were a considerable time of the year at grass, when the proportion of furlough men was usually greater than at other times; but the whole corps assembled at headquarters once a year and were kept together for a couple of months to perfect themselves in its evolutions preparatory to its being reviewed, after which most of the officers were again indulged in leave of absence, many of the men allowed to go on furlough and several troops detached to out quarters. Such a service had many attractions. While detached, everybody had an opportunity of amusing himself; the officers with the gentlemen whose estates were in the neighbourhood, and the men among the farmers, having hardly any duty to do; but when the regiments prepared to be reviewed, then was fashionable dissipation carried to a great length. And even some scenes and theatrical decorations, concerts, private plays, balls, dinners, and suppers in the evening, which gave the country towns where headquarters happened to be, an air of life and gaiety which they only possessed in an inferior degree at other times. The Officers now meeting again, after such long separation from each other, in affluent circumstances, which they had improved while they had lived with their friends, justly looked on the time of the year they were to be reviewed in as the pleasantest season. The mornings were spent at exercise and the remainder of the time in festivity." [1]

A pleasant enough time, no doubt, but scarcely soldiering. From 1784 to 1798 is not a long cry, and we are yet to read of the Rebellion in Ireland, of the state of insubordination and licentiousness of the Army in Ireland, of a regiment of Dragoon Guards behaving infamously in the face of the attack of a French force, of four regiments of Dragoon Guards removed to England to relearn the discipline they had lost in Ireland, and finally, and that which cuts us most to the heart, of the disbanding of the 5th Dragoons. Through small fault of their own were these gallant and well tried regiments brought into disrepute. The foregoing account of dragoon soldiering by Surgeon Smet tells the tale, and in spite of the Cavalry in Ireland in 1784 and 1786, the British Government took no heed; and the natural consequence was the appalling occurences which have been mentioned and of which we are yet to read.

1. Smet's *Historical Record of the 8th Hussars.*

Clonmel. 16 June, 1785.

" That your Memorialist took a field of exercise for the 5th Dragoons during their assembly and review at Cashell, in the last and present months, as there was no common ground near that town. That he was obliged to pay 13 guineas for that field, as he could not get a proper place for less money. Praying to be allowed to charge the said sum to the contingent bill of said Regiment in the usual manner.

Military Memorials Dublin.

(Signed) James Stewart,
Lt.-Col. 5 Dragoons and
Colonel in the Army.

1787 Muster Rolls

Robert Cunningham's troop at Tullamore.
Earl of Errol's " " Tipperary.

1790

The Regiment was in Dublin.

CHAPTER IX.

The Rebellion in Ireland 1798.—*Disbandment of the* 5*th Dragoons* 1799.

With the origin and causes of the disaffection in Ireland which led up to the rebellion of 1798 we have nothing to do; and as regards the rebellion itself, we confine ourselves to the part played by the 5th Dragoons in its suppression. The barbarous murders and massacres enacted during its course will not be related here.

We have no thrilling charges or hard fought battles against an honorable or foreign foe to read of, but merely the heroic struggles of detached parties of the Regiment for the cause of Law and Order in the midst of appalling brutality and crime, and finally, the last sad scenes in the disbandment of the 5th Dragoons. The story is more than a hundred years old, and cannot be read of in this history but with regret. Our regret, however, is greatly tempered by being told of a few of the instances of the loyal part taken by the gallant soldiers of the ROYAL IRISH DRAGOONS throughout the rebellion, up to the unfortunate circumstances which brought to an untimely end the life of a famous Regiment.

At the commencement of 1798 the army in Ireland was undisciplined and demoralised. Even the few regular troops in the country were tainted by the appalling state of insubordination, in their case the result of having for years been scattered in small parties throughout the country for the protection of small towns and the estates of country gentlemen, with but little supervision or control. " It was a trial which would have corrupted the Ironsides under Cromwell himself. "

The French were meditating a descent upon Ireland, and so great was the danger all round, that Sir Ralph Abercromby was sent over from England to the command of the troops.

Abercromby found the state of the army to be so bad, that one of his first acts towards reform was the publishing of the following General Order, dated the 26th of February 1798 :—

"The very disgraceful frequency of courts-martial and the many complaints of irregularities in the conduct of the troops in this Kingdom, have too unfortunately proved the army to be in a state of licentiousness which must render it formidable to every one but the enemy."

The situation he found himself in was, however, impossible. The Lord Lieutenant was adverse to many of his measures for reform, and he obtained but little support from the English Government. Declining to be a cipher or a tool in the hands of the party that governed the country, he resigned his post. The Government found a difficulty in appointing a successor, and were obliged to place the command temporarily in the hands of the senior officer, General Lake. Fearing that he too might perchance actively endeavour to restore the army to its proper state of discipline, the Lord Lieutenant decreed that no General Order should be issued until first submitted to himself. Such a proceeding, with the army in a state of insubordination and licentiousness, and the country itself in a grave state of discontent, was a sure step towards a rebellion in Ireland.

On the 23rd of May, 1798, a general rising throughout the disaffected parts of Ireland was intended by the rebels. Owing to its discovery in many places, however, the movement was not as wide-spread as it might have been. A partial rising took place in Wexford, Waterford, Wicklow, Kildare and other counties.

"On the outbreak of the rebellion, the 5th Dragoons, standing high in the confidence and estimation of the Government, was ordered to march, with all speed, to be contiguous to the capital, and took up its quarters at Lehaunstown Huts, a position within seven miles of Dublin. Although several of the cavalry regiments were quartered in the immediate distance, this Regiment was brought upwards of one hundred miles to act upon a duty, the most important perhaps, that has occurred during the history of Ireland."

Though the first effort of the rebels to rise in Dublin had been defeated by the vigilance of the authorities, it was known that another attempt was to be made, which resulted in the publication of the following notice to the inhabitants of the metropolis on the 24th of May :—

"Lieutenant-General Lake, commanding his Majesty's forces in this Kingdom, having received from His Excellency the Lord Lieutenant full powers to put down the rebellion, and to punish rebels in the most summary manner, according to martial law, does hereby give notice to all His Majesty's subjects, that he is determined to exert the powers entrusted to him in the most vigorous manner, for the immediate suppression of the same ; and that all persons acting in the present rebellion, or in any wise aiding or assisting therein, will be treated by him as rebels, and punished accordingly. And Lieut.-General Lake hereby requires all the inhabitants of the city of Dublin (the great officers of State, Members of the Houses of Parliament, Privy Councillors, Magistrates, and Military persons in uniform excepted) to remain within their respective dwellings from nine o'clock at night till five in the morning, under pain of punishment."

The Viceroy, Lord Camden, received news that a body of rebels had risen at Rathfarnham, a village about three miles from Dublin, with the intention of attacking a small force of yeomanry there. Lieutenant O'Reily with a troop of the 5th Dragoons was on May the 23rd ordered off in pursuit, and with him went the Earl of Roden and Lieut. Col. Puleston as volunteers. On arriving at Rathfarnham, the 5th were informed that the rebels had gone towards Rathcool. They rode towards that place, and on the way met the yeomanry, who were retreating, after having attacked and been repulsed by the rebels.

The troop of Dragoons halted, and it was agreed that Lord Roden should take half the troop up the road to the right, and Lieutenant O'Reily with the other half troop should go to the left, in order to surround the rebels.

Roden's party came up with the rebels at the first turnpike gate on the Rathcool road, and engaging them, drove them back on to O'Reily's half troop ; the latter killed two and wounded a good many more, but the majority managed to escape, the much-enclosed country preventing a pursuit. The bodies of James Byrne and James Keely, two of the leaders who were killed, were afterwards exhibited to public view in the Castle Yard, Dublin.

At Kildare, on the night of the 23rd of May, an old pensioner of the 5th Dragoons named George Crawford, and his grandchild, a girl of fourteen years, were savagely murdered. He and his wife and the girl were stopped by a party of the rebels, one of whom struck the wife with a musket while another stabbed her in the back with a pike. Her husband endeavouring to save her, was knocked down and disabled by repeated blows from a firelock. While the rebels disputed whether they should kill him, his wife stole behind a hedge and hid herself. They then massacred Crawford with pikes, and the grand-daughter, having thrown herself on his body to protect him, received so many wounds that she also expired. "The fidelity of a large dog, belonging to this poor man, deserves to be recorded, as he attacked these sanguinary monsters, and fought bravely in the defence of his master till he fell by his side, perforated with pikes."

On the 29th of May, two dragoons were sent from Dublin with an express to Lord ROSSMORE, the Colonel of the 5th DRAGOONS, at Newtown-Mount-Kennedy. When within about two miles of that place they were fired on by some rebels lining the hedges on the roadside. The dragoon carrying the express was killed, yet his comrade dismounted, and under fire, took the message out of the dead man's pocket, and remounting, galloped off and safely delivered it to Lord ROSSMORE.

On the 2nd of June a large party of rebels were at Ratoath, some eighteen miles from Dublin, and against these went a party of eleven highlanders and four yeomen. On the way this small party met Mr. Frederick Falkiner and eighteen of the 5th DRAGOONS, who joined them, and at Ratoath charged and dispersed the rebels, killing thirty-five of them in the pursuit.

On the 4th of June an action was fought under General Loftus near Gorey in Wexford, against a large body of rebels. Fifty men of the 5th DRAGOONS were on the left of the loyalists. The centre of the force was under the command of Colonel Walpole, whose lack of prudence and military skill resulted in his total defeat; and his neglect to communicate with General Loftus' force on his left, prevented any reinforcements reaching him in time. The defeat of Walpole left Loftus with but the fifty men of the 5th DRAGOONS, commanded by Capt. Corry, and two hundred of the Dunbarton Fencibles. The situation was an alarming one, for the rebel army from Vinegar Hill had formed a junction with that at Ballymore only a couple of miles or so to the front, and amounted to some twenty thousand men.

The small force then moved back on Gorey, which was found to be in possession of the rebels. Loftus was now between two rebel forces, and he quickly made up his mind that the only way to save his small detachment was to endeavour to fight his way through Gorey. He pushed rapidly on to that town under a heavy fire from the surrounding hills, to which he did not reply. On nearing the place, he, Captain Corry, and Colonel Scott reconnoitred it, and discovered it to be so strongly held that an attempt to force a way through appeared hopeless. The only possible way of escape seemed to be to march round the right flank of the rebel position at Gorey, and to endeavour to join Lord Ancram's force at Carnew, some twelve miles away to the north west, which could only be done by moving south west and crossing the Slievebuoy mountain, and then moving north.

Lieut.-Colonel Scott was at once sent off with the Dunbarton Fencibles, while the General remained behind with Corry and the 5th DRAGOONS to watch the enemy. Corry managed to engage the attention of the rebels by pretending to move on Gorey, and so enabled Scott to cross the mountain unmolested. Corry then quickly followed, and Carnew was occupied early the next morning. This action was a spirited performance, and to quote from the historian : " Thus this small body fairly marched round twenty thousand rebels ; and by the good countenance which they kept, and by preserving their fire, though constantly fired at, they escaped without any injury. "

To turn to the doings of the 5th DRAGOONS in another part of Wexford, we find two troops of the Regiment, under the command of Captain Irwine and Captain Ridge, who had been on the march from Tallow in Waterford to join headquarters at Lehaunstown, diverted at Kilkenny to reinforce the garrison at Ross in Wexford, in consequence of the preparations of a large force of rebels to attack that town. On the 2nd of June Irwine and his Dragoons arrived at Ross, where they found the Clare, Donegal and Meath regiments of Militia, detachments of English artillery and Mid Lothian Fencibles ; and the rebels collecting a force some eighteen thousand strong at Carrickbyrne, about five miles south east of Ross.

In consequence of the long and rapid march, many sick and weak horses were left at Waterford, and each troop in the fight which now took place could therefore only mount thirty odd files.

On the 4th of June the County of Dublin regiment marched in to Ross, and on the same day the rebel army moved to Corbet

Hill, an eminence about a mile and a half from the town, having driven in the loyalist outpost.

A person who was forced by the rebels to attend the march, stated afterwards that their army was organised into parishes and baronies, each having a particular standard; and on their way they stopped at a chapel, where mass was said by the priests at the head of each column.

The garrison of Ross consisted of some fourteen hundred men under the command of General Johnson, and remained under arms all night. The infantry and artillery were mostly in a line outside the walls of the town. On the east and south sides the 5th DRAGOONS were on the quay by the river side, and the yeomen infantry on the bridge over the Barrow.

About four o'clock on the morning of the 5th of June, a sentinel on an outpost shot dead a man who was galloping towards him waving a white handkerchief. On his body was found a letter signed by Bagenal Harvey, commanding "the Army of Ireland," and addressed to the Officer commanding "the King of England's Forces at Ross," to the following effect:—

"As a friend to humanity, I request you will surrender the town of Ross to the Wexford forces, now assembled against that town; your resistance will but provoke rapine and plunder, to the ruin of the most innocent. Flushed with victory, the Wexford forces, now innumerable and irresistible will not be controlled, if they meet with resistance. To prevent, therefore, the total ruin of all property in the town, I urge you to a speedy surrender, which you will be forced to in a few hours, with loss and blood-shed, as you are surrounded on all sides. Your answer is required in four hours. Mr Furlong carries this letter, and will bring the answer."

At about five o'clock, the historian relates, "no less than thirty thousand rebels approached the town, with terrific yells, having four pieces of cannon, besides swivels," their priests with vestments on and crucifixes in their hands marching in the ranks.

"They moved with slow but irresistible progress, like an immense body of lava, which issuing from the bowels of Vesuvius, spreads over the plains of Calabria, and from which man alone can escape, and that by flight only."

One column set fire to the suburbs and forced a number of horned cattle before them through the smoke in order to create confusion in the garrison. The first onslaught was met by the Dublin Militia under Lord Mountjoy, supported by the Clare regiment at Three-bullet-gate. A sanguinary struggle now ensued, but the great numbers of the rebels began to tell; the gallant Mountjoy was killed at the head of his regiment, a gun was captured, and the troops forced into the town. A column of rebel pikemen penetrated at another point, and the garrison was thrown into confusion, and retreat became necessary.

Meanwhile Captain Irwine had collected about sixty men of his two troops of 5th DRAGOONS, and forming them up as best he could, was ordered by General Johnson to charge the rebels at the Three-bullet gate, to gain time for the infantry to retire.

To quote the historian's own words: " This was a service replete with danger, as from the situation of the place, and the continual increase of a desperate enemy, a handful of men seemed precluded from every hope of escaping destruction. Notwithstanding, the order was instantly obeyed, and the detachment, not forgetting that they were a portion of the Royal Irish Dragoons, rode to meet bodies of insurgents advancing against them, armed with pikes from ten to twelve feet long. Nor were the rebels inexperienced in the practice of this formidable weapon. Their instructions were to pierce the horse in the flank, and thus obtain an easy conquest over the rider. " A shot from a twelve pounder at the Three-bullet gate killed Captain Irwine's horse, which fell on his leg, and prevented him from moving for some time. Luckily for him, a loose artillery horse passed, and, laying hold of one of the traces, he was dragged clear of the dead horse and after the retreating infantry into the town. Three times did the Dragoons charge with an enormous loss to themselves.

" Such was the lot of these brave devoted men who in the action at Ross have written a testimonial with their blood, that might at least have softened the rigorous order which was impending over their Regiment, and which in reality was excited more by an unlucky occurrence of unfavorable appearances, and indiscreet reports, than by any actual transgression of its own. We are far from calling the propriety of the order in question, because we are convinced that in calmer times, the grounds upon which that order was given would have been leisurely weighed and maturely executed. "

The Quarter Master of the 5th at last found himself in command, and with only nine men, he with difficulty managed to retreat to the opposite side of the river, where the infantry were rapidly forming up.

During the retreat of the infantry the rebels poured into the town in great numbers. General Johnson planted guns at several cross roads, but the rebels, in spite of great slaughter, continued to press on. " One rebel, emboldened by fanaticism and drunkenness, advanced before his comrades, seized a gun, crammed his hat and wig into it, and cried out, ' Come on, boys ! her mouth is stopped. ' At that instant the gunner laid the match to the gun, and blew the unfortunate savage to atoms. This fact has been verified by the affidavit of a person who saw it from a window. "

The gallant Johnson now rallied the troops on the Kilkenny side of the river and persuaded them to recross into Ross.

" He exhorted his troops to recover their ground and concluded by desiring such as were willing to conquer or die, to follow their general. An awful silence prevailed for a few minutes ; after which the Quarter-Master and the few men remaining of the 5th Dragoons who were able to sit on their horses, pressed eagerly forward and exclaimed, ' They were willing to shed the last drop of their blood in support of their General, and to revenge their fallen comrades. ' A shout of applause followed from every part of the column and terminated with three cheers and a general cry of ' God Save the King and success to General Johnson. "

The troops, and with them the small remnant of the 5th Dragoons, returned to the charge ; Ross was retaken, and an immense carnage of the rebels ensued.

The enemy's losses in killed counted on the field was two thousand six hundred, besides numbers they carried off on cars. The losses of the King's troops were, killed, one Colonel, one Cornet, one Ensign, four Sergeants, three drummers, eighty one rank and file and fifty four horses ; the wounded were one captain, one drummer, fifty-four rank and file and five horses ; while missing were one captain, three lieutenants, one ensign, two sergeants, two corporals, seventy-two rank and file and four horses. Of these losses the 5th Dragoons had in killed alone, Cornet Dodwell and twenty-eight men.

On the 5th of June General Needham led a force, amongst them being a detachment of the 5th Dragoons, to recapture the town of Arklow. The dragoons were sent on ahead to reconnoitre, and were able to take possession of the town without meeting any opposition, and next day the main body marched in.

General Needham had under his command in Arklow a force of some seventeen hundred men, amongst them being one officer and eighteen men of the 5th Dragoons, and four officers and twenty-four men of the 4th Dragoon Guards.

On the morning of the 9th of June at about 11 o'clock news arrived of the advance of the enemy. The garrison turned out, and Cavalry patrols were sent off to reconnoitre, and gradually fell back before two strong columns of rebels, amounting to some twenty-five thousand men. A desperate action ensued, in which the rebels made no headway, and finally the cavalry, led by Sir W. W. Wynne, were ordered to charge, which they did with great gallantry. By 8 p.m. the rebel army was in full retreat, with a loss of a thousand killed and a large number of wounded. During the action, a priest named Murphy, to encourage the rebels, took out of his pocket some musket balls, which he said had been fired at him by the enemy, some of which had hit him without wounding; and others he had caught in his hands. He assured them that the balls of the heretics would not hurt them, as they were under the protection of the Almighty, in whose cause they were fighting. Unfortunately for his promises, Murphy was himself killed by a cannon shot later in the battle, " and the fall of this church militant hero had an immediate effect in damping the ardour of the enemy. " The men of the detachment of the 5th Dragoons covered themselves with glory during this desperate action, and worthily upholding the honour of the Regiment they belonged to, were publicly thanked for their services by General Needham.

On the 18th of June General Johnson moved from Ross towards Vinegar Hill, where the rebels had their headquarters. Reinforcements had brought his army up to four thousand men, and he was able to leave a small garrison in Ross. A detachment, consisting of the few survivors of the 5th Dragoons from the battle at Ross, a part of Hompesch's mounted riflemen, and two companies of sharp-shooters, were sent on in advance, with orders to reconnoitre the rebel position and drive in their outposts. This was done so successfully, that the rebels were pursued with considerable slaughter to within musket shot of their own entrenchments.

On the 21st of June was fought the battle of Vinegar Hill.

General Johnson's column moved against the position on the West side, while General Lake advanced from the North East.

On the evening of the 20th, Johnson was opposed by some rebels who moved out of Enniscorthy, a village on the Slaney on the western foot of Vinegar Hill. The rebels posted themselves on an eminence, against which Johnson fired a few shots from his twelve pounders; " and when the balls lodged on the hill, the rebels vied with each other to lay hold of them. " Johnson then fired some shells, which numbers of rebels " having surrounded for the same purpose, exploded, and blew them to atoms. "

Early next morning, having driven the rebels from the high ground into Enniscorthy, Johnson commenced his attack on the town. The place, however, was very strongly held and fortified, and Johnson's storming party of infantry was driven back with the loss of the guns which had accompanied them. The General sent up reinforcements, amongst them being the detachment of the 5th Dragoons, and another attack against Enniscorthy began. " The 5th Dragoons being the eldest regiment present, " took the post of honour and " led the charge into the town. " The rebels were driven out with " considerable carnage, " Enniscorthy was captured, and the lost guns retaken. The infantry were then ordered across the bridge over the Slaney, and pressed up the steep sides of Vinegar Hill, driving the rebels before them.

General Lake had meanwhile commenced his attack on the north east side, and now joined hands with Johnson, and the rebels abandoned the hill and fled in " wild consternation. "

Captain Ledwell, with another detachment of the 5th Dragoons from Lake's force, now joined their comrades with Johnson, and these representatives of the Irish Dragoons vigorously pursued the rebels. " Their intrepidity and zeal was not a little quickened by the sense with which they went into action, of the loss the corps had sustained in the deaths of so many of their brave comrades at Ross. "

The following paper was found in the pocket of a rebel who had been shot by Captain Hugh Moore, of the 5th Dragoons, A.D.C. to General Needham :—

" Jesus I.H.S. Maria
I trust thee

This is measured of the wounds
of the side of our Lord Jesus Christ

> which was brought from Constantinople unto the Emperor Charles, within a gold chest, as a relief most precious to that effect, that no evil or anything might take him which reads it, hears it, wears it, cannot be hurted by any tempest, fire, water, knife, sword lance or bullet. Neither the devil shall hurt him ; he shall be victorious, and never die an unnatural death, and shall be a sure safety to women with child.

At Ballymore Ustace and in other places "the same zeal and esprit-de-corps for which the Royal Irish Dragoons had been so remarkable for more than a century, were singularly conspicuous."

In the midst of barbarous murders and treason, the 5th Dragoons continued to do their duty, while many of the King's regiments were more or less tainted with sedition, and the entire army in Ireland was in an extremely bad state of discipline. On the 27th of August, General Lake, with some seventeen hundred men and eleven guns, in a strong position on the heights of Castlebar, was confidently awaiting the attack of Humbert and his thousand Frenchmen, who had landed a few days before at Killala Bay. The attack was delivered, and the whole of Lake's force fled in a panic. The Longford and Kilkenny Militia, Fraser's Fencibles, Galway Volunteers, and, we regret to read, a regiment of Dragoon Guards who shall be nameless, fled ignominiously before the disciplined troops of the French General. The Artillery, a hundred men of the 6th Foot, and a few of Roden's Fencibles, alone stuck to their ground. The depredations of this force on the march too had "exceeded all description."

To such an ebb had discipline dwindled, that an officer of the Guards writing of his men, stated that if kept in the country another six months he would not answer for their subordination.

Again, in July, 1799, four regiments of Dragoon Guards, amongst them the offenders mentioned above, "were removed to England to relearn the discipline which, through small fault of their own, they had forgotten."

But another fate was reserved for the 5th Dragoons. The Regiment had behaved with great gallantry and much self-sacrifice during the rebellion, but trouble was now to come, and a brave and gallant corps was to be charged with the greatest of crimes a soldier is capable of, that of disloyalty to his King.

A number of recruits had enlisted into the 5th with the intention of acting in concert with a party of rebels in attacking a force of troops engaged in stamping out the last dying embers of the rebellion. The plot was discovered in time, but there had apparently been much desertion in the Regiment, and it was said that there were men in the ranks who had taken the oath of the United Irishmen. In addition to this, the Lord Lieutenant, Lord Cornwallis, had reported the want of discipline in the Corps, and recommended that it should be removed from Ireland. The King, however, took a far stronger view, and ordered its disbandment. The officers of the Regiment petitioned for a Court of Enquiry, but this was refused. How the sentence was carried out will be read in the last pages of this chapter.

The following story of the closing scenes in the life of the 5th Dragoons is taken from an account published in the *British Military Library* of April, 1800 :—

" In reading this brief and unaffected narrative of unquestionable facts, it will not be possible for the man of candour to forbear looking back upon a series of brilliant actions, and after he has passed them in review, to cast an eye of regret upon the fallen state of such an ancient corps.

If being unfortunate constitutes transgression, then the 5th or Royal Irish Dragoons, may truly be said to have been highly culpable ; for never did a more unlucky combination of accidental circumstances, happen to overwhelm a body of men ; nor was that combination rendered less murderous in its effects by the cordial co-operation of *every* person connected with the corps to avert its consequences. We lament to say, not upon vague grounds, but upon strict, honourable and forthcoming evidence, that the resentment of government was, in all probability, not a little increased by personal misrepresentations and private views. We have been credibly informed that it was the wish, because it was the interest, of some individuals, to get the 5th Dragoons removed from the Irish establishment, for the purpose of enhancing the value of their commissions, in the event of its being sent to England. It is possible that under such impressions, certain persons might not have been without hopes that should government be deceived into a belief of the Corps being disaffected, His Majesty might be induced to order it to the Cape of Good Hope or to the East Indies. In either of which cases considerable advantages were looked for by those persons who consulted their own private emolument at the expense of the Corps.

Although we think it right to furnish the public with every authentic particular, which may lead hereafter to a thorough knowledge of one of the most singular instances of censure and disgrace, that has occurred in the military history of Great Britain, nevertheless we are equally aware that nothing should go forth, which may affect the honour and reputation of responsible individuals. Far be it, therefore, from us to insinuate, that anything herein mentioned could possibly allude to the Colonel or Commanding Officer of the Regiment. We will not hesitate to say that they exerted themselves to the utmost, in vindicating the character of their Corps. Nor should it be forgotten, that at this melancholy period, almost every Regiment belonging to the Irish establishment, was more or less tainted by the admission of disaffected persons, who, under the fictitious characters of persecuted Royalists, that had been driven from their homes by the rebels sought refuge in the Army. How the Regiment in question was visited by this insidious evil, will appear by the following statement.

Some time after the 5th Dragoons had arrived at Lehaunstown Huts an order was received that the strength of the Regiment should forthwith march for Dublin. It was, however, directed that a few men from each troop should be left behind to take charge of the heavy baggage, sick men and horses, &c. subsequent to which several small detachments were sent out to different stations. The Officers who commanded them were instructed to receive eligible recruits. Many very fine fellows were accordingly enlisted, particularly at a place called Castle Corner, and were sent to Head Quarters. It is much to be lamented (although in justice to the Officers we acknowledge the moral impossibility of diving into the minds of men) that some precautions had not been taken to ascertain the real character of every individual who offered to be enrolled. Almost all of them, as the event afterwards evinced were rebel partizans, and had insinuated themselves into the 5th Dragoons, agreeable to a preconcerted plan for surprising Lehaunstown ; to which place all the recruits and men unfit for immediate service had been transmitted. A conspiracy was accordingly entered into by these new comers, in concert with the rebels in the adjoining mountains. The design was, that on a certain night an attack should be made on the garrison, whose whole effective strength consisted of about seventy dragoons, many of them invalids, and somewhat more than an equal number of the King's County Militia. The recruits to a man were concerned in this black plot ; and the massacre of every officer and loyal soldier was prevented, only an hour before the scheme was proposed to take place. The conspirators were seized and suffered according to their deserts. The Regiment, however, had the mortification to find it announced in the public papers, that several privates belonging to the 5th or Royal Irish Dragoons, had been found guilty by a general Court Martial of joining the rebels. Some little time after two brothers of the name of Fen'ey, who had deserted from Drogheda (where the Regiment was stationed after the rebellion had been suppressed) and who, under pretence of procuring arms for the rebels, had pillaged several houses in that neighbourhood, were taken by the yeomanry in the act of thieving, and would have been executed on the spot, had they not promised to discover the names of several dragoons, who, they said, were concerned in the same nefarious practices, and were sworn to support the rebel cause.

The influence of scandal is well know to acquire multiplied virulence in its progress. A malignant report, having once obtained circulation, is shortly so much metamorphosed, that the original fabricator can with difficulty recognise his own bantling. The truth of this remark was strongly illustrated in the present instance. It was industriously propagated that a most fortunate discovery had been made relative to the 5th Dragoons. This Regiment was stated by some people, to have been corrupted by the rebels, and to have formed a plot to massacre the yeomanry ; by others it was said, with equal ingenuity, and no less malice, that the 5th Dragoons had organized a conspiracy to murder their own officers, and afterwards to join the rebels with their horses, accoutrements &c. Such were the gossip stories of the day.

Unfortunately for the cause of truth the Commanding Officer was at this time absent from the Regiment, on leave. It is to be feared that reports, scarcely less exaggerated, were from self interested motives communicated to Government by persons of seeming respectability. The Feneys, under a strong escort, were ordered up to Dublin to be tried by a General Court Martial, and every measure was adopted to bring this serious charge to light. A man of the name of M'Nassar was transmitted along with the two brothers. The Feneys were accordingly brought into Drogheda, escorted by the yeomanry under the charge of having confessed a combination between themselves and several others belonging to the 5th Dragoons.

The Commanding Officer of the garrison, and some gentlemen belonging to the 5th Dragoons, attended on this occasion for the purpose of receiving their information. The result of this enquiry terminated in the *important* discovery that *one* man, by name James M'Nassar, constituted the whole of this dreadful plot, as far as it regarded the Royal Irish Dragoons. He was accordingly apprehended. It is here necessary to state that James M'Nassar had, for a considerable time, been looked upon as a very bad character ; he had been repeatedly punished, and was drafted into the wagon corps, who sent him back to the Regiment as unfit to serve in that capacity. A General Court Martial was ordered to sit in Dublin Barracks. James M'Nassar having turned King's Evidence, the veracity of the Feneys, and the character of the 5th Royal Irish Dragoons were at issue. The result of this interesting and important trial was, " *That the two Feneys should suffer death, having been convicted of robbery and other evil practices ; and that James M'Nassar should be transported beyond the seas.* " But not an iota of charge appeared throughout the proceedings that could impeach any other individual belonging to the Regiment, which at the time consisted of more than six hundred men. It is incumbent, however, upon us to declare, that a man named Ryan, a degraded sergeant, and a Corporal Gallagher were suspected by the Commanding Officer, and were confined, but no specific charge of disaffection could be proved against them.

Such were the ostensible marks of criminality under which the 5th Dragoons so unfortunately fell. We have collected the particulars with as strict a regard to truth, as the delicacy of the subject merits, and we make no doubt but we shall stand justified for having so done, in the eyes of every candid and impartial soldier.

On the 10th of April following (1799) the 5th or Royal Irish Dragoons was disbanded at Chatham in conformity to His Majesty's order.

Order for disbanding the 5th Regiment of Dragoons.

Horse Guards, April 8th, 1799.

" His Majesty has taken into his most serious consideration, the representation which has been made by His Excellency the Lord Lieutenant of Ireland, of the conduct of the Fifth, or Royal Irish Regiment of Dragoons; and is of opinion that the insubordination, and the departure from discipline, and principles which have ever distinguished the British Army, therein exhibited, require, especially in these times of warfare and exertion, that they should be marked by a punishment which may be severely felt, and be long remembered by those misguided persons, who have been guilty of the atrocious acts of disobedience, which have brought this indelible stigma on the corps; and may serve as an example to all others, of the consequences of such seditious and outrageous proceedings, and of His Majesty's firm determination to maintain subordination and discipline in his Army, and to support the authority of his Officers in the execution of their duty.

It is on these grounds His Majesty's determination, that the 5th or Royal Irish Regiment of Dragoons, shall be forthwith disbanded.

At the same time that the King judges it requisite for the good of the Service, to make this severe example, His Majesty has graciously condescended to direct that General Lord Rossmore shall be assured, that His Majesty is persuaded of the concern which, as a soldier, his Lordship would feel at such a circumstance occurring in any part of the Army, and is sensible of the particular mortification he must experience in the present instance, from the event of which, however, his Lordship cannot, in the smallest degree, suffer in His Majesty's estimation.

His Majesty is graciously pleased further to express his persuasion, that there are many valuable officers in the Regiment, who have used their best endeavours to restore the order, and to preserve the credit of the corps; and, though in this measure of indispensible severity, it was impossible to make any exceptions, the majority being clearly implicated in the misconduct for which the whole are suffering; yet His Majesty will hereafter make the most pointed descrimination, and those of any rank who are deserving the Royal favour, may rely on His Majesty's disposition to attend to their merits, and to avail himself of their future services. In consideration of the expense to which the officers of the 5th or Royal Irish Regiment of Dragoons have been unavoidably exposed, His Majesty has been graciously pleased to direct, that their full pay shall be continued to them, till the 24th December next, at which period they will be placed on half-pay.

(Signed) Harry Calvert.
Adjutant General."

"The attention which Lord Rossmore uniformly paid to the good order and discipline of the 5th Dragoons, can only be equalled by the heart-felt regret, with which he has been obliged to abandon his trust. His Lordship was, in fact wrapt up in the Regiment he commanded; and every Officer still concurs in expressing the fullest testimony of his unremitting zeal and personal sacrifices for its welfare."

"Previous to this event, the Officers had sent in a Memorial to Government, praying for an investigation of their conduct, and of the character and behaviour of the Regiment, before a Board of General Officers. They failed in their petition, and every attempt at producing an enquiry into the business, was rendered fruitless. We are, however, happy to add from very unquestionable authority, that the Regiment marched above two hundred miles on foot, through this country (England), in perfect good order. It remained some weeks in Chatham before its final dissolution and was publicly thanked by General Fox for its exemplary good behaviour during the march, and its unremitting regularity whilst it was under his command. The conduct of the Regiment was the more praiseworthy on these trying occasions, as it was repeatedly provoked by allusions to the cause of its dishonour, and was frequently reproached by the inhabitants as it marched through the towns, and not a little mortified by the significant whispers of other soldiers. The privates were drafted into other regiments of the line, in which they have since obtained the best testimonies of good behaviour, cleanliness and discipline. Several among them have since been appointed Non-commissioned officers and continue to prove that the disgrace they suffered, has not diminished their zeal and attachment to the Service.

We cannot conclude without expressing our regret that a Regiment which has so frequently deserved well of its country, should have incurred His Majesty's displeasure. Let us hope, that like the Phœnix, it may some time or other rise out of is own ashes, be restored to the Army, and add fresh laurels to those of Ramillies and Hochstet.

Time we trust, and a calm re-consideration of presumed delinquency, will restore this Veteran Regiment to the rank it formerly held with so much satisfaction to the different sovereigns under whose banners it fought, and with so much credit to itself."

Such was the end of the Royal Irish Dragoons. Its most distinguished career was not to save the Regiment from dishonour. The stories of the captured kettledrums at Blenheim, and of the surrendered battalions of Picardie on the field of Ramillies were to pass from the keeping of its sons to the dull pages of the historian, while for years the blank page in the Army List was to remain a standing reproach to the British Army.

CHAPTER X.

The re-raising of the Regiment, 1858.—*Services from* 1859-1884.—*Ireland.*—*Aldershot. The Non-Commissioned Officers' chevrons and the Irish Harp.*—*Hounslow.*—*Light Dragoons to Lancers.*—*Aldershot.*—*Norwich.*—*Indian Service.*—*Lucknow.*—*Sealkote.*—*Hussan Abdul.*—*Homeward bound.*—*Colchester.*—*Aldershot.*—*The chevron Badges.*—*Norwich.*—*The Norwich Cup.*—*Brighton.*—*Woolwich.*—*York.*—*Newbridge.*—*The old battle honours.*—*Dublin.*

For fifty-nine years the figure " 5 " found no place amongst the Cavalry regiments of the Line in the Army List. From the 4th Light Dragoons one turned the pages to the 6th Dragoons. During this period the regiments of the British Army had been gaining fresh laurels in the Peninsula Wars, Waterloo, the Crimea and the Indian Mutiny; but though these distinctions were denied the Royal Irish Dragoons, the names of many members of the old regiment were borne on the Peninsula and Waterloo rolls of other corps. At last, however, the hope of the historian was to be fulfilled: the figure " 5 " was to be restored to its place in the Army List, and the veteran regiment of the 5th Royal Irish Dragoons was to arise, like the Phœnix out of its own ashes, and again stand in line with the famous regiments of the British Cavalry.

In the beginning of the year 1858, in consequence of the strong force of cavalry which it was found necessary to send to our East Indian possessions for the purpose of quelling the mutiny which had broken out in the Native army in the previous year, a considerable augmentation of the existing cavalry regiments was decided on, and, in addition, the 5th Dragoons were ordered to be re-embodied.

On their re-organization the 5th Dragoons were ordered to be clothed and equipped as Lancers, and to be styled the 5th or Royal Irish (Light) Dragoons, (Lancers) and to assume the motto " Quis Separabit " in addition to the Harp and Crown which were formerly borne on the standard and appointments of the 5th Dragoons.

The following is a copy of the original order for the embodiment of the Regiment :—

General Order.

Horse Guards, 9th January, 1858.

His Royal Highness the General Commanding-in-Chief has much pleasure in communicating to the Army the Queen's Command to cancel the Adjutant General's letter dated the 8 April 1799 announcing the Royal determination of His late Majesty George the Third, to disband the 5th Royal Irish Regiment of Dragoons ; that letter is cancelled accordingly.

The Queen commands that the 5th Royal Irish Dragoons be restored to its place among the Cavalry Regiments of the line, and His Royal Highness feels assured that this mark of Her Majesty's grace and favour will be appreciated, and that the 5th Royal Irish Dragoons will emulate other regiments in discipline and loyalty, and vie with them in promoting the glory of the British Arms.

By Order of His Royal Highness
The General Commanding-in-Chief.
(Signed) G. A. Weatherall.
Adjutant General.

Major-General Sir James Charles Chatterton K.H. was appointed Colonel of the Regiment on the 23rd of February, and on the 19th of the same month Brevet Lieutenant-Colonel George Augustus Filmer Sulivan, from the Scots Greys, had been appointed Lieutenant-Colonel and was ordered to proceed to Newbridge in Ireland to commence the organization of the Regiment, the establishment of which was to be in accordance with the following War Office authority :—

War Office, 10th March, 1858.

Sir,

I have the honour to signify to you that the Queen has been pleased to order the establishment of the Regiment under your command to be formed from the 3rd February 1858 so as to consist of the numbers mentioned in the margin hereof.

I have etc.
(Signed) Peel.

For Officer Commanding
5th Dragoons, Lancers.

The establishment shewn in the margin of the letter was as follows :—

LIEUTENANT-GENERAL SIR J. C. CHATTERTON

From a painting in the possession of the CAVALRY CLUB.

8 TROOPS.

1 Colonel,
1 Lieutenant Colonel,
1 Major,
8 Captains,
8 Lieutenants,
8 Cornets,
1 Pay Master,
1 Adjutant,
1 Riding Master,
1 Quarter Master,
1 Surgeon,
1 Assistant Surgeon,
1 Veterinary Surgeon,
1 Regimental Sergeant Major,
8 Troop Sergeant-Majors,
1 Quarter Master Sergeant,
1 Pay Master Sergeant,
1 Armourer Sergeant,
1 School Master (to be appd. by U. S. of S.)
1 Saddler Sergeant,
1 Farrier Sergeant,
1 Hospital Sergeant,
1 Orderly Room Clerk,
24 Sergeants,
1 Trumpet Major,
8 Trumpeters,
32 Corporals,
8 Farriers,
537 Privates,

660 Total.

428 Troop horses.

There was no lack of recruits, for by the 21st of May the Regiment consisted of four hundred and eighty five privates, while the commissioned and non-commissioned ranks were almost complete, and the troop horses numbered two hundred and seventy one.

Of the non-commissioned officers fifty two were transferred from other regiments for the purpose of assisting in the drilling of the recruits, and the men were enlisted in the usual manner in Ireland, England and Scotland, with a few transfers from other corps.

On May the 29th the Regiment was inspected by Lieutenant-General Sir Richard England, G.C.B., commanding the Curragh Division, who wrote to Major-General Sir James Chatterton to the following effect:— "I shall be very glad if you will convey to

Lieutenant-Colonel G. Sulivan my sentiments of approbation at the progress the Regiment has made under his immediate command, which I shall not fail to submit to H. R. H. The General Commanding-in-Chief. "

The following are the names of the Officers appointed on the re-organization of the Regiment :— Colonel Sir J. C. Chatterton Bart, K.H., formerly 12th Light Dragoons.

Rank	Name	From
Lt.-Col.	G. A. F. Sulivan	From Scots Greys.
Major	Robert Portal	„ 4th Light Dragoons.
Captain	G. H. Hillier	„ 80th Foot.
„	W. H. Slade	„ 7th Hussars.
„	H. Timson	„ 6th Dragoons.
„	W. G. Dunham Massy	„ Military Train
„	A. E. McGregor	„ 14th Light Dragoons
„	A. Gamwell	„ 16th Lancers.
„	F. R. Grant	„ 12th Lancers.
Lieutenant	A. Murray	„ 12th Lancers
„	T. W. Vallance	„ 95th Foot.
„	A. Murphy	„ Military Train.
„	G. M. Bright	„ 2nd Foot
„	H. N. Salis	„ 10th Foot.
„	E. F. Weaver	„ 1st Dragoons
„	J. H. Cowan	„ 60th Rifles.
„	F. W. Carden	„ 7th Hussars.
„	W. Edgeworth	„ 8th Foot.
Cornet	R. Mather	„ „
„	R. J. W. Birch	„ „
„	A. Malcolmson	„ „
„	John Chaffey	„ „
„	G. W. V. Cotton	„ „
Pay Master	J. A. Dyer	„ „
Lieut : & Adjt.	E. F. Weaver	„ 1st Dragoons.
Quarter Master	J. Addy	„ Military Train.
Surgeon	Luke Barrow	„ Staff.
Vety. Surgeon	W. C. Lord	„ 14th Light Dragoons.

The dress and appointments of the Regiment at this period were as follows :—

Tunic of blue cloth, double breasted, scarlet facings, two rows of buttons, seven in each row, the distance between the rows being eight inches at the top and four at the bottom, lappels of Regimental facings to be worn doubled back, excepting on the line of march or in bad weather, skirt nine inches in height, and lined with black, collar two inches deep of Regimental facings and rounded in front with the following distinctions of rank.

Colonel :— collar laced all round with gold lace one inch wide. Silver crown and stars embroidered on each side.
Lieut : Colonel :— Laced as above and badge of a crown.
Major :— Laced as above and badge of a star.
Captain :— Collar laced on upper edge with crown and star.
Lieutenant :— Laced as above with badge of star.
Cornet :— Laced as above with badge of star.

The waist of the tunic to be long; plain pointed cuff of the Regimental facings ; on each shoulder a double gold cord, a welt of the Regimental facings down the sleeve and back seams, down the edge and front and round the skirt. Trousers of blue cloth with two gold lace stripes down each outward seam, each stripe ¾ inch wide leaving a light between.

Girdle of gold lace two and a half inches wide, with two crimson silk stripes.

Sword, steel mounted, half basket hilt, blade slightly curved, thirty five and a half inches long, and one and a quarter inches wide. Scabbard steel.

Belts, gold laced, with crimson silk stripes up the centre and edged with scarlet morocco.

Pouch of scarlet morocco with solid silver flap.

Gauntlets, white leather.

Cap, the square top Lancer cap covered with scarlet cloth and trimmed with gold lace, with gold lines to fasten round the neck and green horse hair plume.

Saddlery : Hussar pattern saddle with brass head and cantle. Buckles on bridle and ornaments of brass.

Sheepskin of black Ukraine lambskin with leather seat, trimmed with scarlet cloth.

Shabraque of blue cloth cut round before and behind, three feet eleven inches in length, two feet four inches in depth, trimmed with gold overall lace. Device on the fore corners V. R. and crown in gold, on the hind corners crown over harp within garter, with words " Quis Separabit " round it, and the lances crossed under the garter.

The Regiment were mounted on strong well bred horses of various colours purchased from dealers in Ireland. The average height of the horses was 15 h. 3 in.

The average height of the men was 5 ft 7 inches.

Total average weight of a lancer in marching order was 17 stone 7 lbs.

On the 1st of August 51 men volunteered to the 6th Dragoons for service in India.

In August and November the Regiment was inspected by the Earl of Cardigan and by Sir J. Chatterton, both of which inspections elicited from the Horse Guards extremely favourable reports on the great progress made in its organisation.

In March 1859 the Regiment was again inspected with satisfactory results, and for the first eighteen days in May of the same year was employed in affording aid to the Civil Power in various parts of Ireland during the elections.

On the 23rd and 24th of May, Major-General Parlby, Commanding the Cavalry Brigade of the Dublin Division, inspected the Regiment at Newbridge. On the first day the interior economy of the Regiment was gone into, and on the second day the Regiment, consisting of four strong squadrons, marched to the Curragh in review order. There they proceeded to march past at a walk, trot and gallop, and went through a variety of manœuvres under Lieutenant-Colonel Sulivan. After the inspection the Regiment was formed in close column of squadrons, and addressed by the General in the following terms : " I cannot refrain from addressing you a few words on this occasion. I assure you that I was not at all prepared to see what I have seen this day. I am not only delighted but surprised with the state in which I find the Regiment. You have performed movements to-day which would have done credit to any regiment in the service, and your advance in line was admirable, the oldest regiment could not have done it better. The personal appearance of both men and horses is most remarkable ; in fact I think you are beautifully mounted. What I have seen yesterday and to-day, the almost total absence of crime, the reports that have been made of your excellent behaviour while employed in aid of the civil power during the elections, all this satisfies me that the greatest pains have been taken by your officers, and the greatest attention and zeal have been displayed by yourselves. I shall not fail to mention this in my report. I expect you will shortly be seen elsewhere, but wherever you may go it merely depends upon yourselves, the well earned reputation you have got. I consider you fit to go anywhere and shall be proud to have you in my brigade. "

It should be remembered that the Regiment was only authorized to be raised in February 1858, and had been recruited, mounted and detached on duty in a little more than twelve months.

On the 15th, 16th, 17th and 18th of February, 1860, the Regiment marched by squadrons from Newbridge to Dublin, for embarkation to Liverpool en route to Aldershot.

The Major-General commanding the Cavalry Brigade wrote the following letter on the Regiment leaving his command :—

Officers and non- Commissioned Officers of the Regiment
in Newbridge 1858.

Cavalry Brigade Office, 21st February, 1860.

Sir,

The Fifth Royal Irish Lancers having now embarked for England I consider it my duty, and but justice to that young Regiment, to bring to the favourable notice of his Lordship the General Commanding in Ireland, their highly creditable conduct during this their first march to Dublin, and during the embarkation; nothing could exceed the order and regularity with which all the duties were conducted.

The condition of the horses, the very clean smart appearance of the men and their soldierlike turn out on their march through Dublin, together with the total absence of all misconduct fully bear out the high character earned by the 5th Lancers ever since they were embodied.

On their arrival at Aldershot I am confident that they will justify the high opinion I have always expressed of them at every inspection, and that they will meet with the approbation of His Royal Highness the General Commanding-in-Chief."

(Signed) W. Parlby,
Major-General Commanding Cavalry Brigade.

For Major-General Gascoigne,
Commanding Dublin Division.

On March the 4th the Headquarters of the Regiment arrived at Aldershot, and on the 12th of the same month the Regiment was inspected by the Inspector General of Cavalry (Lord Cardigan), which resulted in a very favourable report to the Horse Guards.

On the 16th of March the following letter was received from Lieutenant-General Chatterton, Colonel of the Regiment:—

"Lieutenant-General Chatterton has received with unmixed gratification the admirable report on his Regiment, their excellent conduct during the late long march and bringing their horses in such a state of condition and efficiency, circumstances almost unknown in the Cavalry service and particularly creditable to a corps so recently raised. Having the greatest interest in the well being and credit of the Regiment, the Lieutenant-General requests Colonel Sulivan, the Officers, Non-commissioned officers and Lancers will accept his best thanks for their keeping up that high state of discipline, the result of unwearied zeal and attention upon the part of the Commanding Officer and those under his command."

Shortly after its arrival at Aldershot Camp the Regiment was inspected in brigade by H. R. H. The Duke of Cambridge, Commander-in-Chief, who expressed himself as being highly pleased with the appearance and discipline of the Regiment.

On March the 20th a petition was forwarded to the Horse

Guards that the Non-commissioned officers of the Regiment might wear the Irish Harp above their chevrons. Permission for this was granted in the following letter to Sir James Chatterton, dated at the Horse Guards on the 2nd April :

"I have the honour to receive and submit to the General Commanding-in-Chief your letter of the 20th ultimo with its enclosure from the Officer Commanding the 5th Lancers, soliciting permission for the Non-commissioned officers of the Regiment to wear the Irish Harp above their chevrons, and I am directed to acquaint you that His Royal Highness has been pleased to approve of the above badge being worn by all non-commissioned officers above the rank of corporal, and I am accordingly to request that you will obtain and send to this Department for sealing a pattern of the Badge in question."

On the 11th of March, 1861 Colonel G. Sulivan resigned the command of the Regiment, retiring from the Service, and was succeeded by Lieutenant-Colonel Robert Portal. The following order was published to the Regiment by Colonel Sulivan :

"Non-Commissioned Officers and men, before resigning command of the 5th Royal Irish Lancers, I beg to thank you for all the support you have given me...... By your conduct you have raised the Regiment to its present high position, and I sincerely trust that the same spirit will ever animate you. Though no longer destined to command, I will ever watch over your future career with the deepest interest."

The Regiment remained at Aldershot Camp until the autumn of 1861, being, together with the other troops at the camp, frequently reviewed by Her Majesty Queen Victoria and other crowned heads.

In August, four troops marched to Hampton Court Palace, Kensington and Hounslow, and in September, two troops to Woolwich, and the remainder and Head Quarters to Hounslow.

Early this year the establishment of the Regiment was reduced from 626 men and 428 horses to 600 men and 400 horses.

In August 1861, the title of "Light Dragoons" was abolished, and the regiments of Light Dragoons were constituted Lancers and Hussars.

On the 13th of August, 1862 the Regiment returned to Aldershot, where they remained until the 3rd September, when Head Quarters and three troops marched to Norwich, three troops to Ipswich, and the remaining two troops to Northampton.

Lieutenant-Colonel Portal resigned the command of the Regiment on the 23rd January, 1863, and retired from the Service. He was succeeded in the command by Lieutenant-Colonel W. H. Slade.

Lieutenant-Colonel Portal, in a farewell order to the Regiment, felt assured that the 5th Lancers would continue to retain the very high character they had secured for themselves, and expressed the deep interest he should always feel in the career of a Regiment which he was satisfied was second to none in Her Majesty's Service.

On the 10th of March, the Regiment formed part of the procession on the occasion of the marriage of H. R. H. The Prince of Wales with H. R. H. the Princess Alexandra.

On the 27th of April, the Regiment was ordered to be held in readiness to proceed to Bengal in relief of the 8th Hussars, and in June marched to Aldershot, where they arrived by the 3rd of July. The dismounted party, with the women and children, were sent to Chichester. The horses were soon after handed over to various regiments, and on the 10th of July the troops marched to Chichester to await the arrival of the vessel at Portsmouth, to convey them to Calcutta.

The Left Wing of the Regiment under the command of Major W. Dunham Massy, consisting of

10 Officers
14 Sergeants
1 Trumpeter
2 Farriers
12 Corporals
220 Rank and File
36 Women and 39 children.

embarked on board the Ship "Newcastle" at Portsmouth on the 22nd of July and sailed next day for Calcutta. The Depôt Troop, to be left at home, marched from Chichester barracks to the Cavalry Depôt at Canterbury on the 25th of July, under the Command of Captain W. L. Brown. Its strength consisted of

3 Officers
10 Sergeants
1 Corporal
1 Farrier
52 Rank and File
18 Women and 35 children.

On the 27th of July the Head Quarters and Right Wing consisting of

16 Officers
18 Sergeants
3 Farriers
6 Trumpeters
16 Corporals
221 Rank and File
33 Women and 54 Children,

under the command of Captain Wyatt, embarked on board the ship "Herefordshire" at Portsmouth, and sailed the same day for Calcutta.

To follow the fortunes of the Left Wing, they arrived at Calcutta on the 16th of November, after a fair passage of one hundred and seventeen days. From there they proceeded next day up the river on flats to Chinsurah, whence they marched on the 24th and 25th to the Hoogley station, and from there by rail and bullock train to Cawnpore, via Benares and Allahabad. At Cawnpore they took over four hundred and eighty one horses from the 8th Hussars, and on the 24th of December marched for Lucknow, leaving a detachment of 1 Officer, 3 sergeants and 24 rank and file, with 272 horses, at Cawnpore to await the arrival of the remainder of the Regiment. The following casualties occurred in the Left Wing from its leaving England to its arrival at Lucknow:—

Died at sea—1 child.
Died at Chinsurah—1 sergeant.
Drowned at Bhangulpore—1 man.

The Headquarters and Right Wing of the Regiment made a longer passage. Sailing from Portsmouth on the 27th of July, they were obliged to put into Mauritius on the 10th of November for provisions and medical comforts. Sailing from there on the 11th, they were again obliged to put into port at Madras on the 18th of December 1864, for further supplies. They sailed from Madras the following day, and arrived at Calcutta on the 19th of January, 1864, after a passage of one hundred and seventy seven days. Landing next day, they were placed under canvas on the glacis of Fort William. On the 22nd, 23rd and 25th of January they went by rail and bullock train via Benares and Allahabad to

1864

Cawnpore, where they arrived on the 1st, 2nd and 3rd of February respectively, and, taking over the horses awaiting them, marched from Cawnpore on the 7th of February, and arrived at Lucknow on the 10th.

The following casualties occurred in the Right Wing from the time of leaving England :—

Died at Sea — 1 man, 1 woman and 2 children.

The Regiment remained at Lucknow until the 25th of November, 1869, when they marched for Campbellpore and Lawrencepore. The destination, however, was altered to Sealkote, which place was reached on the 24th of February, 1870.

1869

1870

The Service troops of the Regiment were reduced by one troop on the 1st of July.

Sabretaches were taken into wear by the Officers of the Regiment on the 1st of January, 1873. The sabretache was of black patent leather, bearing the Regimental Crest.

1873

On the 13th of December 1873, the Regiment with a strength of 19 officers and 339 N.C. officers and men, marched from Sealkote for the camp of exercise at Hussan Abdul. They arrived at the camp on the 31st of December 1873, and took part in all the manœuvres. They left on the 11th of February 1874, arriving at Sealkote on the 25th of Febraury.

1874

In the following April the Regiment was placed under orders for home, and in July volunteers were given to the following corps :—

3rd Hussars	— 10	10 Hussars	— 1
4th "	— 6	13th "	— 7
9th "	— 35	15th "	— 24

The four hundred and eighteen horses of the Regiment remained at Sealkote under the charge of Captain L. M. Carmichael, on the departure of the Regiment, to await the arrival of the 8th Lancers.

The Regiment commenced its move down country for England on the 22nd of October under the command of Lt.-Colonel Dunham Massy, and arrived at Bombay by rail on the 17th of November, and embarked the same day on board H. M's Indian Troopship " Euphrates " for Home.

The Officers who embarked with the Regiment were :—

Lieut.-Colonel W. G. Dunham Massy.
Major W. S. Brown.
Major M. P. Blake.
Capt. Major C. H. Chichester. Bt.
„ J. M. McNair.
Captain J. S. Benyon.
„ G. R. R. Poole.
Lieut : R. Chambers.
„ A. B. Harvey.
„ A. C. Spencer.
Sug : Lieut : J. F. L. Russell.
„ M. Beaumont.
Adjt. C. W. B. Bell.
Rdg. Mr. T. Fletcher.
Qr. Mr. D. Shawe.
Surg. Major J. Atkinson.
Surgeon W. Cherry.
„ P. Shepherd.
Vety. Surgeon W. B. Edmonds.

and three hundred and fifty four Non-Commissioned Officers and Men.

The last annual inspection of the Regiment at the close of its Indian career was made by Major-General Macdonel, Commanding the troops at Sealkote, on the 4th and 5th of March, 1874. In connection with the inspection report the Commander-in-Chief in India made the following remarks in a letter to the G. O. C. Rawal Pindi Division, dated 24th of June, 1874: " I am directed to request you will convey to Lieut.-Colonel Massy and the Officers under his command an expression of His Excellency's satisfaction at receiving so creditable a report, which shows that the Regiment is in a high state of efficiency in every detail at the close of its Indian service. "

The Regiment disembarked at Portsmouth, and proceeded at once to Colchester, where it arrived on the 24th of December.

On the 26th, two new troops, " B and H, " were organised out of the Depôt troop from Canterbury.

On the 28th, the Regiment took over the horses of the 9th Lancers, each troop taking over the horses of the corresponding troop of the 9th.

The strength of the Regiment now fixed on the Home establishment was 28 Officers, 506 Non-Commissioned Officers and men, and 320 horses.

1875

On the 29th of March 1875, in a letter from the Horse Guards, His Royal Highness the Field Marshal Commanding-in-Chief approved of the plume for the Officers' dress caps to be made of cock's feathers. The cock's feather plume was ordered to be worn at Levees and on full dress parades in Review order; upon other occasions when plumes were worn, the regulation horse hair plume was to be worn.

On the 18th of June 1875, the Regiment left Colchester by march route for Aldershot, where it arrived on the 24th, and went under canvas.

The marching strength was 26 Officers, 37 Sergeants, and 366 rank and file.

The Regiment took part in the Aldershot summer manœuvres, and moved into the West Cavalry Barracks on the 1st of September.

H. R. H. The Commander-in-Chief ordered that the silver badges worn by the Non-commissioned Officers of the Regiment should be worn upon, instead of above the chevrons.

From the date of the return of the Regiment to England, till October 1875, one hundred and forty recruits had been enlisted for the 5th Lancers in the following districts :—

Regimental Head Quarters	47
London	55
Portsmouth	3
Liverpool	5
Belfast	13
Dorchester	1
Bristol	6
Birmingham	4
Halifax	1
Hull.	5

The Regiment was inspected by Major-General Sir Thos. McMahon, Inspector General of Cavalry, on the 11th of October, and on that officer's report H. R. H. The Field Marshal Commanding-in-Chief wrote the following letter :—

"The most creditable and highly satisfactory report on the 5th Lancers has been perused with the greatest pleasure by His Royal Highness, who agrees with Sir Thomas McMahon that no regiment ever returned from

India in better order or more generally efficient, and you will be good enough to convey to this corps, and to Lieut.-Colonel Massy its commanding officer, His Royal Highness' appreciation of this highly satisfactory state of things. (W. O. 7 April, 1876).

1876 On the 3rd and 4th of July, 1876 the Regiment, with a strength of 24 Officers, 28 Sergeants, 336 Rank and File and 291 horses, under Lieut.-Colonel Dunham Massy, marched from Aldershot to Lewes Camp, where it arrived on the 6th of July. There the Regiment took part in the summer manœuvres, and returned to Aldershot on the 19th and 20th of July.

The Regiment was again inspected on the 3rd and 4th of October, and again earned a very flattering report from H. R. H. The Commander-in-Chief.

1877 The 9th of April, 1877, saw the Regiment under the command of Lt.-Colonel Dunham Massy on the march from Aldershot to Norwich, in relief of the 1st Royal Dragoons ; a detachment of two troops under Brevet-Major Chichester being sent to Ipswich. The Regiment arrived at Norwich on the 18th of April.

The marching strength was as follows :—

19	Officers	53	Chargers
48	Sergeants	314	Troop horses
8	Trumpeters	48	Women
427	Rank and file	97	Children.

The annual inspection in August was again most satisfactory to H. R. H. the Commander-in-Chief.

1878 In May, the Regiment was raised to War Establishment with a personnel of 652 and 480 horses. At the same time the Regiment was opened for recruiting, and in less than five weeks one hundred and forty seven recruits were enlisted in the following districts :—

Regimental Head Quarters	46
London	37
Liverpool	9
Belfast	6
Dublin	5
Leicester	6
Nottingham	3
Shrewsbury	3
Reading	3
22 other districts	29

The Regiment marched out of Norwich on the 23rd of June, 1878, for Brighton, to relieve the 20th Hussars, a detachment of one troop under Brevet-Major Vandeleur being sent to Shorncliffe. The following was the marching out strength :—

26	Officers	54	Chargers
51	Sergeants	314	Troop horses
8	Trumpeters	47	Women
537	R. and F.	66	Children.

On the departure of the Regiment from Norwich, a motion was set on foot by the citizens to present the Regiment with a testimonial, in recognition of its exemplary conduct and popularity, and of the assistance rendered by the Officers and men in giving several performances in aid of the City's Institutions. A subscription headed by the Mayor, the Bishop, and all the most influential citizens of Norwich, and a number of the country gentlemen, was collected for the purpose.

The result was that a magnificent silver-gilt cup, of considerable value, was presented to Colonel Dunham Massy and the Officers in the Council Chamber of Guildhall at 1 p.m. on the 24th of June, 1878.

The following account of the ceremony is taken from a leading paper of the time :—

"The Mayor rose, and amid cheers said :—

Fellow Citizens, the inhabitants of Norwich have rather got the start of us to-day, but I do not regret their having done so, a portion of the Royal Irish Lancers left here this morning at half-past nine, amid such a hearty demonstration that, so far as my experience goes, was never before given to a regiment on its departure from the city.

For some weeks past there has been a very wide-spread desire on the part of the inhabitants of Norwich, to give some expression of their feeling and respect to Colonel Massy and the Regiment, of which he has the command. Everyone must have noticed this fact, and efforts have been made to bring it to a satisfactory conclusion.

As soon as a decision was come to, to make a presentation to Colonel Massy and the Regiment, we took steps to carry out the desire of the citizens. The inscription on the plate has been written by one of the Committee who has taken a deep interest in the movement, and we were not able to alter a single word. If I were to stand before you and speak for an hour I could not better express the feelings of the citizens than does the inscription which is as follows :—

"Presented by the Mayor, the Sheriff, and the Citizens of Norwich, and the Residents of Norfolk to Colonel Dunham Massy and the Officers of the 5th Royal Irish Lancers in appreciation of the exemplary conduct of the Regiment during its stay among them, and in grateful remembrance of the many benefits conferred by it upon the City and its institutions."

Ladies and gentlemen, that is the inscription, and I for one feel that it expresses all we desire to say. I am delighted to find this presentation is made to this Regiment, to the entire Regiment, for though it will be presented to Colonel Massy and the Officers, we include the entire body. I have an unspeakable pleasure in informing this meeting that during the stay of the 5th Royal Irish Lancers in Norwich, there has not been a single instance in which the Military and Civilians have come into collision. That is a grand thing to say of a number of men located in a city like Norwich, when we know what human nature is. It is a remarkable thing that I, as Mayor of Norwich, should be able to stand up and say that a regiment has conducted itself in a manner to gain the esteem of the whole city. I am old enough to remember the time when the sword room has been occupied with enquiries with regard to collisions between the military and civilians; and, so remembering the past, it gives me the greater pleasure to state there has been no single instance during the stay of this Regiment in Norwich in which any of the men have come into collision with the Citizens. May I ask you, Colonel Massy, and in addressing you I also address your brother Officers, to accept this piece of plate. The plate itself is of little value compared with the feeling that has prompted the presentation. You sir, as Commanding Officer of the Regiment, have carried out the duties of your position so as to gain the respect and esteem of all who know you. There is an old adage, and a true one, that good masters make good servants. The remarks I have been able to make with regard to your Regiment, Colonel Massy, are to be traced mainly to the example and to the character of yourself and your brother Officers."

In his Observations on the excellent Inspection report on the Regiment for 1878, H. R. H. The Commander-in-Chief noted that its conduct whilst stationed at Norwich was very commendable.

While at Brighton the Regiment gave an Assault at Arms during the week ending the 22nd of February, 1879. The amount realised from the Assault at Arms, £150, together with an additional £50, was distributed amongst the Brighton charities: and the following acknowledgment was received by the Regiment:—

"At a meeting of the Town Council of the Borough of Brighton held at the Town Hall on the 5th day of March 1879.

It was resolved—

That the letter addressed to the Mayor by Colonel Massy, commanding 5th Royal Irish Lancers be entered on the minutes of the Council's proceedings, and that the Council do hereby tender to Colonel Massy and the Officers and men of the Regiment their warm thanks for the munificent contribution of a sum of £200 amongst the Charities of the Borough, that the Council at the same time desire to express their appreciation of the good conduct and behaviour of the men under his command during the time they have been stationed at Brighton, and their regret at the departure from the Borough before the expiration of the usual term, of a Regiment that has made itself so deservedly popular.

It was resolved—

That the Town Council do forward a copy of the foregoing resolution to the War Office."

Colonel W. G. Dunham-Massy retired on half-pay on the 15th of March 1879, and was succeeded in the command of the Regiment by Lt.-Colonel W. St. Browne.

The Regiment left Brighton for Woolwich on the 7th of July, arriving there on the 9th.

On the 17th of June 1880, the Regiment marched to Aldershot for the summer drills, a small depôt being left at Woolwich, and on the 3rd and 4th and 11th of August proceeded by march route to York, where they arrived on the 19th and 20th of August, being preceded by the depôt from Woolwich. 1880

A dismounted party went by rail from Aldershot to Birmingham on the 13th of August.

The Regiment remained at York until February 1881, when it moved to Newbridge in Ireland. The Left Wing left York on the 22nd of February, and the Right Wing two days later. The marching out strength shewed 17 Officers, 449 N. C. O. and men, 52 chargers, 309 troop horses, 44 women and 36 children. 1881

On the 1st of July of this year the establishment of the Regiment was fixed at 24 Officers, 469 N. C. O. and men, and 300 troop horses.

The following letter was received by the Officer commanding on the 4th May, 1882:—

Horse Guards, War Office, 29 April, 1882.

Sir,

By desire of H. R. H. the Field Marshal Commander-in-Chief, I have the honour to acquaint you that Her Majesty has been graciously pleased to command that the victories of "Blenheim," "Ramillies," "Oudenarde," and "Malplaquet" shall be inscribed on the appointments of the 5th Lancers, and the same will be recorded in the next issue by the Queen's Regulations and Orders for the Army in addition to the present achievements,

I have etc.

(Sgd.) G. J. Wolseley.
Adjutant General.

The following was published in Brigade Orders at the Curragh Camp on the 3rd of October, 1883 :—

" Major General C.C. Fraser V.C., C.B., Inspector General of Cavalry in Ireland, desires before the departure of the Regiment from Newbridge to record his appreciation of the high standard of efficiency in all departments, arrived at by the 5th Royal Irish Lancers.

The Regiment is soldierlike in bearing, steady and quick in the field, expert in the use of weapons, and extremely well conducted in quarters.

The Riding, the training, and the horses in the School, in the field, and in leaping fences of different descriptions are excellent.

The systems of Interior Economy and of Stable management are very good and are well carried out.

The 5th Lancers excel in the accomplishments required of a Cavalry Regiment.

In sketching 233 N. C. O. and men are proficient, and 56 are prominently good.

In Reconnaissance the duties are well understood and the reports sent in exhibit great intelligence.

The Signalling and Pioneering are of the highest order.

Within the past two years, the Non Commissioned Officers of the Regiment have won the Inter-Regimental prize for good shooting.

The Regiment stands first in signalling in all the Cavalry.

Further, the recent General Order shows that the Regiment has taken 13 First Class Certificates of Education, which places it by a long way at the top of the list of Cavalry Regiments.

The Inspector General of Cavalry in Ireland compliments Col. W. St. Browne, commanding the 5th Royal Irish Lancers, on the success of his endeavours to extend, with the aid of his Officers, the cultivation of these high attainments throughout the ranks of the Regiment. "

The following observation was made by His Royal Highness the Field Marshal Commander-in-Chief on the Report of the Inspection in Dublin of the 25th of April, 1884, by Major General Lord Clarina.

" The condition of this fine Regiment appears to His Royal Highness to be in all respects very satisfactory and creditable to all concerned. "

Officers of the Regiment at Island Bridge Barracks,
Dublin 1884.

CHAPTER XI.

The Nile Expedition 1884-1885.—*Abu Klea.*—*Abu Kru.*—
Fall of Khartoum.

In September 1884, the Regiment was called upon to furnish 2 Officers and 43 N.C.O. and men to join the Heavy division of the Camel Corps being formed for service with the Nile Expedition.

The detachment was selected from volunteers, and left Dublin on the 19th of September, under the command of Major L. M. Carmichael and Lieut. H. Costello, for Aldershot, whence they proceeded to Southampton, and embarked on the 26th of September for Egypt.

This sudden despatch of troops to the Nile was due to the Government having at last determined that it was necessary to rescue General Gordon from his perilous position at Khartoum.

General Lord Wolseley was to concentrate his troops for the attempt at Korti on the Nile. On looking at the map of the Nile it will be seen that at Korti the river makes an enormous bend to the North between that place and Khartoum. For this reason it was determined that from Korti a double expedition should be despatched, the Desert Column commanded by General Sir Herbert Stewart to move straight across the desert, following a line of wells to the Nile at Metammeh, and a Nile Column commanded by General Earle to follow the river to Aber Hamed and Berber, with a view to co-operating with the Desert Column for the relief of Gordon at Khartoum.

For Sir Herbert Stewart's Desert Column a Camel Corps drawn from the Cavalry and Infantry was to be organised. The Cavalry portion was to be composed of detachments from the Cavalry Regiments in Great Britain, subdivided into "Heavies" and "Lights;" while the Infantry part was drawn from the Brigade of Guards, the Royal Marines and mounted infantry of regiments in Egypt.

The following was the eventual composition of the Heavy Camel Regiment of the Corps,

STAFF.

Lt.-Col. the Hon. R. Talbot, First Life Guards, Commanding.
Capt. Lord St. Vincent, 16th Lancers, Adjutant.
Surgeon Falvery, A.M.D., Surgeon.
Lieut. G. Leigh, First Life Guards, Acting Quartermaster.

O. C. DETACHMENTS	SUBALTERNS	DRAWN FROM.
Major the Hon. C. Byng	Lt. Lord Rodney	1st Life Guards
Capt. Lord Cochrane	Lt. R. J. Beech	2nd " "
Major Lord A. Somerset	Lt. Lord Binning	Royal Horse "
Capt. A. L. Gould	Lt. R. F. Hibbert	2nd Dragoon "
Capt. J. W. Darley	Lt. C. W. Law	4th " "
Major W. H. Atherton	Lt. St. J. Gore	5th " "
Major W. Gough	Lt. J. F. Burn-Murdoch.	1st Royal Dragoons
Capt. W. H. Hippisley	Lt. R. Wolfe	2nd (Scots Greys) Dragoons
Major L. Carmichael	Lt. H. Costello	5th Lancers
Major T. Davison	Lt. W. B. Browne	16th "

Each detachment consisting of 43 N.C.O. and men.

The detachments making up the Light Camel Regiment were drawn from the 3rd., 4th., 7th., 10th., 11th., 15th., 18th., 20th., and 21st. Hussars. Each detachment also of 43 N.C.O. and men.

The Guards' Camel Regiment was formed from the Guards' Brigade and a detachment from the Royal Marines ; and the Mounted Infantry Camel Regiment, as we have seen before, from Infantry battalions in Egypt.

The men selected for this service had all to be marksmen or first-class shots, not under twenty-two years of age, medically fit, and good characters. They were armed with a rifle and sword bayonet, and wore a bandolier over the left shoulder holding fifty cartridges. The dress was a Khaki serge jumper, cord breeches, puttees, ankle boots and a pith sun helmet.

The Heavies and the Guards detachments embarked on the 26th. of September in the P.O. steamer "Deccan" and sailed the same day for Alexandria, arriving there on the 7th. of October. From there they went to Cairo and then on to camp at the Pyramids. On the 13th. the Heavies moved by train to Assiot and thence by steamer up the Nile to Assouan. Here they were fitted out with camels and marched for Korti, where Lord Wolseley was concentrating the two expeditions for Khartoum. It was a weary, trying march up to Korti, and many a soldier was sadly learning that the camel is not a pleasant beast to ride. There was a man, however,—a sailor—in the Nile Expedition, who admired a camel he rode exceedingly, because, being pitched up out of his saddle incessantly, and caught dexterously as he descended, "the camel had only missed him twice throughout an afternoon."

The troops working up the river in their barges were also experiencing a trying time.

The following general order shews the anxiety of the commander to accelerate the passage of the expedition.

"To the sailors, soldiers, and marines of the Nile Expedition.

The relief of General Gordon and his garrison, so long besieged in Khartoum, is the glorious mission which the Queen has intrusted to us. It is an enterprise that will stir the heart of every soldier and sailor fortunate enough to have been selected to share in it, and the very magnitude of its difficulty only stimulates us to increased exertions. We are all proud of General Gordon and his gallant and self-sacrificing defence of Khartoum, which has added, if possible, to his already high reputation. He cannot hold out many months longer, and he now calls upon us to save his garrison. His heroism and his patriotism are household words wherever our language is spoken; and not only has his safety become a matter of national importance, but the knowledge that our brave comrade needs help, urges us to push forward with redoubled energy. Neither he nor his garrison can be allowed to meet the sad fate which befell his gallant companion in arms, Colonel Stewart, who, when endeavouring to carry out an enterprise of unusual danger and folly, was treacherously murdered by his captors. We can and with God's help will — save General Gordon from such a death. The labour of working up this river is immense, and to bear it uncomplainingly demands the highest soldier-like qualities, that contempt for danger, and that determination to overcome difficulty which in previous campaigns have so distinguished all ranks of Her Majesty's Army and Navy. The physical obstacles that impede our rapid progress are considerable: but who cares for them when it is remembered that General Gordon and his garrison are in danger? Under God their safety is now in our hands, and come what may we must save them. It is needless to say more to British soldiers and sailors."

The middle of December found Lord Wolseley and his troops collected at Korti, and the final arrangements for the Desert and the River columns being rapidly completed.

The camp at Korti was a pleasant place after the long and toilsome journey by boat and camel. The tents of the Camel Corps were pitched under spreading groves of trees extending to the banks of the river, which there took a winding course in broad still reaches. The days were occupied with field days on camel-back in the desert, and by "stables" three times a day, this latter duty chiefly consisting in picking the ticks out of the camel's hide.

Meanwhile affairs at Khartoum were evidently approaching a crisis. The Mahdi had occupied Omdurman, only a few miles from Khartoum, and had summoned Gordon to surrender the city. The answer was :— "If you are the real Mahdi, dry up the Nile and come over and I'll surrender;" whereupon, rumour has it, that the Mahdi accepted Gordon's challenge, and having collected his forces and chanted some spells, sent them all into the river, with the result that enormous numbers were drowned, and the remainder scrambled back half dead.

After this, severe fighting was continually going on round Khartoum, but the gallant Gordon and his garrison were successfully holding out; but news was difficult to get, and the spies of the Intelligence Department showed a certain reluctance in braving the dangers of the desert and the vigilance of the Mahdi's followers; the news that they did bring in, too, was somewhat untrustworthy.

Briefly stated, the plan of the approaching campaign was that the greater portion of the mounted troops under Sir Herbert Stewart was to advance across the desert from Korti to Metammeh, establishing fortified posts at the wells along the route. Sir Charles Wilson was to go with Stewart, and at Metammeh was to proceed in Gordon's steamers to Khartoum ; and having communicated with Gordon, to return to Metammeh to report the result to Sir Herbert Stewart. Simultaneously with the despatch of the Desert Column, a force under General Earle was to be sent up the river to punish the murderers of Colonel Stewart and the Consuls, and then to advance to Berber to co-operate with Stewart's force in an attack on the Mahdi before Khartoum, under the personal command of Lord Wolseley, who was to have joined Stewart with the remainder of the mounted troops and a force of infantry.

On the 30th. of December a messenger, who had been sent to Khartoum on October the 29th., returned, and brought into

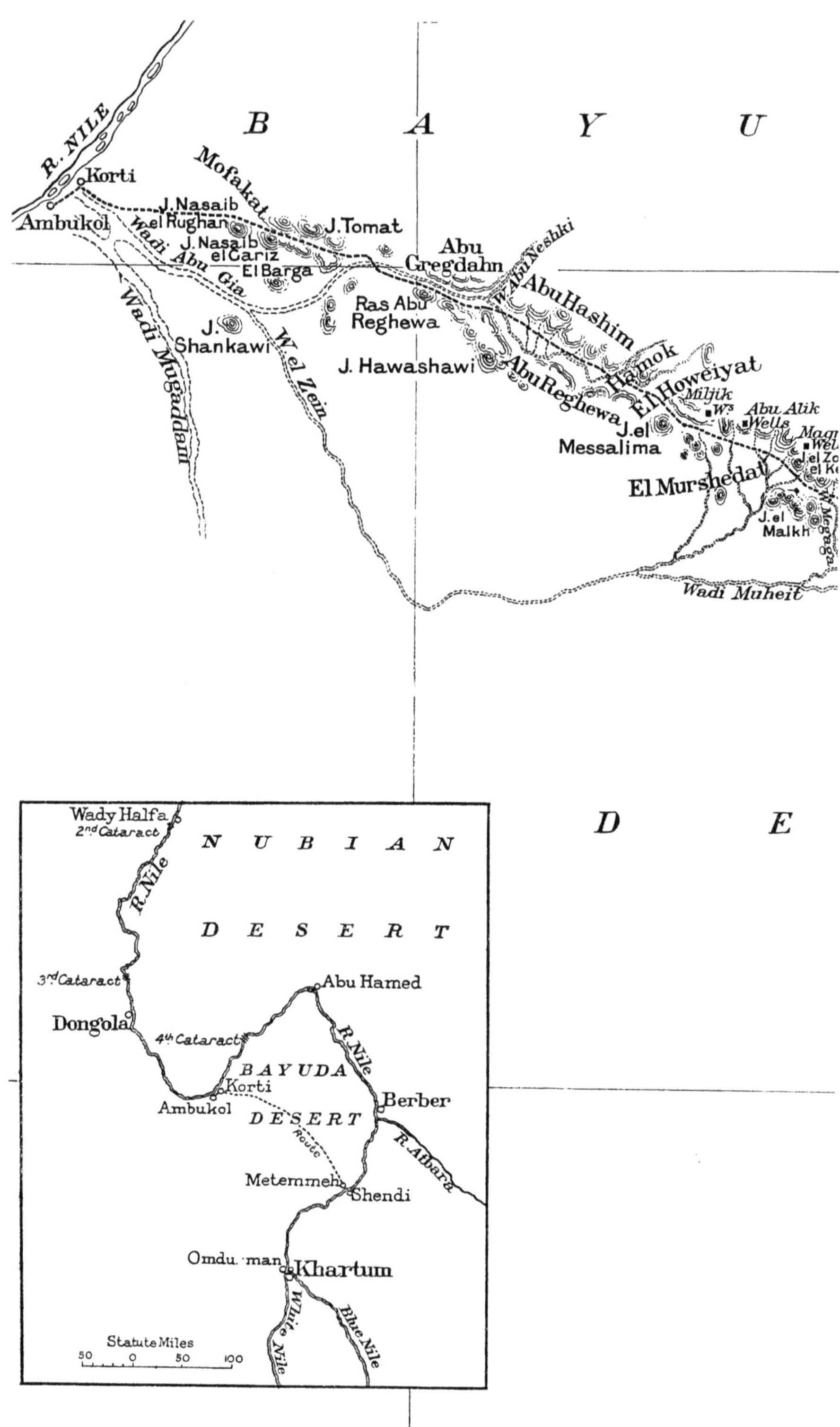
R. NILE
Korti
Ambukol
B A Y U
Mofakat
J. Nasaib
W. el Rughan
J. Nasaib
el Gariz
El Barga
J. Tomat
Abu Gregdahn
Wadi Abu Gia
Ras Abu Reghewa
Wadi Mugaddam
J. Shankawi
W. el Zein
J. Hawashawi
Abu Hashim
Abu Reghewa
Hamok
El Howeiyat
J. el Messalima
El Murshedat
Abu Alik Wells
J. el Malkh
Wadi Muheit
D E
Wady Halfa
2nd Cataract
N U B I A N
D E S E R T
R. Nile
3rd Cataract
Dongola
4th Cataract
Abu Hamed
R. Nile
B A Y U D A
Korti
Ambukol
D E S E R T
Route
Berber
R. Atbara
Metemmeh
Shendi
Omdurman
Khartum
White Nile
Blue Nile
Statute Miles
50 0 50 100
32

DESERT MARCH OF THE NILE EXPEDITION

1884-5

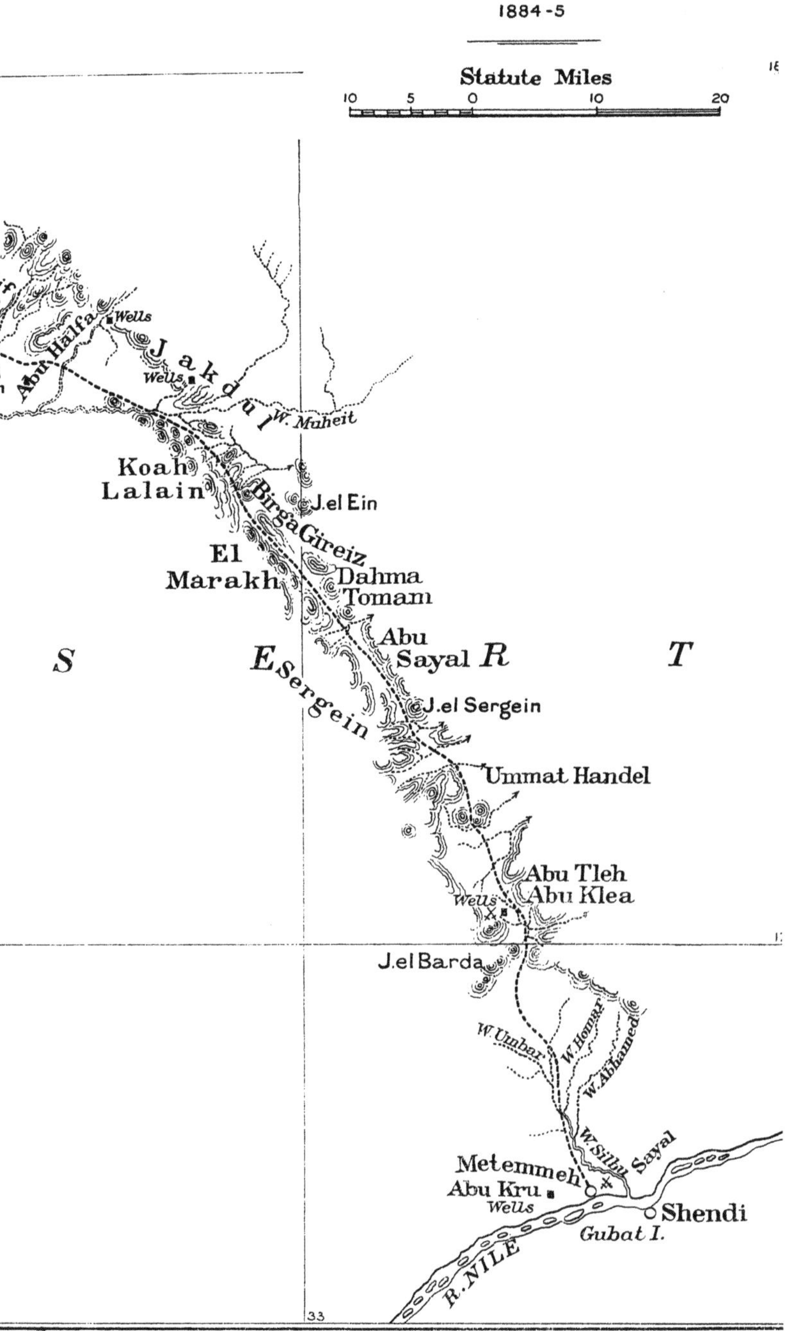

A. Doubleday & Co., Ltd.

Korti a piece of paper the size of a postage stamp, on which was written " Khartoum all right. " It was signed C. G. Gordon, and dated 14th. of Dec. 1884; and the messenger said he was told to deliver a verbal message to the effect that food in Khartoum was running short, and the troops suffering from want of provisions; he went on to say, " We want you to come quickly. Do not scatter your troops, the enemy is numerous; bring plenty of troops if you can; " and the advice was to come only by way of Metammeh or Berber, and to do so without letting rumours of the advance spread abroad.

On the 30th of December, Sir Herbert Stewart started from Korti with part of the Desert Column. The column was preceded by thirty-four scouts of the 19th. Hussars, who were followed by the Guards Camel Regiment, and 650 camels belonging to the Heavy and Light regiments. The men of these two divisions of the corps remained behind, their camels, as well as 500 transport camels, being loaded with provisions and stores. The mounted infantry brought up the rear.

The Guards were dismounted at Gakdul, and remained in charge of the stores, while their camels, together with those of the other Camel Corps, returned to Korti, when the " Heavies " and " Lights " were remounted, and the Guards' and transport camels loaded up again with stores and provisions, and, accompanied by the Sussex regiment, returned to Gakdul. On the 12th of January this second column had reached Gakdul.

Leaving 400 men of the Sussex Regiment to garrison Gakdul, Sir Herbert Stewart marched off from Gakdul on the 13th. The numbers of the force which paraded at 2 p. m., outside the hills on the plain were,

Naval Brigade, with one Gardner gun.	about	30	of all	ranks.
Heavy Camel Regiment	"	380	"	"
Three troops of 19th Hussars	"	90	"	"
Half-battery, Royal Artillery, with three 7-pounder screw guns.	"	30	"	"
Royal Engineers	"	25	"	"
Guards' Camel Regiment	"	367	"	"
Mounted Infantry Camel Regiment	"	360	"	"
Sussex Regiment	"	100	"	"
Medical and Commissariat Staff	"	45	"	"
Native drivers	"	120	"	"

A total of about 1500 men, 90 horses, and 2300 camels; a mere handful of men to be exposed to the savage onslaught of some 12,000 of the Mahdi's fanatics. With the exception of the 19th Hussars on their horses, and the natives who walked, the whole of the force was mounted on camels. Three powerful camels carried a screw gun and its ammunition between them; 100 rounds of ammunition were taken per gun. A number of camels were fitted with litters for the sick and wounded. A train of some 500 camels carried stores of all sorts to form a depôt at Metammeh.

With the scouts of the 19th in front, the "Heavies" led the way, followed by the Guards, then the baggage and stores, and the Mounted Infantry bringing up the rear.

Soon after starting, a Remington rifle was picked up on the rocks. This, and the marks of recent horse tracks, rather pointed to the fact that the advance was known to the enemy. The first afternoon's march was over a vast gravelly plain, with gentle undulations. After covering about ten miles, during which fifteen camels succumbed, the column halted for the night. Reveille next morning sounded at 3, and before 5 the force was again on the move. This day's march was over a tract of loose sand, and at 10 o'clock a two hours' halt was called, and the stragglers brought up. Recent horse tracks were seen, shewing that the enemy's scouts were about, but no messages had been received from the Hussars' scouts.

Many of the camels were falling from want of food and from overwork. If a camel dropped, his load and saddle were taken off him and placed upon another, already loaded, for the spare camels had all been used up. The poor brutes toiled on in an extraordinary way. A camel would be seen going slower and until the tail of the animal in front, to which he was tied, looked like coming off; then he would stop for a second, give a mighty shiver, and drop down stone dead.

After passing Jebel-el-Nus the scenery changed to a broad grassy valley with some trees, and rocky hills all round. At 5 p.m. the column camped for the night at Jebel Sergain, where the camels were tied down and preparations made for a probable attack, so that the wretched beasts could not move to get a meal of the savas grass which grew plentifully in the neighbourhood.

Next morning the column again started before daylight, and at daybreak the hills of Abu Klea could be seen in the distance, and the 19th Hussars were ordered to push on and occupy the wells.

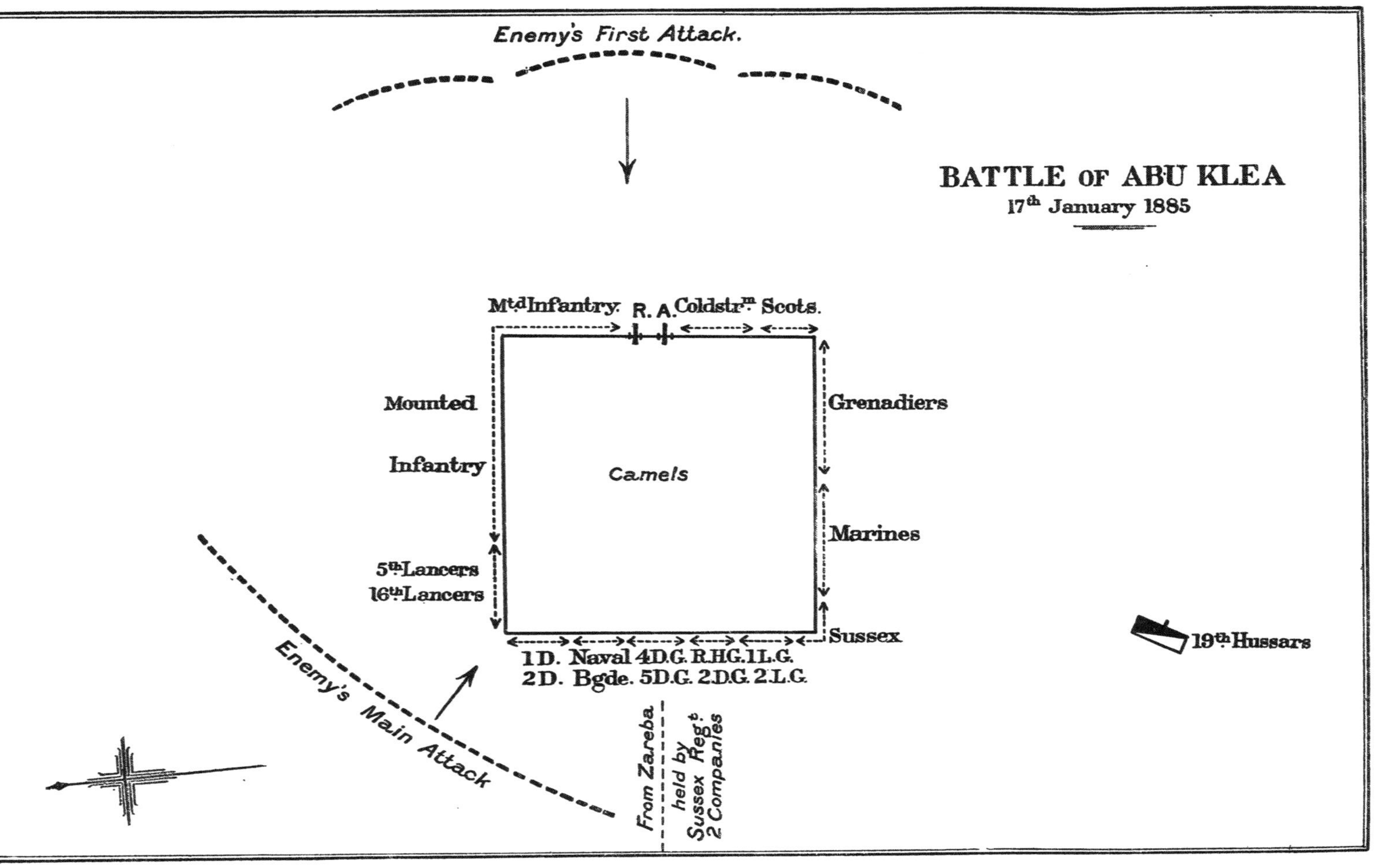

HISTORY OF THE FIFTH LANCERS. A. DOUBLEDAY & CO., LTD.

Between ten and eleven the force halted, and shortly after Major Barrow of the 19th came in to report that he had found the enemy in force between the main column and the wells. With three or four men he had pursued some of the Arab scouts into the Abu Klea valley, but had been forced to retire. The route from the halting place of the column to the Abu Klea wells was through a pass of the mountain in front; and when this was ascended, detached bodies of the enemy could be seen on the hill tops ahead. Sir Herbert Stewart and Sir Charles Wilson went on to reconnoitre the position of the enemy; the former returning to select a place at which to halt the convoy, while the latter went on and joined the Hussars down the valley, whence could be seen a long line of banners, and puffs of white smoke from the rifles of the enemy being fired at the advanced party of the Hussars, though at too great a distance for the bullets to reach them. Sir Charles Wilson returned to report that there was a large force in front, part of which must belong to the Mahdi's army, and that a serious encounter might be looked for. A couple of thousand yards to the right of the British column swarms of the enemy began to wave their spears and to execute a wild dance, after which they commenced firing.

Sir Herbert Stewart had halted the column on a stony plateau, where a zariba was quickly built of stone and the thorny branches of the mimosa. Pickets were sent out to occupy two hills on the left, where the mounted infantry built a small fort. It was getting late and the General had decided not to attack until the morning. Emboldened by the growing darkness, the enemy's riflemen were creeping up on the right of the advanced post, and commenced a long range fire from a hill, which became so heavy that the picket had to be withdrawn; the enemy still creeping round the right, the cavalry vedettes were also withdrawn.

The firing lasted all night, while the men lay in the zariba waiting for an attack which was not made; though sometimes the beating of tom-toms sounded quite close. And in the dense darkness, if anyone struck a match to light a pipe, or if, in the hospital, where there were already some wounded, a light was shewn for an instant, it always drew a bullet. However, there was not much damage done, and the morning of the 17th of January broke at last. The enemy's fire became hotter, and a few officers and men were hit. Some of their horsemen rode down to the right of the zariba, but were dispersed by a few rounds of shell. Sir Herbert Stewart determined to march out and give battle, leaving a force to hold the zariba; and, after having breakfast under a brisk fire from the enemy, the men were delighted to get the order to form square and advance.

The square was formed by the Guards and Mounted Infantry in front; the rear face by four companies of the Heavy camel regiment, with its fifth company round the angle, and on the left face of the square; the detachment of the Sussex Regiment on the right face towards the rear; the Naval Brigade and the Gardner gun between the third and fourth companies of the Heavies; and in the centre some thirty camels for carrying water, ammunition, and wounded men, driven by natives. Some of the Sussex, and the baggage guards, were to remain in charge of the zariba, and the 19th. Hussars were to operate on the left of the square, the front and flanks of which were covered by skirmishers.

The square was rapidly in order, and the troops marched down the valley towards the row of flags which stretched across it, while the 19th. moved off to the left. Several times the square halted, and returned the fire of the enemy who moved along parallel with it. Men were falling fast, and the whistling of bullets overhead was incessant. Sir Charles Wilson has given a graphic description of the advance, and the sudden appearance of the enemy, in *Egypt and the Soudan.*

"When the skirmishers got within about 200 yards of the flags the square was halted for the rear to close up, and at this moment the enemy rose from the ravine in which they were hidden in the most perfect order...........They advanced at a quick even pace, as if on parade, and our skirmishers had only just time to get into the square before they were upon us; one poor fellow who lagged behind was caught and speared at once... I could not have believed beforehand that men in close formation would have been able to advance for 200 to 400 yards over bare ground in the face of the Martini-Henrys..............When they got within eighty yards the fire of the Guards and mounted Infantry began to take good effect, and a huge pile of dead rose in front of them. Then, to my astonishment, the enemy took ground to the right as if on parade, so as to envelope the rear of the square. I remember thinking 'By Jove, they will be into the square!' and almost the next moment I saw a fine old sheikh on horseback plant his banner in the centre of the square behind the camels. He was at once shot down, falling on his banner...........I had noticed him in the advance with his banner in one hand and a book of prayers in the other, and never saw anything finer. The old man never swerved to the right or left and never ceased chanting his prayers until he had planted his banner in our square......directly the sheikh fell the Arabs began running in under the camels to the front part of the square. Some of the rear rank now faced about and began firing. By the fire Herbert Stewart's horse was shot............Almost immediately afterwards the enemy retired, and loud and long cheering broke from the square. They retired slowly, and for a short time hesitated in the valley before they made their final bolt. During this period of excitement groups of three to five Arabs, who had feigned death, would start up from the slain and rush wildly at the square. They were met by a heavy fire but so badly directed that some of them got right up to the bayonets.........There was one strange incident. An unwounded Arab armed with a spear, jumped up and charged an officer. The officer grasped the spear with the left hand, and with his right ran his sword through the Arab's body ; and there for a few seconds they stood, the officer being unable to withdraw his sword until a man ran up and shot the Arab."

There is little doubt that when the Arabs got into the square, some officers and men fell by the rifles of the ranks who turned about and fired into the square. This, it was surmised, caused the death of Carmichael of the 5th. Lancers, and of Gough of the Royals.

Many of the rifles were rendered useless by the cartridges jamming, and the men had to take to the bayonets. The Gardner gun jammed at the tenth round.

In *The Nineteenth Century*, Lieut.-Colonel the Hon. R. Talbot, commanding the Heavy Camel Regiment, gives an account of the battle.

The total strength of the Heavies was 390.

RIGHT WING.

1st. Company—1st. and 2nd. Life Guards.
2nd. „ —The Blues and The Queen's Bays.

LEFT WING.

3rd. Company—4th. and 5th. Dragoon Guards.
4th. „ —Royal Dragoons and Scots Greys.
5th. „ —5th. and 16th. Lancers.

By the time the attack took place, the company of the Royals and Greys had been partly moved from the rear to the left face, to fill up the gap caused by the gradual lengthening out of the sides of the square, due to the impossibility of keeping up the strings of camels. By this movement the rear of the square was considerably weakened.

The route taken was parallel to, and a few hundred yards from, the wady, or shallow ravine, that ran on the left to the wells of Abu Klea, in which grew stunted trees and thick grass concealing deep water courses, giving admirable cover for the enemy ; and the course of the march was commanded by hills to the right and rear occupied by Arab riflemen.

Directly the square started from the zariba, many men and some officers were hit ; amongst them being Lord St. Vincent and Major Dickson. Skirmishers from the Heavies were sent out to the rear and rear flanks to silence the fire of the enemy. After a slow march of some two miles, a large force of Arabs at about 500 to 700 yards' distance sprang up and advanced as if to attack the left leading corner of the square. The square was halted, and moved to the right on to a slight elevation, a simple movement for men but difficult for camels, some of which were left outside. There was a gap on the left face of the square through which the Gardner gun was taken into action until it jammed. A solid column of the enemy was seen to be advancing from the wady on the left, but the fire of the rear face of the square had to be reserved, being masked by the skirmishers who were still out ; the last of whom was overtaken and speared. Close upon the heels of the skirmishers came the great body of Arabs, led by their chiefs on horseback. Until that moment the Heavy Camel Regiment had withheld their fire, which was then delivered at the advancing column. Taking advantage of the opening in the square, the Arabs hurled themselves with terrific rapidity and fury upon it. The company of the 4th. and 5th. Dragoon Guards had a few moments before been wheeled outwards by Colonel Burnaby, with the intention, as Lt.-Colonel Talbot understood, of bringing their fire to bear ; "but no sooner did he see that not only on the flanks but on the rear, the attack was being developed, than he rode in front of the company and shouted to the men to wheel back. The order was obeyed, the men stepping steadily backwards. Before they had got back into their original place, the Arabs were in through the interval thus created, and through the gap already existing at the left rear corner of the square. Burnaby, whose horse had fallen, was one of the first to be attacked, and as he lay on the ground he received a mortal wound in the neck from a sword cut."

The Royals, Greys, and 5th. Lancers, upon whose rear the

camels pressed, hampering their free movement, were now attacked in rear by those of the enemy who had succeeded in passing the 4th. and 5th. Dragoon Guards, coming through and under the camels, at the time they were engaged with the enemy in front. "A severe hand-to-hand fight ensued, in which the strength and determination of our men told, and not an Arab escaped alive. The affair was a matter of moments, and from first to last not more than five minutes elapsed. The fire of the mounted infantry principally, and of the Guards Camel Regiment (who faced their rear rank about), of the detachment of the Sussex and of the right wing of the Heavy Camel Regiment, prevented the Arabs from re-inforcing their attacking column; but the brunt of the fight, the hand-to-hand encounter, was borne by the left wing of the last-named regiment. No men could have fought better, and although two detachments lost their officers, their places were at once assumed by the non-commissioned officers. It was an Inkerman on a small scale—a soldiers' battle; strength, determination, steadiness, and unflinching courage alone could have stemmed the onslaught."

It was at this point that Major Carmichael of the 5th Lancers was killed, and his subaltern, Lieutenant Costello, wounded.

The British losses were: —

	KILLED.	WOUNDED.
Officers	9	9
N. C. O. and men	65	85

The brunt of the attack fell upon the left wing of the Heavy Camel Regiment and the Naval Brigade, and it was at that corner of the square that all the Officers were killed, and the majority of the casualties occurred. Seven of the officers killed belonged to the Heavies, and the other two to the Naval Brigade.

The following were the losses of the Detachment of the 5th Lancers.

KILLED.

Major L. M. Carmichael.
No. 2344 Corporal C. Percival.
" 2333 Lance Corporal S. Parker
" 2085 Private C. Peters.

No. 1897 Private E. Bell.
" 2155 " J. McGrath.
" 2129 " A. Russell.

WOUNDED.

Lieutenant H. Costello and 2 N. C. O. and men.

For the bravery he displayed during the engagement, No. 2151, Private G. Austin, 5th Lancers, was eventually awarded the medal for "Distinguished Conduct in the Field."

It was now getting on for five o'clock, and the 19th Hussars who had been sent on to look for the Wells, sent back word that they had found and occupied them. The square slowly moved towards the wells, where the troops bivouacked in square for the night. A party of three hundred volunteers from the Heavies, Guards and Mounted Infantry, went back to the zariba for the baggage camels, whence they returned at seven o'clock next morning, and the troops were able to get their first meal since noon of the 16th,—and this was the 18th.

The enemy's losses must have been very great, for a staff officer counted some eleven hundred bodies on the ground round the place which had been occupied by the square.

A small garrison of the Sussex Regiment, the wounded, and the stores, were left behind in occupation of the Wells, and at 3.30 p. m. on the 18th the Column continued its march to the Nile. After a terribly hard night, during which more than once the direction was lost and at one time the confusion amongst the baggage camels was terrible, a halt was called just before daybreak, and the guide, escorted by some hussars, went on to reconnoitre. The Officer in charge of the party soon reported that he had seen Metammeh, and that the enemy were moving from it towards the column.

About 7 a.m. Sir Herbert Stewart, seeing that the force would have to fight its way to the river, some five miles distant, decided to breakfast the men, then close up the transport and march for the river bank, with the fighting men going between the transport and the town. A zariba was then commenced of boxes, camel saddles, sand etc., and during its construction the enemy's sharp-shooters commenced firing with their Remingtons. A great number of camels, which were tied down in the centre, were shot. The attack grew hotter as the parapet in front of the men gradually rose, and a company of guardsmen were extended along a low ridge fifty yards in front of the zariba to endeavour to keep down the fire.

Inside the zariba men were falling fast, and at a little after 10 a.m. Sir Herbert Stewart was mortally wounded with a bullet in the groin, and Sir Charles Wilson took over the command.

The zariba and two small redoubts having at length reached a condition to resist an attack, it was decided to leave in it a garrison consisting of half of the Heavies, the 19th, Naval Brigade, guns, and baggage, and that the remainder should fight their way to the river.

A square was formed with the guards and mounted infantry in front and on its flanks, and with the remaining half of the Heavies and Sussex in the rear. Some camels for wounded, water, etc. were in the centre, and small reserves of dismounted hussars and sappers were at the corners of the square.

The square was formed under a heavy fire, and as the men of each corps forming it came up, they lay down on the ground in proper position. At length, at about half-past two in the afternoon, the advance commenced, and under a continuous and heavy fire the square slowly moved forward. At an occasional halt, volleys were fired wherever the enemy's smoke appeared thickest, while the guns from the zariba shelled them when they could get a target. The march was a terrible one and the men were falling quickly. Matters were becoming extremely critical. To go on looked like fighting to the last man; to retreat meant being utterly destroyed. Suddenly the Arabs began to collect in large bodies in front, and the long wished for moment had arrived. "Thank God! They're going to charge!" and on they came towards the left hand corner of the square. As our men halted to receive the charge they gave a wild cheer. "Cease fire" was ordered, and instantly obeyed, and then, with the enemy at 300 yards distance, "commence fire." The soldiers aimed low, and fired their volleys as steadily as on an Aldershot field day. The advancing host of savages seemed to melt away before the continuous roar of musketry. Not an Arab got within eighty yards of the square, and in a few minutes the front ranks were swept away, and the Arabs, brave as they were, wavered, scattered and bolted towards Metammeh. Three ringing cheers arose from the parched throats in the square, and the battle of Abu Kru was won. No casualties occurred during the charge.

The work, however, was not yet over ; the Nile must be reached before nightfall. After a short halt, the square moved on again, and eventually, half an hour after dark, the river was reached.

The wounded were first taken to drink. Perfect discipline was observed, and not a man left his place in the ranks until his company was marched up to take its fill. The front face having drunk its fill was marched back to relieve the rear face, and so on, in order that no flank should be left undefended in case of attack.

A zariba of bush was formed, sentries and pickets posted, and in a few minutes the exhausted force was asleep.

Twenty-one officers and men had been killed, and eighty-six wounded.

Before daybreak next morning the force stood to arms in expectation of an attack, which, however, was not delivered. A position was then selected at the deserted village of Abu Kru. It was soon placed in a condition of defence, and garrisoned by a hundred men of the Heavies, and in it were placed the wounded. Leaving a detachment of the Sussex Regiment to keep a watch on Metammeh, the remainder of the force marched back to the camel zariba, and brought up the troops and stores which had been left there.

The next morning the Heavies, the Guards, and the Mounted Infantry moved out to the attack of Metammeh, but the place was found too strong, and the force retired. During the advance, Gordon's promised steamers from Khartoum arrived and promptly landed a party of Egyptian soldiers and four guns, which soon came into action. With the steamers came a message from Gordon that Khartoum was all right, though woefully short of provisions.

On the evening of the 23rd a convoy was despatched to Gakdul to fetch provisions, and the next morning Sir Charles Wilson started in Gordon's steamers for Khartoum, taking with him twenty men of the Sussex Regiment and some 150 of Gordon's soldiers. Gordon in his messages had asked for a few red-coats to be sent, as their presence would do wonders in Khartoum ; the Sussex men were accordingly rigged out in red serge jumpers.

On the 31st of January the Convoy returned from Gakdul with stores and three guns, but with no fresh camels.

An officer with the force at Gubat (Abu Kru) thus describes the morning of the 1st February. " No member of our small force as long as he lives will ever forget this morning, " writes Lieutenant Dawson in his journal. " Just at dawn I was awoke by some one outside our hut calling for Boscawen. I jumped up and went out to see who it was, and then made out, to my surprise, Stuart-Wortley, whom we all thought at Khartoum. I looked towards the river, expecting in the faint light to see the steamers ; then seeing nothing, and observing by his face that there was something wrong, I said, ' Why, good heavens, where are the steamers—what is the news ? ' He said, ' The very worst. ' Then it all came out. " Sir Charles Wilson and his steamers had got close to Khartoum, where they were received with a tremendous fire, and found the place in possession of the enemy. They returned with difficulty, and on going down the Shabluka cataract, his steamers were wrecked. The troops were landed on an island and entrenched themselves, whilst Stuart-Wortley was sent down river in a nugger for assistance. To continue quoting from Dawson's journal : " The first necessity was of course to get Sir Charles Wilson off the island, and also to be ready at any moment for an overwhelming force coming down from Khartoum and cutting us off. The Mahdi was now free to move his whole force, numbers impossible to estimate, and besides was largely re-inforced by guns...... and 15,000 stands of rifles. "

Sir Charles Beresford and his blue jackets were sent up in a steamer to rescue Wilson, which, under great difficulties, they succeeded in doing; and another convoy, escorted by 300 Heavies, Guards, and Mounted Infantry, started off for Gakdul for stores and ammunition, and took with them some 110 sick and wounded. On the 6th of February Sir Charles Wilson left Abu Kru for Korti, with an escort of the Guards Camel Regiment, and on the 11th of February the convoy from Gakdul arrived, bringing with them six companies of the Royal Irish Regiment, and General Butler.

It was evident that for the small force at Abu Kru to attempt to take and hold Metammeh, or to hold on to its camp on the Nile bank, would be futile now that Khartoum had fallen, and the Mahdi's followers were swarming down. Butler therefore retired from Abu Kru to Abu Klea. The convoy of wounded was sent away previously, and near Abu Klea was attacked by a large force of the enemy, who, however, retired on the arrival of the Light Camel Regiment from Gakdul.

From Abu Klea Butler finally retired to Gakdul, where his force arrived on the 26th of February. Sir Herbert Stewart died of his wound on the 16th of February.

From Gakdul, the force marched to Korti, having lost during the expedition 30 officers and 450 men out of its strength of 2000.

The River Column, under Major-General Brackenbury, returned to Korti on the 8th of March.

"This then was the end of the double expedition which it was hoped would meet to co-operate for the relief of Khartoum and the rescue of General Gordon, of whose death there was now little doubt on the part of the Officers in command, although in England it was not till some time afterwards that the hope was abandoned that he had escaped to the Equatorial provinces, or had found protection with some friendly tribe, or was even kept in durance by the Mahdi himself." The rebels had entered Khartoum at daylight on the 26th of January, through the treachery of Farag Pasha, who opened the gates in the South wall. Nearly all the native stories agree that Gordon, on learning that he was betrayed, made a rush for the magazine in the Catholic mission building. Finding that the building was in possession of the enemy, Gordon returned to Government House and was killed while trying to re-enter it. Some say that he was shot; others that he was stabbed.

Writing of the departure of the troops of the Desert Column from Korti down the Nile, the author of "With the Camel Corps up the Nile" says: "The first to move were the Heavies, who, after having been in turn Cavalry, Infantry, and Mounted Infantry, now made their *dèbut* as boatmen. With about seventeen men and their kits to one boat, they paddled off on the afternoon of the 11th of March. Eight boats took the whole lot, a sadly reduced remnant of the magnificent corps which had started across the Bayuda." The corps had suffered greater losses than the other camel regiments. Out of a total of 20 officers and 390 men, only 8 officers and 210 men were returning.

The Camel Corps returned slowly down the Nile, and eventually embarked at Alexandria on board the P. & O. s.s. "Australia," and sailed for home on the 4th of July.

On the 27th of May, during the journey down the Nile,

Lieutenant Costello of the 5th Lancers died of enteric fever at Abu Fatmeh.

The following were the losses of the Camel Corps during the campaign :

HEAVY CAMEL REGIMENT.

	OFFICERS.			MEN			TOTAL LOSS	
	K.	W.	Died.	K.	W.	Died.	Officers	Men.
1st Life Guards	0	0	0	2	0	2	0	4
2nd " "	0	1	0	2	0	3	0	5
Royal Horse Guards	0	1	0	1	4	4	0	5
2nd Dragoon Guards	0	1	0	5	1	2	0	7
4th " "	2	0	0	7	5	1	2	8
5th " "	1	0	0	10	7	1	1	11
1st Dragoons	1	0	0	12	4	3	1	15
2nd "	1	0	0	11	5	4	1	15
5th Lancers	1	1	1	5	4	5	2	10
16th "	1	0	1	4	1	3	2	7
Total :	7.	4.	2	59.	31.	28	9.	87

GUARDS CAMEL REGIMENT.

	K.	W.	Died.
Officers	0.	3.	0
Men	26.	38.	23

MOUNTED INFANTRY CAMEL REGIMENT.

	K.	W.	Died.
Officers	0.	5.	1
Men	11.	67.	?

Upon arrival in England, the 5th Lancers detachment were present at the inspection of the Camel Corps by Her Majesty the Queen at Osborne.

For their services in the Soudan, the Detachment received the Egyptian Medal and two clasps for " Nile 1884-85, " and " Abu Klea. "

Major A. G. Spencer and Lieutenant Ayrton were sent out to the Nile to take over the command of the 5th Lancers detachment, *vice* Carmichael and Costello.

The following are the names of the Officers, Non-Commissioned Officers and Men of the 5th Lancers Detachment of the Camel Corps who fell during the Nile Campaign.

KILLED.

Major L. M. Carmichael.
No 2344 Corporal C. Percival.
No 2333 Lance Corporal S. Parker
No 2085 Private C. Peters.
No 1897 „ E. Bell.
No 2155 „ J. McGrath.
No 2129 A. Russell.

DIED.

Lieutenant H. Costello.
No 2425 Private C. Rayman.
No 2445 „ J. Harvey.
No 2240 „ G. Watson.
No 2207 „ E. Marks.

CHAPTER XII.

Suakim, 1885.—*Hasheen.*—*McNeill's Zareba.*—*Tamai.*

Meanwhile an expedition was being organised for Suakim, to oppose the gathering hordes of savage tribesmen who were being collected by Osman Digna against that place. In February 1885, the Regiment was called upon to furnish two squadrons for service with the expedition, which was to be commanded by Lieut.-General Sir Gerald Graham V.C.

The Squadrons were under the command of Lieut.-Colonel Chichester, and with him went the following Officers,

Major A. B. Harvey.
Captain E. C. Gilborne.
Captain L. H. Jones.
Captain J. Sinclair.
Lieutenant B. Mundy.
Lieutenant W. H. Goodair.
Lieutenant J. H. Rennie.
Lieutenant M. McNeill.
Lieutenant M. B. Doyne.
Lieutenant J. Richardson.

and 249 Non-commissioned Officers and men, and 200 horses.

The Detachment sailed from Kingstown on board the S.S. "Lydian Monarch" on the 20th of February, and on the 24th the Headquarters of the Regiment, under Lt.-Colonel Vandeleur, moved from Dublin to Dundalk.

Upon arrival at Suez, Lieut.-Colonel Chichester was ordered to Cairo, and Major Harvey took command of the service squadrons. The "Lydian Monarch" arrived at Suakim on the 13th of March, and disembarked the squadrons, with whom was Sir Gerald Graham.

On the night of the 11th-12th, a determined attack had

been made on the British camp, and successfully driven off. Small parties of the enemy continued nightly to get into camp, and on one or two occasions they succeeded in getting into a tent and stabbing the men. "The nearly naked Hadendowas, with bare feet and greased skins, as dusky as the night, crept and glided on their faces along every hollow and gully, carefully taking advantage of each bush that could screen their approach, and, if alarmed, lying perfectly still, after casting the sand with a rapid noiseless motion over their prostrate bodies, so that the keenest eye could hardly detect them from a stone. When they wished to make a sign to each other they imitated the cry of the desert birds with marvellous fidelity, and often has this low plaintive cry been the signal for their onslaught. Sometimes it was a volley followed by a rush with swords and spears, but more often a dark figure would seem to rise out of the very ground at the sentry's feet and stab him in the back ; or if it was impossible to get sufficiently near to him unperceived, they would wait until he moved away on his beat, knowing well his exact position by the crunching of his heavy ammunition boots on the gravel, and, wriggling past like serpents, slip among the tents ; then would follow the death scream, the rush of feet, and fierce volleys poured in rapid succession into the night after a few shadowy forms disappearing into the darkness, content at having achieved their work of murder and mutilation. This would be the signal for a general alarm along the whole line ; the Arabs further out in the desert would open fire with their Remingtons, bullets would come whizzing into the camp in all directions, and the force be kept on the alert until the long hours of the night passed away and the sun rose on another day of incessant work at the wharves and trenches for men who had enjoyed no sleep." And in addition to these trials there were the sandstorm, the great heat and the flies to endure.

On the 16th, the laying of the Suakim-Berber Railway was commenced.

On the 19th of March the Cavalry Brigade made a reconnaissance to Hasheen. Early in the morning the brigade, consisting of the 5th Lancers, 20th Hussars, 9th Bengal Cavalry,[1] R.H.A., and Mounted Infantry, was turned out. The English Cavalry were thrown forward, and gradually spread themselves

1 The 9th Bengal Cavalry, a sword regiment, had been armed with lances before leaving India.

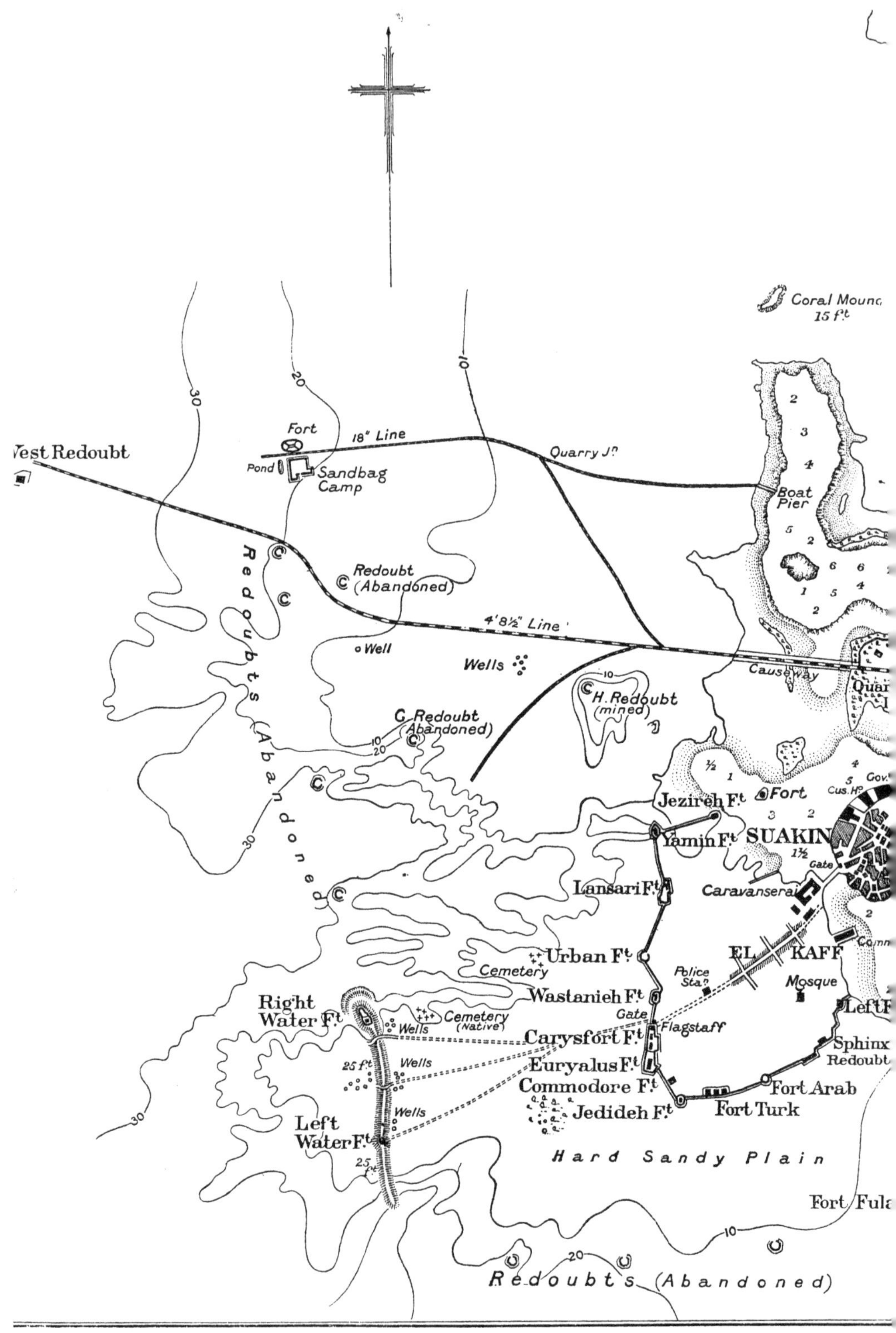

West Redoubt
Fort
18" Line
Pond
Sandbag Camp
Quarry Jn
Coral Mound 15 ft
Boat Pier
Redoubts (Abandoned)
Redoubt (Abandoned)
4'8½" Line
Well
Wells
Causeway
H. Redoubt (mined)
Redoubt (Abandoned)
Jezireh Ft
Fort
Yamin Ft
SUAKIN
Gate
Lansari Ft
Caravanserai
Urban Ft
Cemetery
EL KAFF
Police Stan
Mosque
Right Water Ft
Cemetery (Native)
Wells
Wastanieh Ft
Gate
Flagstaff
Carysfort Ft
Euryalus Ft
Commodore Ft
Jedideh Ft
Sphinx Redoubt
Fort Arab
Fort Turk
Left Water Ft
Hard Sandy Plain
Fort Fula
Redoubts (Abandoned)

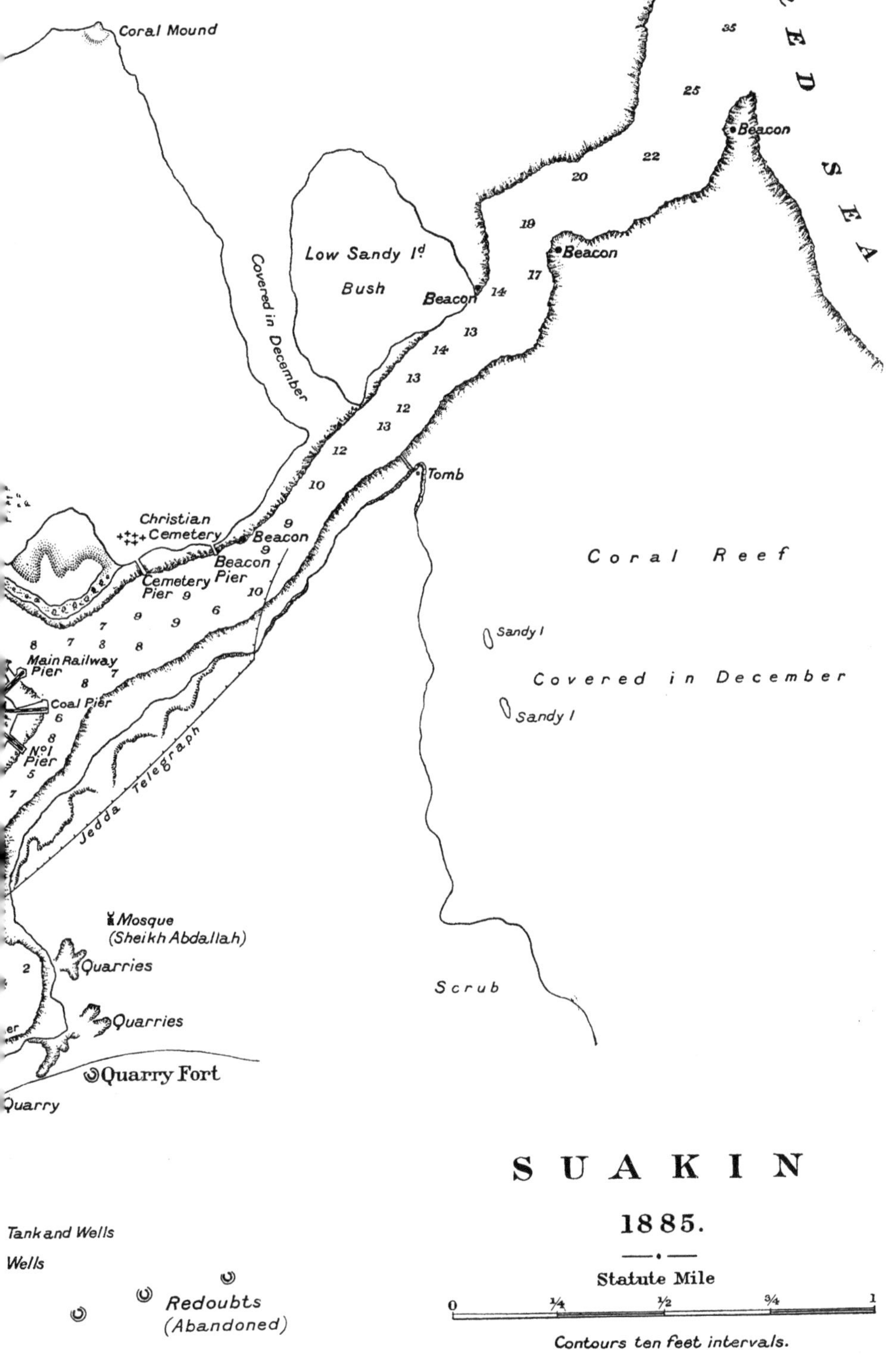

A. DOUBLEDAY & CO., LTD.

over the plain like a huge fan, while the Indian cavalry, guns, and Mounted Infantry followed in support. An eye witness writes: "I do not think I ever witnessed a more imposing spectacle than was presented by the beautiful working of this cavalry force, as they gradually felt their way across the plain towards the mountains in the direction of Hasheen." The village, however, was practically deserted, and the force returned to camp by one o'clock. Some prisoners were taken, and the British loss was one man killed, and an officer and a man wounded. Before retiring, a letter addressed to Osman Digna was left stuck on a cleft stick near the village. It was a reply to a boastful letter sent by him to our camp a fortnight before, in which he referred to the power and success of the Mahdi, and exhorted our Generals to submit and become Mahommedans. In reply he was reminded of the inability of the Mahdi's troops to withstand the British, and he in turn was exhorted to surrender. This letter was next day found trampled in the dirt.

The advance from Suakim to Hasheen was ordered for the next day, much to the delight of the camp. The men, weary with daily toil, sick of remaining within the limited space of the camp, and under extremely unhealthy conditions, and harassed by night attacks and alarms, were only too glad at the prospect of an active engagement with the enemy. At 5.30 a.m. the whole force moved out, with the exception of the 53rd (Shropshire) which regiment remained behind to guard the camp,

The formation was an open square. The front face was composed of the battalions of the 2nd Brigade under Sir John McNeill, the 49th, 70th, and Royal Marines. On the right face were the Guards Brigade, under General Freemantle; and the left face was formed by the Indian Brigade, consisting of the 15th Sikhs and 17th and 28th Bombay Native Infantry under General Hudson. Inside the square were the artillery, a small battery of six Gardner guns, the ambulance and the transport. The cavalry, consisting of two squadrons of 5th Lancers, two of 20th Hussars, and four of 9th B.C. and the greater part of the Mounted Infantry, were opened out in advance of the square, and scouting every yard of ground. At 8.30 a.m. the square had reached the first ridge of hills, the enemy retiring before the advance. The 70th (East Surrey) Regiment set to work to construct four sand-bag redoubts, with zaribas on the summits, on the left of the line of advance.

About 9 a.m. the 49th (Berkshire) Regiment reached the foot of Dihilbat Hill, which rose sheer above them like a wall, the enemy clustering on its summit. Supported by the Marines, the men of the Berkshires, without firing a shot, swarmed up the side of the mountain in attack formation, and under a heavy fire from the enemy. On reaching a ledge halfway up, they opened fire. The enemy returned the fire for a few minutes, until the Marines, advancing to enfilade them, they wavered and ran ; whereupon the Berkshire men crossed a ravine in front of their spur, and rushing on, gained the summit. The Indian infantry deployed and advanced upon the village of Hasheen, the Guards' Brigade, formed in square, covering their rear ; while the Mounted Infantry acted on the extreme right ; and the artillery came into action.

Meanwhile a large body of the enemy, driven from the hills on the left by the advance of the Berkshires and Marines, descended into the plain on the other side, and a squadron of the Bengal Cavalry were sent to intercept them. The enemy, however, charged the 9th B.C. who were obliged to fall back, losing four of their men, whose horses were hamstrung by the enemy. When the collision came, the Arabs threw themselves on the ground, slashing at the horses' legs with their swords; and, with surprising agility, they sped over the ground after the retreating horsemen.

The Indian Infantry, formed in two sides of a square, had now reached Hasheen, with the Guards in complete square covering their rear. A party of the enemy of about a hundred and fifty strong suddenly appeared from behind a hill within three hundred yards of the Guards, and actually charged the brigade, but were met with such a withering fire from the square, that they turned and fled.

On the extreme right rear of the British force the 5th Lancers now got their opportunity. A number of the enemy tried to break through in the direction of the redoubts which the 70th were constructing, and Major Harvey charged the Arabs on the flank, going right through them, and then, wheeling about, riding them down a second time. The Arabs practised their usual tactics, and lay flat on the ground when they saw the Lancers approaching, doing their best to hamstring the horses as they passed. But on this occasion they had the Queen of weapons to reckon with, and the lance put an end to many a Mahdist before he could put his plan into execution. The leader of this dashing charge was himself wounded by one of the lances of the Bengal Cavalry, with which an Arab had armed himself. Kneeling on the ground, the fellow kept himself in front of Harvey, who was somewhat perplexed to know what he was going to do, so he went straight at him with his drawn sword. The Arab suddenly jumped on one side, and as Harvey passed him, endeavoured to run him through with the lance. So quick was the Arab, that the sword was too late to parry the point, and the spear was lodged deeply in the rider's thigh, so deeply indeed, as to wrench it from the Arab's grasp. With the bridle in one hand and his sword in the other, there was no possibility of Harvey withdrawing the lance, which caught in a bush and nearly unhorsed the gallant major. Another officer of the Fifth laid four of the enemy low before he emptied his revolver, and such execution did the lancers do, that only seven of the enemy made their escape. The casualties amongst the Fifth were, Troop Sergeant-Major Nicholls and four men killed, Major Harvey and seven men wounded.

While this charge was being made, the artillery opened fire upon two large bodies of the enemy. One of these, about 2000 strong, was retreating in front. The other, quite double that strength, which had come from Tamai, was on the British left rear. About 1 p.m., the second, and the Indian brigades were ordered to fall back upon the guards, and at 2 p.m. the whole force began their march back to the hill occupied by the 70th Regiment. The Indians led, followed by the 46th and Marines, while the Guards' brigade, still in square, with the artillery, transport etc. in the centre, brought up the rear. The Guards were under a heavy fire, and halted every two hundred yards or so to fire volleys into the scrub. The ridge where the 70th had now completed the redoubts and zariba, was reached at 3 p.m. and a halt was called. Leaving the 70th with two guns and four Gardners, and five days provisions and water, to hold the redoubts and zariba, the remainder of the force set off on their return march to Suakim, reaching camp about seven in the evening. The action of Hasheen cost the British force some 22 officers and men killed and 43 wounded. The enemy must have lost very heavily. On the 25th March, the 70th were withdrawn to Suakim, and the zariba and redoubts they had built destroyed. Major Harvey was soon after invalided home, and Captain Gilborne assumed command of the two squadrons.

It was now determined to march on Tamai without further delay, and as the distance from Suakim was too great for troops to make the journey and return to Suakim in one day, it was decided to establish two zaribas on the line of advance, one four miles, and the other eight miles from Suakim. At 7 a. m. on Sunday the 22nd of March, the advance on Tamai began. One squadron of 5th Lancers scouted in front, followed by the British square consisting of 30 of the Naval Brigade with four Gardner guns, a detachment of Royal Engineers with Field Telegraph, the telegraph uncoiling its wire as they went, a battalion of the Berkshire Regiment, and another of the Royal Marine Light Infantry. The second square followed on the right rear, and consisted of the Indian infantry brigade. In its centre was a vast and unwieldy column of transport, consisting of 580 camels with 11,500 gallons of water, 500 camels with supplies, and about 400 pack-mules, draft-horses, and baggage camels with commissariat, water tanks, ammunition, and ambulance : a total of 1500 baggage animals. The force was under the command of Sir John McNeill, with General Hudson in charge of the Indian contingent.

It was terrible work getting the vast convoy of animals along in the midst of the thorn bushes, and it soon became evident that the original plan of forming a zariba eight miles out could not be carried out in time for the Indian brigade to return. As the force advanced, the Lancers began to report that parties of the enemy were hovering about. Sir John McNeill therefore determined to halt at six miles and there to form one zariba. At 10.30 a.m. the force emerged on a fairly open space of sandy and gravelly ground, some 300 to 400 yards across, dotted with thorn bushes and surrounded by thick bush and scrub, from which material was cut to form a zariba, consisting, as usual, of a hedge about four feet high, with a two foot ditch behind. It was a triple zariba, three enclosures of diamond shape standing corner to corner. That nearest Suakim was for the Royal Marines and half the Naval Brigade ; the centre one was for the camels ; and that nearer Tamai for the Berkshires, two Gardner guns, and the other half of the sailors. The work of building the zariba began. Infantry pickets were thrown out 150 yards to the front. A quarter of a mile or so further, in front of these pickets, the squadron of 5th Lancers, under Captain L. H. Jones, covered some three miles of frontage, with Cossack posts and a support nearer the zariba. Every precaution was adopted which the small numbers at his disposal would admit of. In proportion to the ground to be covered, it must be admitted that the exterior line was very weak. Owing to the scarcity of mounted men, it is evident that it was impossible to extend the radius of observation, as by so doing the distances between the posts would have been increased; while to patrol the bush, or to scout two or three miles from the main force was practically impossible to such a handful of troopers.

About 2.30 p. m. when the zariba was nearing completion, a Lancer rode in and reported to Sir John McNeill that the enemy was gathering in front and advancing rapidly, and was almost immediately followed by another Lancer, with similar intelligence. Orders were at once given for the working and covering parties to be called in, and while these instructions were being carried into effect, the Cavalry and Infantry pickets were to be seen coming in on every side, with the Arabs close on their heels.

The Arabs were now surging onwards, chiefly South and West, in one vast impetuous mass, enveloped in clouds of dust, filling the air with a pandemonium of shouts and yells, and making frantic efforts to storm the position. The Infantry faced the fierce rush with undaunted bravery, one battalion of natives only quailing before the shock, and these had been thrown into partial disorder by the stampede of the transport animals.

The Arabs now crowded in by the uncompleted salient, killing some of the sailors who were gallantly defending their charge. The rear-rank of the Berkshire half-battalion, then engaged in defending the western face of the redoubt, were now ordered to face about and occupy the vacated position through which the Arabs were pouring. This they at once did, and meeting the enemy half-way, quickly despatched every Arab who had entered, one hundred and twelve dead bodies being afterwards counted as having fallen within the limits of this redoubt alone.

The first wild rush of the enemy checked, the position was safe, and the Arabs were being repelled at every point. No sound could now be heard save that of the steady and sustained roar of musketry. With unflinching firmness the British and Indian troops settled to their work, and volley after volley was repeated, with all the rapidity and precision of clockwork. The fire was too terrible to be withstood, and the enemy fell back and disappeared among the bushes. The whole affair lasted but twenty minutes.

Colonel Way, an eye-witness, thus describes the battle. " Everything seemed to come at once, camels, transport of all kinds, including water carts, ammunition, mules, native infantry, Madras sappers, sick bearers, transport corps, Cavalry, and Arabs fighting in the midst. All these passed close by me and went out at the other side of the zariba...... The dust raised by this crowd was so great that I could not see anything beyond our zariba for a minute or two, and it was impossible to see who was standing or what was likely to happen. The men behaved splendidly, and stood quite still. It was about the highest test of discipline I shall ever see, as in my opinion nothing could beat it. " At the moment of the Arab onslaught, the camels and transport were being marshalled outside the zariba preparatory to returning to Suakim, and had stampeded.

About 3.30 p.m. a large force of the enemy appeared to the south-east of the zaribas, but these were dispersed by a counter-attack delivered by the Marines, led by Sir John McNeill in person.

A squadron of the 20th Hussars and another of the 9th B. C. patrolling from Suakim, reached the zariba about 4 p.m. On the way they had had a skirmish with the enemy, and on being joined by Lieutenant Goodair's troop of 5th Lancers, had pursued the Arabs, and prevented an attempt of the enemy to creep round by the sea-coast and turn their flank, and were able to push the Arabs back towards the zariba, where, being enfiladed by the fire of our men, they dispersed towards the sea.

The zariba was now completed and strengthened, and every preparation made for the possibility of a night attack.

When darkness came on the men were ranged all round the defences two deep, as if on parade. One rank lay down and slept, while the other, with bayonets fixed, kept two hours watch, fully armed and ready. Absolute silence reigned, a profound but watchful silence, not a light was shown, and nothing was permitted that might attract attention.

A sufficient garrison was left to defend the zariba, and the squadron of the 5th Lancers returned to Suakim with the remainder of the force.

In a telegraphic despatch on the battle dated Suakim, March 23rd, 1885, Lt.-General Graham stated: " The cavalry, 5th Lancers, did their best to give information, but the ground being covered with bush, it was impossible to see any distance. "

The British Losses were :—

	K.	W.	M.
Officers	4	8	1
N. O. C. & Men	66	125	35
Followers	34	18	122

A large number of camels and other transport animals were killed.

The casualties in the 5th Lancers squadron were Lieut. Richardson and 4 N.C.O. and men missing (killed) ; and 1 man killed.

Early in the day during the advance from Suakim, Lieutenant Richardson of the 5th Lancers, with four of his men, had been sent on an officer's patrol to the left of the line of advance. His horse getting knocked up after being out some hours, Richardson rode into Suakim about midday, and, on a fresh horse, returned to his patrol duties. Neither he nor his men were ever heard of again. They were no doubt cut off and slain by the Arabs during the course of the afternoon. Richardson's silver whistle was found some time afterwards at a spot in the bush, some six miles from Suakim and a couple of miles south of the zariba. The whistle was distinctly marked with a spear thrust, and encrusted with blood; and it is only too evident that the patrol had been caught in the great Arab wave sweeping down from Tamai on the British force, and killed to a man. The only other traces found of the patrol were at the capture of Tamai, on the 3rd of April, when a Lancer's scabbard and saddle were discovered.

The next few days the Regiment was employed on convoy duty, and on the 2nd of April was part of the force which marched for Tamai, where, on the next day, the enemy made some slight opposition. The British losses amounted to 1 man killed; and 1 officer, 14 men and 1 follower wounded. Tamai was destroyed, and on the 4th the force returned to Suakim. The Australian contingent—which arrived at Suakim on the 29th of March—took part in this advance. There is no need to follow the campaign further. The Regiment was employed on convoy duty, patrolling etc. with some small fighting, until by the 16th of May the campaign was at an end and the " construction of the railway to any considerable distance, postponed. "

On the 20th of May the 5th Lancer squadrons embarked at Suakim on board the S.S. "Lydian Monarch" for home. At Suez Captain Sinclair's squadron, with Lieutenants Rennie and McNeill, was landed to take over the horses of the 20th Hussars, for the purpose of bringing them to England.

The first squadron disembarked at Portsmouth on the 12th of June, and joined Regimental Head-Quarters now quartered at Brighton.

The second squadron, after staying at Suez a fortnight, was railed with the 20th Hussars to Alexandria, where they embarked on board the S.S. "Oregon," and landed at Portsmouth on the 27th of June. They went to Aldershot, where the 20th Hussars horses were handed over to various regiments, and then rejoined Head Quarters at Brighton.

In recognition of their services during the campaign, Major Harvey was promoted to Brevet Lieutenant-Colonel, and Captains Gilborne and Little, (the latter had been employed on the staff) each received a Brevet Majority; and in addition to these Officers, Captain Jones was mentioned in despatches. For their services the two squadrons received the Egyptian Medal with clasps for "Suakim 1885" and "Tofrek."

The following Officers, N.C.O. and men fell in the campaign.

KILLED.

Lieut. J. Richardson.
No. 1486 Troop Sergeant Major C. Nicholls.
2294 Corporal G. Pell.
2388 Lance Corporal J. Blood.
2621 Private J. Edwards.
1993 " J. Rose.
1999 " J. Shaughnessy.
2366 " J. Wilson.
1770 " J. Howard.

DIED.

1500 Sergeant C. Yorke.
2079 Private F. Turner.
2629 " M. Ryan.

In 1886, the Past and Present Regiment erected a memorial in St Martin's Church, Brighton, to the memory of the Officers,

Non-Commissioned Officers, and men of the Regiment who fell in the Nile and Suakim Expeditions.

Lieut.-Col. Vandeleur had resigned the command of the regiment on the 25th of May and was succeeded by Lieut.-Colonel Ward Bennitt, who was promoted from the 6th Inniskilling Dragoons on the 29th of July, Lt.-Col. Harvey having in the meantime been temporarily in command.

CHAPTER XIII.

Brighton.—Aldershot.—Shorncliffe.—Indian Sercice, 1888 to 1898.—Mhow.—Fever Epidemic.—Meerut and Alighar camp.—Muttra.—Commander-in-Chief's Cup.—Lieut.-General Dunham Massy appointed Colonel.—Cholera.—The Queen's Cup.—Regiment embarks for South Africa, 1898.—Natal.—Ladysmith.—Pietermaritzburg.—Colonel Scott Chisholme's farewell.

In General Order 10 of 1886, Her Majesty the Queen was graciously pleased to approve of the Regiment, amongst others engaged in the operations in the vicinity of Suakim during the year 1885, bearing the words "Suakim 1885" on their appointments.

On the 21st of June the Regiment moved to Aldershot, and encamped on Cove Common, and took part in the summer drills, at the conclusion of which, in August, Head Quarters and five troops returned to Brighton, and three troops to Hilsea, where 1887 they remained until the 14th of April 1887, when the Regiment marched to Aldershot, and moved into the South Cavalry Barracks.

The Commander-in-Chief made the following remark on the Inspection Report on the Regiment by Major-General Sir Drury Lowe K.C.B.

"His Royal Highness read the report of this admirable Regiment with great satisfaction."

1888 The Regiment remained at Aldershot until the 18th of July 1888, when it marched to Shorncliffe Camp, handing the South Cavalry Barracks over to the 16th Lancers.

On the 19th of September, "G" troop, with a strength of 50 N. C. O. and men, under the command of Major Fletcher and Lieut. Daniel, embarked in H. M. Troopship "Crocodile" at Portsmouth for Bombay, en route to Mhow. On arrival at Mhow this advanced party took over the horses of the 6th Dragoon Guards.

The Regiment was now under Orders for India, and on the

Lieutenant-Colonel D. MASSY

1876

13th of October His Royal Highness the Commander-in-Chief made a farewell inspection, and expressed himself highly pleased with everything he saw, and with the smart appearance of the Regiment.

At daybreak on the 20th of November 1888, the Regiment moved by rail from Shorncliffe to Portsmouth, and embarked on board H. M. Troopship "Euphrates." Early next morning the "Euphrates" sailed for Bombay; a depôt troop being sent to the Cavalry Depôt at Canterbury from Shorncliffe.

The following Officers embarked in the "Euphrates" for India.

Colonel Ward-Bennit (commanding)
Lt.-Col. A. B. Harvey.
Major C. E. Johnstone-Douglas.
" A. Weston.
Captain J. M. Fawcett.
" F. Ayrton.
" H. J. McLaughlin.
" J. H. W. Rennie.
Lieut. M. B. Doyne.
" W. E. R. Collis.
Lieutenant and Adjutant J. R. Harvey.
Lieut. H. V. Platt.
" H. V. Bailey.
2nd Lieut. E. B. Wilson.
" M. P. R. Oakes.
" J. B. Scriven.
" A. V. L. Wood.
" W. J. F. Horwood.
Hon. Lieut. and Quarter-Master G. Waterman.
" " and Riding Master H. Payne.
Attached Major and Pay Master Ternan.
and
2 Warrant Officers.
540 Non-Commissioned Officers and Men.
42 Women.
46 Children.

The "Euphrates" arrived at Bombay on the 18th of December, whence the Regiment was railed to Mhow via Deolali, arriving there on the 24th of December.

1889

At the end of January 1889, the Regiment was inspected by Brigadier-General George Luck, C.B., Inspector General of Cavalry

AA

in India, and in February by Major-General His Royal Highness The Duke of Connaught. His Royal Highness and General Luck both expressed their satisfaction at the appearance and state of the Regiment.

During their stay in Mhow the 5th Lancers suffered severely from enteric fever ; in a couple of months the epidemic carried off some fifty N. C. O. and men.

In October the Regiment was under orders to move to Meerut, in relief of the 8th Hussars, whose horses were to be taken over, and the advanced party left Mhow by rail on the 8th of that month, the Regiment following on the 16th and 17th. The horses were left in Mhow for the 18th Hussars, who were coming out from England.

Before the Regiment left Mhow, Major-General R. Gillespie C.B. issued the following order : "The General Officer is exceedingly sorry the Regiment is leaving his command. As regards the discipline of the Regiment he has had no trouble with either Non-commissioned officers or men. The manner in which the Regiment settled itself down on arrival was very remarkable, and it would have been a great pleasure to him to have retained it at Mhow. The Colonel and the Regiment have his most hearty and sincere good wishes and he will always be glad to hear of their welfare."

1890 The following is an extract from a letter from Sir Frederick Roberts, Commander-in-Chief in India : "I inspected the 5th Lancers at Meerut last December and was well satisfied with the turn-out and general appearance......The men are a fine body and well set up, and ride well. Their conduct has been very good. The horses are nearly all country-bred, they are of a good stamp and I think will give satisfaction. There has been a great advance in musketry, attributed by the A.A.G. for Musketry to the increased attention paid to the drills and shooting by the Officers and non-commissioned Officers.

1891 On the 18th of November 1891 the Regiment marched to Alighar, and took part in the manœuvres of the Cavalry Division at that camp of exercise during November and December. The Regiment was posted to the 2nd Brigade of the Southern Force, under Colonel T. A. Cooke, during the manœuvres from Alighar to Meerut. Meerut was reached on the 17th of December, and after a march past the Commander-in-Chief on the 19th, the regiments of the Division marched off to their respective stations ; the 5th remaining at Meerut.

1892

This year the Colonel of the Regiment, Lieut-General The Hon. J. G. Calthorpe, was transferred to the 5th Dragoon Guards, being succeeded by Lieut.-General the Hon. C. W. Thesiger. In connection with Lieut.-General Calthorpe's transfer the following Regimental Order was published—"Her Majesty the Queen having been graciously pleased to transfer Lieut.-General the Hon. S. J. G. Calthorpe from the Colonelcy of the 5th Lancers to that of the 5th Dragoon Guards, he is permitted by the Commanding Officer to express through the medium of the Regimental Orders to all ranks of the Regiment his regret at being no longer on the roll of the 5th Lancers, and in bidding them farewell and God speed can assure them he will always take a deep interest in the fortunes of the Regiment and look back with pride to the time he had the honour of being at the head of so distinguished a corps."

1893

At the end of 1892 and commencement of 1893 the Regiment took part in the District concentration of troops at Meerut.

During this year the Regiment was inspected by the Inspector General of Cavalry and by Major-General C.E. Nairne C.B., and both Officers expressed their complete satisfaction at everything they saw ; the good condition of the horses being specially commented upon.

Early in November the Regiment moved into camp, being relieved by the 5th Dragoon Guards from England, who took over the barracks.

The Regiment took part in the District manœuvres from the 10th of November to 10th of December, and on the 11th left by march route for Muttra, after a stay of five years in Meerut.

The Regiment arrived at Muttra on the 4th of December. A somewhat remarkable incident happened at Jeyt, the last camp before Muttra. During the evening stable hour, a herd of black buck was observed immediately in front of the camp. After some hesitation they commenced to move round the flank of the camp, and ultimately made a dash between the flank and the place where the transport animals were picketed. They were at once pursued by the transport drivers, who hurled every available missile at them, and headed them off. The herd then rushed through the camp, and passed so close to the rear guard, that the sentry on duty made a lunge at a fine buck with his lance, and brought him down.

1894 Lieut.-Colonel Johnstone-Douglas, Commanding Officer, died at Solon, Punjaub, on the 11th of August 1894 of abcess on the liver. Lt.-Colonel Johnston-Douglas was succeeded in the command by Lt.-Colonel Scott Chisholme.

The inspections of the Regiment in 1894 produced very favourable reports, and H.R.H. the Commander-in-Chief made the following observation ; " This Regiment appears to His Royal Highness to be in a high state of efficiency. "

1895 At his inspection in March 1895, the General Officer commanding recorded that he considered " the 5th Lancers to be in very good order and particularly fit for active service. The men are well inured to India, their conduct and health are good. They are a smart Regiment with very good Non-commissioned Officers. "

His Excellency the Commander-in-Chief made the following remark on the Inspection reports on the Regiment for 1895. " The reports show the 5th Lancers to be in a very efficient state. It is the best shooting Cavalry Regiment in India, and has proved its superiority not only in the annual course, but also in the competitions for prizes open to the British Army in Bengal. "

This year the Regiment won the Commander-in-Chief's Musketry Prize, being at the head of the list of all British Cavalry and Infantry in India. The following were the team :—

S. S. M. Owen.	L.-Cpl Hamilton
„ Lay.	„ Beadle.
S. Q. M. S. Perrin.	„ Wing.
Sergeant King.	Private Fitzgerald.
L.-Sergeant Myhill	„ Grain.

On the 22nd of October, a party consisting of 3 Officers, 1 Sqdn. Sergeant Major, 2 Sergeants, 1 Trumpeter and 56 rank and file marched to Agra, and formed the escort to His Excellency the Viceroy of India (The Earl of Elgin) on the occasion of his meeting the Princes of India.

1896 During January 1896, the Regiment marched to Alighar and was there brigaded with the Imperial service Cavalry of the Rajahs of Bhurtpore and Rampur, during a camp of instruction under Lieut.-Colonel Scott Chisholme. At the conclusion of the camp the 5th returned to Muttra.

1896 saw another change in the Colonelcy of the Regiment. Lieut-General the Hon. C. W. Thesiger was transferred to the

14th Hussars, and was succeeded by Lieut.-General W. G. Dunham-Massy C.B. as from 4th of October.

The inspections of this year produced extremely favourable reports.

1897

Lieut.-General Sir Baker Russell K.C.B. Commanding the Forces in Bengal, inspected the Regiment in February 1897, and stated his appreciation of the appearance and turn out of the Regiment. "The drill" he reported, "was very good, and there was an absence of all unnecessary noise." He was also much pleased "with the knowledge of the country round Muttra shown by all ranks, men as well as Officers and N.C.O."

The magnificent state of the Regiment at this period can be gathered from the following incident. At the conclusion of the last parade of the drill season of 1896-1897, "Jabber" Chisholme wheeled the Regiment into line and sounded the "Officers" call. They galloped out to him, when, returning their salutes, he said, "Gentlemen, I have called you out to look at such a regiment of cavalry as you are unlikely ever to see again. Turn about and look at the Regiment". They turned and looked, and the youngest amongst them knew that the Colonel was right, as with pride they gazed on that long line of *five* strong squadrons, standing motionless under the Indian sun; not a horse out of its place, and intervals and distances correct enough to more than satisfy the keenest of critics. Their Indian soldiering was drawing to a close, and such regiments, they knew, did not exist out of India.

During the summer of 1897, a severe outbreak of cholera occurred in the Regiment, on which subject General Sir Baker Russell wrote to the Officer commanding. "I very much admire the tone and spirit of the report, and am full of admiration for the grand coolness and pluck and discipline shown by the 5th Lancers under your command. I shall be much obliged if you will convey my sentiments to the Officers, Non-Commissioned Officers and private soldiers of the Irish Lancers and express to them my condolence on the loss of their comrades."

The 5th Lancers were now under orders for South Africa, and Major-General Sanford, commanding the Meerut District wrote, "I am sorry not to be able to see the Regiment again before it goes to South Africa to bid it 'Good-bye'; but it has my best wishes, and more than this, my firm conviction that it will do right well and maintain its own excellent reputation wherever it may go, and whatever it may be called upon to do."

On the report of the inspections by Major-General Grant (Inspector General of Cavalry) and Major-General Sanford (commanding the Meerut District), H.E. the C. in C. observes—"Quite an exceptionally good report."

During November the Regiment won the "Queens Cup" (Cavalry) of the Army Rifle Association, the following being the team:—

Lieut. C. Arkwright.	L.-Sergt.	Myhill
S. S. M. Owen	"	Elworthy
Sergeant King	L.-Cpl.	Hamilton.
L.-Sergt. Burkinsher.	"	Simpson.

1898 The Regiment being under orders for South Africa, on the 29th of January 1898 handed over the horses etc. to a detachment of the 9th Lancers from Natal. The same day the advanced party left for Bombay with the heavy baggage.

On the following day the Regiment left Muttra by train, after a stay in that cantonment of over four years. A halt was made the same night at Jhansi, and on the following nights at Hoshungabad Khandwa, and Deolali rest camps. Deolali was not left until the 8th of February, when the journey was continued to Bombay, and the Regiment embarked the same evening on board the Royal Indian Marine Ship "Clive."

On leaving the Bengal command the General Officer commanding, Sir Baker Russell, made the following remarks on the 5th Lancers.

"A first rate Regiment in first rate order. I never saw a better. It is in every way fit for active service. I am sorry to lose it from my command."

The following was the strength of the Regiment embarking in the "Clive."

18 Officers.
522 N.C.O. and Men.
3 Officers' wives.
4 " children.
19 Women.
27 Children.

Officers of the Regiment in Natal, 1899.

The Regiment on sailing from Bombay had been a little over nine years in India, during which time it had earned the highest opinion, in quarters, and at the various camps of instruction, of every one with whom it had come in contact.

Roll of Officers on the strength of the Regiment at the time of leaving India.

Lieut.-Col. J. J. Scott Chisholme (commanding).
Major J. F. M. Fawcett.
" H. J. McLaughlin.
" A. C. King.
" R. J. Spurrell.
Captain H. W. Graham, D.S.O.
" E. O. Wathen.
" A. Parker.
" W. A. Adams.
Lieutenant M. P. R. Oakes.
" J. B. Scriven.
" J. B. Jardine.
" R. C. Browne-Clayton.
" W. T. Willcox.
" M. Caillard.
" C. Arkwright.
" M. F. McTaggart.
2nd Lieut. H. F. Fraser.
" W. H. T. Hill.
" R. G. Hooper.
" S. S. Williams.
" K. D. McKenzie.
" J. Bruce.
Lieut. and Adjutant H. H. Hulse.
Hon. Lieut. & Quarter Master G. Waterman.
Hon. Lieut. & Riding Master Payne.
Canterbury Depot. Capt. E. B. Wilson.
Lieut. R. P. J. Gwyn.

After an uneventful voyage the "Clive" arrived off Durban, Natal, on the 24th of February, but owing to a case of small-pox amongst the native crew, she was placed in quarantine, and anchored at the mouth of the harbour.

On the 5th of March the Regiment disembarked, and proceeded by rail to Pinetown, remaining there under canvas until the 10th. Entraining at 1.30 p.m. a halt of an hour in the evening was made at Pietermaritzburg, where all ranks were entertained by the 7th Hussars. Ladysmith was reached at 5 a.m. on the 11th. The Regiment marched to the Camp and took over the horses and quarters of the 9th Lancers, who left the same day, to embark at Durban for Muttra in India.

From the date of disembarkation the Regiment was reduced from four to three service squadrons and a Reserve troop, with a depôt troop in England.

The establishment of a service squadron was 131 of all ranks, and 105 horses; that of the Reserve troop was 104 of all ranks and 45 horses; a total of 497 Officers, N.C.O. and men, and 360 horses abroad with Head Quarters.

While at Ladysmith, the Regiment, with the 7th Hussars, formed the cavalry of the force engaged in the Biggarsberg manœuvres.

On the 24th of October 1898 the Regiment left Ladysmith by march route for Maritzburg, being relieved by the 18th Hussars from India. The capital of the colony was reached on the 29th, where the quarters of the 7th Hussars in Fort Napier were taken over, that regiment leaving for home.

1899 Owing to the prevalence of horse-sickness in Pietermaritzburg during April 1899, the 5th Lancers, with the Mounted Infantry and Artillery quartered there, marched to the Nottingham Road Camp. While there, the force camped at Nottingham Road were engaged in operations with the troops at Mooi River, the latter having come to the high ground from Ladysmith on account of horse sickness.

The Regiment returned to Fort Napier on the 8th of May, and on the 10th it was on duty upon the occasion of the opening of the Natal Parliament by His Excellency the Governor, Sir Walter Hely-Hutchinson.

On the 12th of August 1899 Colonel Scott Chisholme was placed on half pay on the termination of his period of command, and was succeeded by Lieut.-Colonel J. F. M. Fawcett.

On giving up the command of the Regiment, Colonel Scott Chisholme personally bid the men farewell in their barrack rooms, and subsequently he published the following Regimental Order.

"I cannot express to you all how much I feel leaving the Regiment. I thank you that you have made the five years of my command so exceptionally happy. I shall ever feel proud that

Colonel J. J. SCOTT CHISHOLME

1896

I have been your Colonel, and am confident that you will uphold the high esteem in which you are held. Comrades Farewell."
To the great regret of the whole regiment Colonel Scott Chisholme's period of command expired, as after events proved, upon the eve of War.

CHAPTER XIV.

South African War 1899.—*Elandslaagte.*—*Rietfontein.*—
The return of the Dundee Column.—
The battle of Ladysmith (Lombard's Kop or Modder Spruit, Nicholson's Nek.)

"Soon after our return to Maritzburg from the Nottingham Road Camp," writes an Officer in his diary, "the excitement of a probable Boer War began. Then followed the Bloemfontein Conference, and I remember the day the Colonel came into the ante-room and announced its failure. 'Who-whoop' everyone shouted, and the Colonel walking out of the room laughingly said 'You are a bloodthisty lot of devils." The majority of us sadly thought it most improbable that the Boers would show fight, in spite of the fact that relations between Great Britain and the Transvaal were evidently getting more strained every day. We were then told that the possibility of war was not to be discussed, and the men were forbidden to speak of it."

By August there was a subdued excitement in Natal, and though soldiers held their peace, civilians were constantly to be seen standing in little groups in the streets discussing the situation. Everywhere one heard the question "Will they climb down?" During August, too, suspicious looking characters began to appear in Fort Napier; men with imaginations voted them Boer spies. Were they? Who knows? And the provost sergeant kept his eye on them.

During this month, too, officers of the Natal garrison were sent off to various parts of the Colony sketching towns, districts, and possibly, future battlefields. In connection with these duties an amusing paragraph appeared in a local paper. The correspondent writes, "Some of the inhabitants of this town were alarmed when an Officer and his servant, in their uniforms of the 5th Lancers, made their appearance. All sorts of wild rumours were current, among them that the officer (Lieut. . . . . .) had come to select a camping ground for a regiment of cavalry. I understand, however, that he was only on the usual road survey that the military authorities make occasionally. It is evident that the majority of inhabitants in this small town know very little of uniforms of the British Army, for some of them called the soldier a "Dragoon," others "Hussar," and one man I heard declare that he belonged to the Royal Engineers. One young schoolboy asked a resident what corps the soldier belonged to, and when told the "Horse Marines", took it in, and was quite satisfied. The Lieutenant has now left Greytown, and gone on to Middle Drift."

The powers that be were evidently preparing for war, for the barrack square was being filled with mobilization carts and wagons, and everybody was being examined by the doctors. And away at the end of the town, in the safely guarded enclosure of the Agricultural Show grounds, an irregular corps of horse, which was destined in the near future to gain exceptional honours at Elandslaagte, Ladysmith and other hard fought fields, was being rapidly and secretly raised. So secret was the organization of the regiment that Maritzburg knew but little of it. Parties of twenty men, seldom more, left Johannesburg during the exodus, being quietly and unostentatiously met at Maritzburg, and equally quietly taken to the Agricultural Show grounds. Here also occasional horses appeared, and boxes of equipment, saddlery and rifles; mostly at the expense of the loyal members of the Johannesburg community. The raising of this corps was entrusted to Colonel Scott Chisholme, who, with the assistance of an officer of the 5th Lancers as adjutant, and some N. C. O. of the same regiment, rapidly had the continuous stream of new arrivals sworn in, dressed, equipped, drilled, and ready to take the field. It was a proud day for the officers of this new corps when Her Majesty was graciously pleased to order that the force was to be called "The Imperial Light Horse."

In the early days of September, while manœuvring near Maritzburg, some of the men of the Regiment rode into some cultivated ground, a most reprehensible proceeding, which so annoyed an old lady who happened to be there, that she complained bitterly; whereupon her husband remarked to her "Shut up, they will be fighting for you soon." There was no doubt about it, in spite of people endeavouring to think the contrary, war was swiftly and surely approaching.

In a local paper appeared the following paragraph. "In

HISTORY OF THE FIFTH LANCERS.

A. DOUBLEDAY & CO., LTD.

appreciation of the Transvaal's assurance that they are desirous of a peaceful solution of the crisis, the 5th Lancers spent Sunday morning at the camp in sharpening their lances and sabres."

On the following Sunday night, the 24th September 1899, the Officers of the Regiment were sitting down to dinner when they were joined by Major General Sir Penn-Symons, commanding the troops in Natal, who promptly ordered that a squadron of the Regiment should leave by rail next day for Ladysmith, and that the other two and Head-Quarters were to follow by march route on the day following. The Ladysmith garrison he remarked, was to move up to Dundee on to morrow.

The "C" squadron was ordered to go by rail, and entrained on the following afternoon. With the squadron went Captain M. P. Oakes, Lieutenants R. P. Gwyn, W. T. Willcox, and W. T. Hill, and the maxim gun detachment. The horses went in one train and the men in another, the latter travelling in open trucks. The 2nd King's Royal Rifles left the same afternoon. The excitement in Maritzburg was immense. Pipes and tobacco were showered into the men's trucks by the excited crowd, and during daylight at every station the trains met with a huge ovation. People begged the cap badges of the men, and after the relief of Ladysmith an officer was told at Howick that a prominent inhabitant of that village had bought a lance flag from one of the men for £5.

The squadron arrived at Ladysmith at 3 a.m., and at once detrained and rode up to the camp, and went into the quarters of 18th Hussars, who had left the previous day for Dundee. There was no news in Ladysmith beyond the rumours of large gatherings of Boers on the Drakensberg. Until the arrival of Head Quarters the squadron was employed patrolling.

Meanwhile Head Quarters and "A" and "D" Squadrons marched from Maritzburg on the morning of the 26th September for Ladysmith via Dargle Road, Nottingham Road, Mooi River, Estcourt, and Colenso, and arrived at Ladysmith on the 2nd of October. The Reserve troop with 2nd Lieutenant Hooper and Riding Master Payne remained behind in Maritzburg. The following Officers marched with the Squadrons.

Major A. C. King — in temporary command
Lieut. H. H. Hulse — Adjutant.

"A" SQUADRON.

Lieut. R. C. Browne-Clayton.
2nd Lieut. H. F. Fraser.

"D" SQUADRON.

Captain A. Parker.
Lieut. A. V. L. Wood.
„ C. Arkwright.
„ M. M'Taggart.

Hony. Capt. G. Waterman — Quarter-Master.
Major Holyoake R. A. M. C. — Medical Officer.
Vety. Lieut. Coley. A. V. D. — Veterinary Officer.

The Natal Volunteers were now called out and posted at various places on the railway. The Boers were massing along the Natal Frontier, at Wakkerstroom, Sand River, and along the Drakensberg. The Regiment was employed patrolling night and day.

At 2.30 a.m. on the 10th, information reached Ladysmith that 1200 Boers had reached Colworth from the Free State, and that a squadron of Natal Carbineers was retiring before them. "C" and "D" squadrons of 5th Lancers were promptly turned out and rode to Besters, where they were joined by an armoured train and a company of the Liverpool Regiment. The Natal Carbineers were found to be all right, and the information a false alarm; the squadrons returned to Ladysmith.

Sir George White, commanding the Natal Field Force, now arrived in Ladysmith; Sir Penn-Symons going on to command the troops at Dundee.

With the policy of holding on to Dundee we have here nothing to do, but having decided on this line of action, Sir George White concentrated his few available troops at Ladysmith and proceeded to accumulate there the large stores of supplies which enabled the town to hold out for the long four months against the Boers.

The original Natal garrison was now divided between Dundee and Ladysmith, and to the latter place were coming the contingent from India. Local Natal forces were at Colenso, Estcourt, Helpmakaar, and at points along the line of railway. By the 11th of October there were fully 20,000 Boers on the Natal frontier.

On the 12th of October 1899 the Regiment was informed of the declaration of War in the following Regimental Order of that date.

"Extract from Garrison Orders.
16 War.
The Secretary of State for War telegraphs from London that War has been declared by the South African Republic, and the British Agent at Pretoria has been instructed to ask for his credentials and withdraw to British territory."

The troops marked their appreciation of the situation by cheering.

For more than a week the assembled Burghers on the frontier had been eagerly waiting to move down into Natal. They looked forward to shooting the red-coats. They had beaten the English Army before, and they were going to beat it again. A prominent member of the Volksraad, Mr Botha, had wired to President Kruger: "May the Vierkleur soon wave over a free harbour." At Albertina on the 11th they had stopped the mail train going to Natal, and now War was at last declared. The average burgher was somewhat troubled at the rumours of lyddite, armoured trains, and War balloons, but after all these were perhaps much exaggerated. In Pretoria it was feared that while the Boers were advancing into Natal, the British would make a raid upon Pretoria in war balloons. "Instructions were given to all telegraph stations to report any balloons seen, with the result that for the first fortnight balloons, singly or in coveys, and usually provided with powerful coloured searchlights, were daily reported from every quarter of the Transvaal."

On October the 12th the Transvaal forces moved towards Penn-Symons at Dundee, while those of the Orange Free State came down the Van Reenan and Tintwa passes towards Ladysmith.

The next morning Sir George White moved the major part of the Ladysmith garrison nine miles out to the west, and the Dublin Fusiliers were brought by rail from Dundee. The Free Staters, however, were not met, and the force returned to Ladysmith, and the Dublins were sent back to Dundee. "D" Squadron of the 5th Lancers continued on towards Mount Tintwa, and returned on the heels of the column.

At 5.30 a.m. on the 16th, the 5th Lancers and some Mounted

Infantry rode towards Acton Holmes, and the 19th Hussars towards Blaaubank. From Acton Holmes the 5th continued on for another seven miles to the westward, "A" Squadron branching off S.W. to Bethany, where they got a good view of the enemy's camp near Bezuidenhout. The Regiment bivouacked at Clydesdale for the night, and next morning "A" and "C" Squadrons returned to Ladysmith, while "D" Squadron reconnoitred towards Mount Tintwa, where they were shelled by a large force of Boers. The shells, however, did not burst.

The Boer commandos were gradually closing in on Dundee and Ladysmith, and on the 19th a portion of General Kock's commando, moving rapidly through the Biggarsberg, seized the coal-fields and railway station at Elandslaagte, and so severed White's communication with Penn-Symons; they also captured a supply train on its way to the Dundee force.

The Tin camp in Ladysmith being in an exposed position, was vacated on the 19th, the troops moving into the town. The 5th Lancers moved into camp on the S. side of the town, and between it and the Klip river. On this day a 5th Lancer patrol was in touch with the enemy near Besters. Meanwhile the Indian contingent were arriving, and some squadrons of the 5th Dragoon Guards and 19th Hussars had already reached Ladysmith.

In these early days of the War the Regiment was continuously reconnoitring to the N; Boers were generally seen in small bodies, and they generally retired.

On the 20th of October the Dundee troops fought the action at Talana Hill. The British Infantry drove the enemy from the position they had taken up during the night on the hill, but at a tremendous cost, amongst the British losses being the gallant Penn-Symons who had fallen mortally wounded.

Meanwhile, at 11 a.m. on the 20th, General French moved out of Ladysmith with the 5th Lancers and the Natal Carbineers to Modder Spruit, some twelve miles along the Dundee road. Halting at Modder Spruit the General sent "C" Squadron of the 5th Lancers on towards Elandslaagte. 2nd Lieutenant Hill's patrol succeeded in capturing two patrols of Boers, from whom it was gathered that Elandslaagte was not held in any great force by the enemy. One of the captured Boers boasted that some six weeks before he had sold a public house in Ireland to come out and fight for the Boers. Unfortunately for the boaster, his captor,

Sergeant Ryan, was also an Irishman, and took infinite pains to impress on his erring countryman the folly of his ways, before handing him over. The force returned to Ladysmith.

Sir George White now decided to clear the Boers from Elandslaagte and to cover and repair the railway and telegraph, and re-open the direct communication with Dundee. To this end General French was ordered to leave Ladysmith early on the morning of Saturday the 21st of October 1899 with the Imperial Light Horse and a battery of the Natal Volunteer Artillery. These were to go by road, while a half battalion of the Manchester Regiment, and the Engineers were to follow by train. The I.L.H. and the battery started at 4 a.m., and by 8.30 a.m. were within a mile of Elandslaagte station. The Natal gunners opened fire on the station buildings and the Boers promptly replied. The Boer artillery was a revelation, better guns and smokeless powder, and good shooting. Seeing that his force was too small to attempt occupying Elandslaagte, General French withdrew to a position which was out of range and covered the armoured train, which had arrived, and telephoned the situation to Ladysmith.

Sir George White promptly commenced to despatch as large a force as could be spared to reinforce General French. "D" Squadron of the 5th Lancers under Captain Parker joined a squadron of 5th Dragoon Guards and the Battery of Royal Field Artillery at Limit Hill, just outside Ladysmith, whence they rode hard to Modder Spruit, to which place French had meanwhile retired. The force joined French about 11 a.m., and the infantry, consisting of the 1st Devonshire Regiment and the 2nd Gordon Highlanders, commenced to arrive by train. Meanwhile, on their arrival, "D" Squadron of the 5th Lancers, and four squadrons of Imperial Light Horse were sent to clear a ridge running parallel to the Boer main position, and about two and a half miles to the south west of it, which was intended to be the starting point of the infantry attack. The Boers, under Pienaar, fell back towards their main position.

Meanwhile, in Ladysmith, shortly after noon, "C" Squadron 5th Lancers under Captain Oakes was ordered to escort the 21st Battery R.F.A. to Modder Spruit, going as quickly as they could, the battery with double teams. At Modder Spruit the squadron joined that of the 5th Dragoon Guards, who were manœuvring against Schiel's German corps and a party of Boers to the North of the railway. On the enemy falling back, these two squadrons cut the wire fences and crossed the railway to the East, eventually halting in the vicinity of Elandslaagte station.

By 3 p.m. all the re-inforcements had arrived, and General French determined to attack. The infantry were sent on to the ridge which "D" Squadron and the I.L.H. had earlier in the day cleared. The Devons marched along its northern end, while the Manchesters moved further to the right, and the Gordon Highlanders were in the centre. The Boer guns opened on the infantry as they appeared, and at 4 p.m. the 21st Battery R.F.A. came into action in the open, on the left of the Manchesters, against the enemy's artillery at 4000 yards range. The 42nd Battery then coming up, the Boer guns ceased firing. "D" Squadron 5th Lancers, supported by the Imperial Light Horse, now gave chase to a party of the enemy who appeared to be galloping south. The squadron pursued to within a short distance of the Boer main position, and retired at a gallop under a heavy fire in open order on the I.L.H., who had dismounted.

About 4.30 p. m. the infantry attack had commenced. The Devons, advancing across the open, made a frontal attack against the Boer position, while the Gordons and Manchesters were working round to the right against the Boer left flank. The Devons, slowly and steadily advancing, walked through the artillery and mauser fire to within 800 yards of the summit of the Boer hill. Here they lay under an appalling fire for over half-an-hour, waiting for the flank attack to develop, and the order to advance.

Meanwhile on the right flank the Manchesters and Gordons were pressing on. On the ridge already mentioned they were joined by the Imperial Light Horse, who, dismounting, doubled out and extended on the right of the Manchesters; while still further to the right "D" Squadron of 5th Lancers were awaiting their opportunity. The flank attack now had half a mile of open slope to cover to reach the main ridge, and in spite of enormous losses they pressed on. At last from the valley below the Devonshire bugles rang out the "charge;" the call spread along the line, and with a cheer and a great rush, the bayonets of the frontal and flank attacks were in possession of the hill, while the Boers were streaming away down the reverse slopes. The "Cease fire" now sounded and the action was apparently over. Suddenly, however, a party of Boers headed by General Kock delivered a counter attack against the soldiers crowded together on the ridge, which went very near to being successful. "The men, perplexed by the 'cease fire' and staggered by the sudden fury of the attack, fell back a hundred yards, uncovering the captured guns, which the leading Boers laid hold of." The Imperial Light Horse on the right again rushed to the charge, but not before they had lost their gallant Colonel, Scott Chisholme. He was binding up a wounded trooper when the counter attack took place, and was shot in the ankle. Waving his men on with his old regimental scarf, he called out "My boys are doing well," and immediately after got a bullet through his lungs, while a third bullet piered his brain. His gallant corps, however, leapt forward, the infantry rallied, and once more the Khaki clad line advanced. Four companies of the Devons stormed the detached hill on the left of the position, and in a few minutes Devons, Manchesters, Highlanders and Light Horsemen were rushing for the laager below, and 'cease fire' again sounded. It was now after half past five, and already getting dusk. During the infantry attack "C" Squadron 5th Lancers and "C" Squadron 5th Dragoon Guards had been lying in a fold of the ground on the left, and a couple of patrols of the Lancers had captured some twenty Boers. But now the opportunity for which they had waited had come. The Boers were streaming off their main position in a Northerly direction, and with the 5th Lancer Squadron under Captain Oakes on the right, and that of the 5th Dragoon Guards under Major Gore on the left, the Cavalry extended, and were let go. As they topped the rise which had concealed them, they found the Boers crossing their front at a distance of a few hundred yards. The Boers endeavoured to get away, but in spite of a donga to be crossed, and the bad rockey going, the big English and Wales horses of the Lancers and Dragoons were soon amongst their little ponies. For over a mile did the two British squadrons ride through the enemy, spearing some forty of them. Then rallying, the troopers wheeled about and galloped back again through the still streaming crowds of fugitives. Many Boers endeavoured to fire their mausers from the saddle, but after the first onset of the cavalry, the Burghers were straining every nerve to gallop away from those terrible lance points.

"Then the scattered troopers were again rallied. The men fell in and cheered madly. There was something awful in the dramatic setting of the scene. The wild troopers forming in the

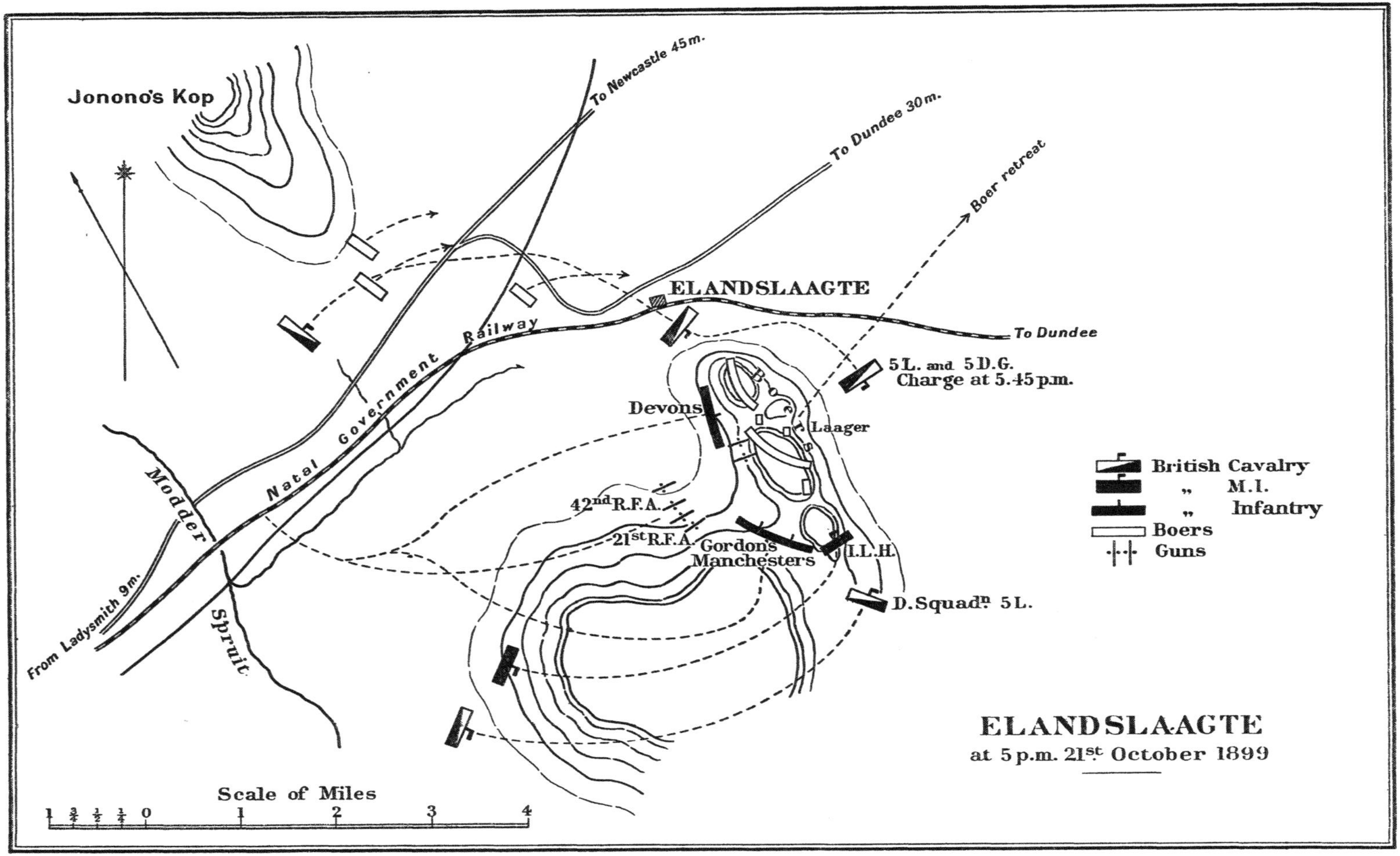
Jonono's Kop
To Newcastle 45m.
To Dundee 30m.
Boer retreat
ELANDSLAAGTE
To Dundee
Railway
Natal Government
Modder
Spruit
From Ladysmith 9m.
5 L. and 5 D.G.
Charge at 5.45 p.m.
Devons
Boers
Laager
42nd R.F.A.
21st R.F.A.
Gordon's
Manchesters
I.L.H.
D. Squadn 5 L.
British Cavalry
" M.I.
" Infantry
Boers
Guns
ELANDSLAAGTE
at 5 p.m. 21st October 1899
Scale of Miles
1 3/4 1/2 1/4 0 1 2 3 4

HISTORY OF THE FIFTH LANCERS. A. DOUBLEDAY & CO., LTD.

thickening darkness, with their reeking weapons bare ; the little knot of prisoners, with faces blanched in fear, herded together at the lance point ; the dim patches on the veldt, which denoted the destruction which had been dealt, and the spasmodic popping of rifles from remote portions of the field as the fighting died out with the last light of day, or as the wounded tried to attract attention."

"This charge created the greatest terror and resentment among the Boers, who vowed at the time that they would destroy all Lancers they captured." On the other hand, an Officer who rode with the squadron stated that in the return gallop he repeatedly saw Boers throw up their hands in token of surrender, and as the lance point was turned and the lancer passed, a rifle would be treacherously fired by the Boer at his generous enemy. A somewhat extraordinary occurrence was the act of Lance Corporal Kelly of the 5th Lancers, who speared two Boers riding on one pony with his lance. A little more light, and few of the Boers would have got away from Elandslaagte.

The troops bivouacked on the field, and a cold drizzling rain made matters worse for those wounded who were not found by the search parties.

"The victory was complete. The enemy had been driven off a strong position of their own choosing and their retreat converted into a demoralised rout." The British losses were 5 Officers and 50 men killed, and 30 Officers and 175 men wounded. The casualties occurred chiefly amongst the Infantry regiments when they stormed the position.

The losses in the Regiment were :—

Killed—No 4600 Private Kinsey, and 2 horses
(In the charge)
Wounded—No 4206 Private Trowbridge
" 4876 " Adams
and some horses.

The Regiment had suffered but little during the action. The lance had struck greater terror into the hearts of the enemy than the rifle, and the Boers themselves are first rate marksmen. The feeling of exultation amongst the troops was excusable, but as they lay endeavouring to snatch fitful sleep or stood guarding their prisoners all through that wet and cheerless night, there was not a man in the Regiment, but bemoaned the death of the gallant Scott Chisholme.

" Killed in action; "—a soldier's death, as he himself used to say: but surely no officer was ever more sincerely mourned on the battlefield by his men, than was " Jabber " by those who had served him so well in India and South Africa. And the hard luck of it all was that fate had decreed he was not to lead the men he had trained so well into their first fight, but at the last moment and on the eve of war, had severed his connection with his Regiment. Some few days after Elandslaagte an officer from another Regiment said to an officer of the Lancers, " I am told that just before your charge your men had heard of Scott Chisholme's death and went into it like madmen killing and slaying all they knew. " The speaker was incorrectly informed, for their former Colonel's death was not known to the men, and the killing and slaying was no more, and probably a great deal less, than can be reasonably expected in a cavalry charge.

" A " Squadron of the Regiment, under Captain Scriven, had left Ladysmith in the early morning and remained out all day to the West of French, and had not been actively engaged.

The Boer losses were some 60 killed, 150 wounded and 200 Prisoners, and 2 guns captured in 1896 from Dr Jamieson, and a large quantity of stores, rifles etc.

Next morning at 3 a.m. the mounted troops started back to Ladysmith, " C " and " D " Squadrons of the Regiment escorting the Artillery; and during the day " A " squadron returned to the town. The Infantry and prisoners soon after daylight entrained for Ladysmith. This retirement from Elandslaagte was somewhat hurried by the movements of the Freestaters on the other side of Ladysmith.

The following Officers of the Regiment were present at Elandslaagte :—

Major A. C. King (commanding).
Lieutenant H. H. Hulse (Adjutant).

" A " Squadron.

Captain J. B. Scriven.
Lieutenant R. Browne-Clayton.
2nd Lieutenant H. F. Fraser.

"C" SQUADRON.

Captain M. P. Oakes.
Lieutenant J. Gwyn.
2nd Lieutenant W. T. Hill.

"D" SQUADRON.

Captain A. Parker.
Lieutenants A Wood.
" J. B. Jardine.
" C. Arkwright.
" M. McTaggart.

The afternoon of the 23rd found the Regiment reconnoitring towards Tinta Inyoni. "C" and "D" Squadrons returned to Ladysmith at dusk, leaving "A" Squadron out for the night.

Meanwhile the force at Dundee was in a difficult position. With diminished numbers in the British Camp, the Boers on their side were adding to their strength, and had placed a 6 inch Creusot gun ("Long Tom") in position on Impati, whence they shelled the camp. General Yule, who had succeeded Penn-Symons in the command, at length decided that the only course left to him to save his force from disaster was to leave all standing in his camp, abandon his stores etc., and slip away at night for Ladysmith. Accordingly at 9.30 p.m. on the 22nd, the retirement commenced. The camp was left standing and the wounded left in the hospital. "So quietly was the manœuvre carried out that the townspeople of Dundee slept through it all, and awoke in the morning in the belief, shared by the enemy, that the camp was still occupied." The column trudged by forced marches down the Helpmakaar road to the Waschbank. The Free State Boers had re-occupied Elandslaagte, and a large force of Boers was known to be in the neighbourhood of Jonono's Kop and on the hills of Rietfontein Farm. To prevent the enemy from attacking the retreating column on its exposed flank, Sir George White decided that by demonstrating in force in the neighbourhood of Rietfontein Farm, some 8 miles out of Ladysmith, he would keep the Boers too fully occupied to annoy the Dundee Column.

About 4.30 a.m. on the 24th of October, the following force marched out of Ladysmith:— the 5th Lancers, 19th Hussars, I. L. H., some Natal Mounted Volunteers under General French, the 42nd and 53rd Field, and 10th Mountain batteries, and an Infantry brigade of Gloucesters, Devons, Liverpools and 2nd 60th. Rifles, under Ian Hamilton. The cavalry pushed on ahead of the main body, which moved in column of route along the Newcastle road. "A" Squadron, who had been out all night, were picked up on the road. When near Modder Spruit, the left flanking patrols of the Regiment came in contact with the Boer outposts on the low spurs of Tinta Inyoni, while from the mountain itself a Boer gun opened fire on the main body. The Regiment drove the enemy off the low spurs, which they occupied themselves. The Boer position was along the lofty ridge of Tinta Inyoni, between which and the road and railway, and running parallel, was another and lower ridge. On to this ridge Sir George White moved his infantry and artillery. His flanks were protected by the mounted troops. The 5th Lancers remained all day on the foot hills of Tinta Inyoni opposite the Boer left, and by dismounted fire were a check on the enemy reinforcing his right. Part of I. L. H. and M. I. were in rear of the 5th Lancers, and the 19th Hussars were further to the right, at Modder Spruit. The left flank was watched by the Natal Volunteers.

About 8 a.m. the infantry deployed along the ridge, and, under a long range rifle fire, slowly advanced to the enemy's side of it, where they remained for the rest of the action, with the exception of the Glosters, who, owing to some mistake, made a short advance down the face of the ridge, which cost them their Colonel and six men killed and forty wounded. On the British left the Boers attempted a flank attack, but were held in check by the Natal Volunteers.

At midday heliographic communication was obtained with the Dundee Column at the Waschbank, and Sir George White, considering he had attained his object, commenced to withdraw his infantry under the cover of the guns.

The Regiment retired by a succession of squadrons from the kopjes they had been holding. As soon as they left the cover of the kopjes they came in for a severe fire, which dropped a few men and horses. During the return march to Ladysmith the Regiment acted as rear guard to the Column.

The British casualties at Rietfontein amounted to 1 Officer and 11 men killed; 6 Officers and 98 men wounded. Of the 5th Lancers the following were wounded :—

No 3789 L. Cpl. Kelly, No 4101 Pte Monahan,
No 4672 Pte. Jack and No 4595 Pte. Swan.

The Boers on their side only admitted to 9 men killed and 21 wounded.

Next morning the Regiment marched out along the Helpmakaar road to the Modder Spruit just East of Mbulwana. From there "A" Squadron was sent on towards the retiring Dundee Column, whom they met in the afternoon at Sunday's River. The Lancers were met with loud cheers and great enthusiasm from Yule's tired soldiers. At 6 p.m. the Column again started for another night march, the squadron of Lancers acting as its rear guard. This last march of the retiring column was a frightful ordeal to the hungry and weary soldiers from Dundee; streaming torrents of rain all night, and inches of mud to toil through. Oxen and horses fell in their traces and were left to die. "It sometimes took hours to cover a hundred yards." However, the men and animals crawled steadily on through mud and rain, and at daybreak reached the Modder Spruit, where the remainder of the 5th Lancers were in bivouack awaiting them. From here a short march brought the Column into Ladysmith.

Next day the men were delighted at receiving a cable from Lieutenant-General Dunham Massy. "Well done, Fifth," he wired from England.

The Boers were now rapidly moving down to Ladysmith. The Free Staters were already a few miles to the North of the town, and by the 27th the Transvaal commandos were laagered along the eastern and north eastern front of Ladysmith.

"Dusty little Ladysmith with her 12,000 troops" was now "the centre of all that the British Empire could do for the immediate defence of Natal." It is a quaint place, and a dusty place, with its little tin houses and parallel and right angled streets, all nestling below a kopje or two, and on the banks of an extremely muddy stream. An unknown outpost of the British Empire, she was destined very shortly to suffer and survive a great siege, as did her godmother, the fair Spanish lady rescued from the siege and sack of Badajoz by the future Governor at the Cape, Sir Harry Smith; important only up to this time for being the junction of the rail from Harrismith in the Free State and that from the Transvaal.

On Sunday the 29th of October the Boers cut the Ladysmith water supply, and further employed themselves in mounting a "Long Tom" on Pepworth Hill, which is some 7000 yards to the North of the Town. The enemy were quickly closing in from the North, and Sir George White decided on a general action for the next day.

The supposition was that the Orange Free State commandos were moving South past Ladysmith, and the idea was to "roll up the Transvaal force from its left flank." The first objective was Long Hill, and an infantry brigade consisting of the 1st and 2nd King's Royal Rifles, the Liverpools, the Leicesters, and the Dublin Fusiliers under Col. Grimwood, was to carry the hill in flank, supported by four batteries. The right flank of this attack was to be covered by General French with the 5th Lancers, 19th Hussars and the Natal Mounted Volunteers, from the direction of Lombards Nek. An infantry brigade of Devons, Manchesters and Gordon Highlanders (joined later by the 2nd Rifle Brigade) under Ian Hamilton, with three batteries, 5th Dragoon Guards, 18th Hussars and I.L.H. were to take up position under the cover of Limit Hill. When Long Hill was carried, both the aforementioned infantry brigades would storm Pepworth's Hill and capture the guns. The cavalry would then pursue by Bell's Spruit and Tinta Inyoni. Seven miles out of Ladysmith this road passes through the defile of Nicholson's Nek, where the pursuit might be easily checked. To prevent this and to keep the pass open, and to occupy a proportion of the enemy, a column consisting of the Gloucestershire Regiment, Royal Irish Fusiliers and 15th Mountain battery was to be sent to occupy the Nek.

The Boer forces north of Ladysmith extended along an irregular crescent from the Harrismith railway on the north west to the laager near Farquhar's Farm just east of Long Hill on the East.

A little before 11 p.m. on the Sunday night the infantry columns moved out.

Day broke crisp and clear on "Mournful Monday," the 30th of October. It is not our intention to detail the events of this unfortunate day. Long Hill was found to be practically unoccupied, and the brigade originally destined for its capture was swung round to face an attack from its right.

The Cavalry moved out at about 3 a.m., part going towards Limit Hill as ordered, and the main body under French along the Helpmakaar road to Lombard's Kop. Here French halted his regiments. About 6.30 a.m. the 5th Lancers and the 19th Hussars rode forward. "Two squadrons of the Lancers rode up to a small kopje forming part of Grimwood's position opposite Long Hill, while the Hussars went further to the right, to what was then Grimwood's right rear. The Lancers, who expected to be fired at, if at all, from their left, received a heavy fire in front and on their right flank as well." The 19th met with an equally warm reception. The men were dismounted and lined the nearest ridge, but soon after mounted and rode back under a heavy pom-pom fire to the nek north of Lombard's Kop, where, taking up a line on the Kopjes north east of Lombard's Kop, French prolonged the line from Grimwood's right.

By 8 a.m. the cavalry on the right could do but little more than hold their own. They and Grimwood's infantry were drawn up on a line some four miles long facing east. Sir George White now sent the 5th Dragoon Guards, 18th Hussars and 21st and 69th Field Batteries to French on the right.

Meanwhile news from the detached column for Nicholson's Nek showed that something was wrong in that quarter, and a helio message to retire was sent, but was not got through.

The Boers were now steadily reinforcing their left and making a bold bid to turn Grimwood's position.

White now determined to abandon the contest, and at midday commenced the withdrawal of his infantry. The retreat into Ladysmith, which was commenced under great difficulties, was materially helped by the "Powerful's" Naval Brigade, who, arriving by rail from Durban, promptly got their 12 pounders into action and engaged the Boer "Long Toms" at their own ranges.

Following the return of the British force into Ladysmith came the news of the surrender of the Nicholson's Nek detachment, and if any succession of unfortunate events was calculated to throw a place into gloomy foreboding that of "Mournful Monday" should have had such an effect on the garrison of Ladysmith. That such was not the case is proved by subsequent events.

The British losses were, 6 Officers and 57 men killed, 13 officers and 245 men wounded, and 43 officers and 925 men reported missing. The Boers on their side must have lost heavily, but no reliable report was obtainable. Beyond some horses killed the Regiment had no casualties.

The next two days the Naval guns and the Boer "Long Tom"

on Pepworth were engaged in a long range artillery duel, while the Boers were rapidly getting their big guns into position and preparing for the investment of Ladysmith, the British being equally busy in their defence preparations. The camp of the cavalry brigade was now moved to below the 'Princess Victoria Battery' (a Naval 4.7 gun) on Cove Redoubt. About noon on the 2nd of November the last train, in which was General French, left Ladysmith for the South! The train was fired on near Pieters but succeeded in running the gauntlet, and a few minutes later the Boers tore up the rails. About 2.30 a.m. on the 2nd, the 5th Lancers with the 5th Dragoon Guards and 18th Hussars, a battery of R.F.A. and some Natal Volunteers made a reconnaissance towards Besters and shelled a Free State laager. "A" Squadron of the Regiment escorted the guns, and came under a heavy shell fire while covering the retirement of the battery, No. 4488 Private Stapleton and No. 4808 Private Butler of the 5th Lancers being wounded. On reaching Ladysmith the force found the Town and Camps being shelled by the enemy from Pepworth and Mbulwana, a high flat topped hill commanding Ladysmith some 6000 yards to the S. E. A sympathiser of the enemy writes of this action,

"On the 2nd November, a large force of Lancers and other cavalry advanced as far as Tatham's Farm near Besters where they encountered the Free Staters...... A brief but fierce fight ensued, the Free Staters crying "No quarter to the butchers of Elandslaagte!" as they shot down their foes."

CHAPTER XV.

The Defence of Ladysmith.

With the rail and telegraph destroyed, and the enemy in possession of all the points surrounding and commanding Ladysmith, the isolation of Sir George White's force was complete, and the investment of Ladysmith an accomplished fact.

The officers of the 5th Lancers present with the Regiment at the commencement of the siege were :—

Lieutenant-Colonel	J. M. Fawcett—Commanding.
Major	A. C. King—2nd in Command.
Lieutenant	H. H. Hulse—Adjutant.
Capt. & Quarter^mr^.	W. Waterman.

"A" SQUADRON.

Captain	J. B. Scriven.
Lieutenant	R. C. Browne-Clayton.
2nd Lieutenant	H. F. Fraser.

"C" SQUADRON.

Captain	E. O. Wathen.
"	M. P. R. Oakes.
Lieutenant	R. P. J. Gwyn.
"	W. T. Willcox.
2nd Lieutenant	W. H. T. Hill.

"D" SQUADRON.

Captain	A. Parker.
"	W. A. Adams.
Lieutenant	A. V. L. Wood.
"	J. B. Jardine.
"	C. Arkwright.
"	M. McTaggart.

ATTACHED.

Major	Holyoake R.A.M.C.—Medical Officer.
Lieutenant	Coley A.V.D.—Veterinary Officer.

STRENGTH OF THE REGIMENT.

18 Officers.
450 N. C. O. & Men.
384 Horses.

The troops in Ladysmith were :—

NAVAL BRIGADE of H. M. S. Powerful : two 4.7. inch, four 12 prs.

CAVALRY BRIGADE:—	5th Dragoon Guards,
	5th Lancers,
	18th and 19th Hussars.
Brigadier—	Colonel Brocklehurst.
ROYAL ARTILLERY:—	13th, 21st, 42nd, 53rd, 67th, 69th.
	Batteries R. F. A. two howitzers of obsolete pattern and the two guns captured at Elandslaagte.
INFANTRY:—	1st Liverpool,
	1st Devon,
	1st Gloucester (four companies),
	1st Manchester,
	1st and 2nd K. R. R.,
	R. Irish Fusiliers (2 companies),
	1st Leicester,
	2nd Gordons,
	2nd Rifle Brigade,
	2nd Dublin Fusiliers (½ company).
R. ENGINEERS:—	23rd Field Company,
	Balloon section,
	Telegraph section.
ARMY SERVICE CORPS and,	
ARMY ORDINANCE DEPT.	
MOUNTED TROOPS:—	Imperial Light Horse,
	Natal Carbineers,
	Natal Mounted Rifles,
	Border Mounted Rifles and a few Natal Mounted police.
LADYSMITH TOWN GUARD.	

A total of some 12,000 men, the British Regiments of which, having just arrived from India or having been quartered in South Africa before the declaration of War, were seasoned hard men.

The town of Ladysmith lies in the N. W. angle of a plain some five miles in length extending from the N. W. to the S. E.

The plain is bounded on the N. and W. by a complex of broken ridges and kopjes varying from 100 to 200 feet in height, which separate it on the N. from the mountainous country of Pepworth and Surprise Hills, and on the W. and S. W. from the great rolling plain which extends to the Tugela. In its S. W. corner stretches the plateau of Caesar's Camp, some 300 feet above the plain. Bester's valley, 2000 yards across, divides Caesar's Camp from the hills which rise in successive tiers on the S. to the main mass of Onderbrock and Grobelaar's hills before sinking down to the Tugela at Colenso. The rest of the S. and S. E. border of the plain is formed by the ridges and kopjes beyond which is Nelthorpe railway station. On the E. is the Mbulwana plateau, two miles long and rising abruptly over 400 feet above the plain, and Gun Hill and Lombard's Kop, while at the N. E. corner beyond the low ridge above the town is a bare space extending to Limit and Flag Staff hills. From N. W. to S. E. the Klip river winds its way through the plain, enclosing the town of Ladysmith in its first great loop.

For the defence of the town a series of field works was rapidly put into execution, and the perimeter was divided into four sections. A proportion of the Artillery was allotted to each section, and the Naval guns mounted along the northern portion of the "A" section. The total length of the perimeter was about 15 miles.

Joubert's force of some 22,000 Boers was equally busy in making preparations for the siege. A continuous bombardment, they hoped, would bring about a surrender of the Ladysmith garrison in a few days. The Boer guns consisted of two 6 inch Creusot Long Toms, firing a 96 lb shell, with an effective range of 10,000 yds; four 4. 7 inch howitzers throwing a 34 lb. projectile a range of over 6000 yards, and some sixteen smaller guns.

The bombardment practically began on the 2nd of November, when Egerton of the Naval Brigade was mortally wounded by a shell from the Boer gun on Pepworth hill. "That's spoilt my cricket" was his remark as he was carried away. On the 3rd, the 5th Lancers rode out at 4 a.m. to Observation Hill, a low ridge to the N. of Cove Redoubt, and remained there on outpost duty until 7 p.m. A certain amount of sniping was indulged in during the day by the enemy on Stone Wall Hill, but with no result. The Imperial Light Horse made a reconnaissance towards

Mounted Infantry and Middle hills, and their retirement had to be covered by the 5th Dragoon Guards with some losses. During the afternoon a shell dropped into a bell tent, in the 5th Lancers camp, in which three men were sleeping, but not bursting, the occupants of the tent escaped with a fright. Next day the Regiment marched before daybreak down to the Klip river, where the horses remained saddled up all day, the men " standing to, " hidden from the Boer artillerymen by the Kopjes on either side of the river. Orders were issued to dye all white horses.

The 5th of November was spent by the Regiment at outpost duty on Observation Hill, while the 5th Dragoon Guards took their turn of " standing to " below the Klip river Kopjes. The enemy were rapidly getting the guns into position and the bombardment was getting more severe. White now opened negotiations on the subject of the removal of non-combatants and women and children to some place of greater safety, with the result that Joubert agreed to the formation of a neutral camp for sick, wounded and non-combatants at the Intombi Spruit below the Boer position on Mbulwana. Ladysmith was to feed the camp, and sick and wounded sent to Intombi were not to return to duty in the garrison upon recovery, and one train a day was to run from the beleaguered town to the neutral camp every morning, returning in the evening. The next day 2nd Lieutenant R. G. Hooper of the 5th Lancers arrived with despatches for Sir George White from the British lines at Estcourt. This Officer had originally been left in Maritzburg with the Reserve troop of the Regiment, and being anxious to rejoin, he obtained permission to endeavour to break through the investing army into Ladysmith. With a trooper of the Natal Police he rode at night from the advanced force of the British Army at Estcourt to Colenso, which place was in possession of the enemy. Cautiously approaching the only bridge across the Tugela, they quickly rode over, and before the Boer sentries on the bridge, which was a long one, were alive to the fact, they were galloping over the veldt for Onderbrook Spruit. Before daylight they arrived at a mission house, and letting their ponies go, were taken in by the hospitable parson and his wife and there remained hidden during the day. The Boers at Colenso had meanwhile discovered their mistake, and were diligently searching the country. Twice during the day parties of Boers called at the mission house, and on each occasion did the good lady stoutly deny the existence of the two Englishmen in her house, and with gifts of bread got rid of her dangerous visitors. At night, leaving the policeman at the mission house, Hooper left with a staunch Kaffir guide belonging to his host, and after some hairbreadth escapes while passing through the investing Boers, crept into Ladysmith at 3 a.m.

The Regiment spent the 7th on Observation Hill, and the 8th "standing to" down by the river. The Cavalry Brigade camp was badly shelled on the former day, but as the men and horses were away no casualities occurred, while on the latter date the Boers dragged a second "Long Tom" on to Mbulwana, which joined in the bombardment of the town at a range of 7500 yards.

A week of steady bombardment had now gone by, and yet Ladysmith showed no signs of surrendering. The enemy were getting impatient, and on the 8th Joubert called a Council of War, at which an assault on the defences was decided on for the next day.

The 9th of November was the turn of the 5th Lancers for outpost duty on Observation Hill. At 3 a.m. the squadrons rode out to take up their usual positions on the hill, "A" squadron on the left, "C" in the centre and "D" on the right. On appoach-ing their positions, advanced scouts were as usual sent on ahead, and day was just breaking as these men reached the top and discovered the Boers running across the valley from the dongas at the foot of Surprise Hill. A number had actually reached the railway wire fencing at the foot of Observation Hill West, and another five minutes would have found them in possession of the ridge. The scouts of "A" Squadron however promptly opened fire on them and checked their rush, while the squadron, rushing up with their carbines, soon drove them back to the shelter of the nearest dongas. It was a close thing, but all three squadrons were now in position. Covered by a heavy long range fire from Stone Wall Hill (or Bell's Kopje), the Boer supports were now coming along in considerable numbers from the dongas of Surprise Hill and Bells' Kopje. The attack was mainly directed against the "A" squadron position on Observation Hill West, and one troop of "C" Squadron was sent galloping across to reinforce "A." This troop, leaving their horses at the foot, ran up the Kopje, and on reaching the top rushed through an enfilading fire from the enemy on Stone Wall Hill across an open bit some fifty yards wide to the shelter of some rocks. The fire the troops had just run through was pretty hot, but nothing compared to the next open space of some sixty or seventy yards to the "A" Squadron firing line, across which the men had to face a very severe frontal in addition to the enfilading fire. There were three troops of "A" and one of "C" now in the firing line, sheltered by a low loosely built stone wall some eighteen inches high, from which the Lancers kept up a slow and steady carbine fire at the dongas in the valley below, and checked numerous attempts of the enemy to rush across the open. The low stone wall was now the mark of every Boer rifle on Stone Wall Hill and in the Surprise Hill dongas, and the rifle fire directed on it was tremendous. However, the training of the past few years was not thrown away, and the men knew how to take every advantage of cover. Not a man was hit, and this cavalry squadron kept a large contingent of Boers armed with a better weapon at bay. As the light got better, the enemy's howitzer on Surprise Hill, and two guns up Bell's Spruit, opened fire on the low stone wall, and gave the Lancers steady doses of shrapnel, but with no effect ; lying low behind the wall, with a steady and well aimed fire from their carbines, they prevented every attempt of the Boers to come out of their dongas.

About 9 a.m. two companies of the Rifle Brigade were sent up in support from Leicester Post, and soon after the Boer Long Tom on Mbulwana joined in the bombardment of Observation Hill West. This gun shelled the hill in rear, and caused some casualties amongst the led horses, which were moved round a projecting spur, and necessitated the whole of the supporting riflemen to be moved up into the more sheltered firing line of the 5th Lancers. The din on the hill was appalling, and about 11 o'clock shells were breaking on the Kopje almost without intermission, while the rifle fire was incessant. A little before noon there was a lull in the firing, but at 12 o'clock the Sailors' guns in the Naval batteries got wonderfully energetic, planting a succession of shells in the Boer batteries, and as suddenly ceased. This was followed by distant cheering, and for a moment some on Observation Hill thought the enemy had got in at some point, but as the cheering was taken up and carried on from battery to battery and post to post round the Ladysmith defences, one soon

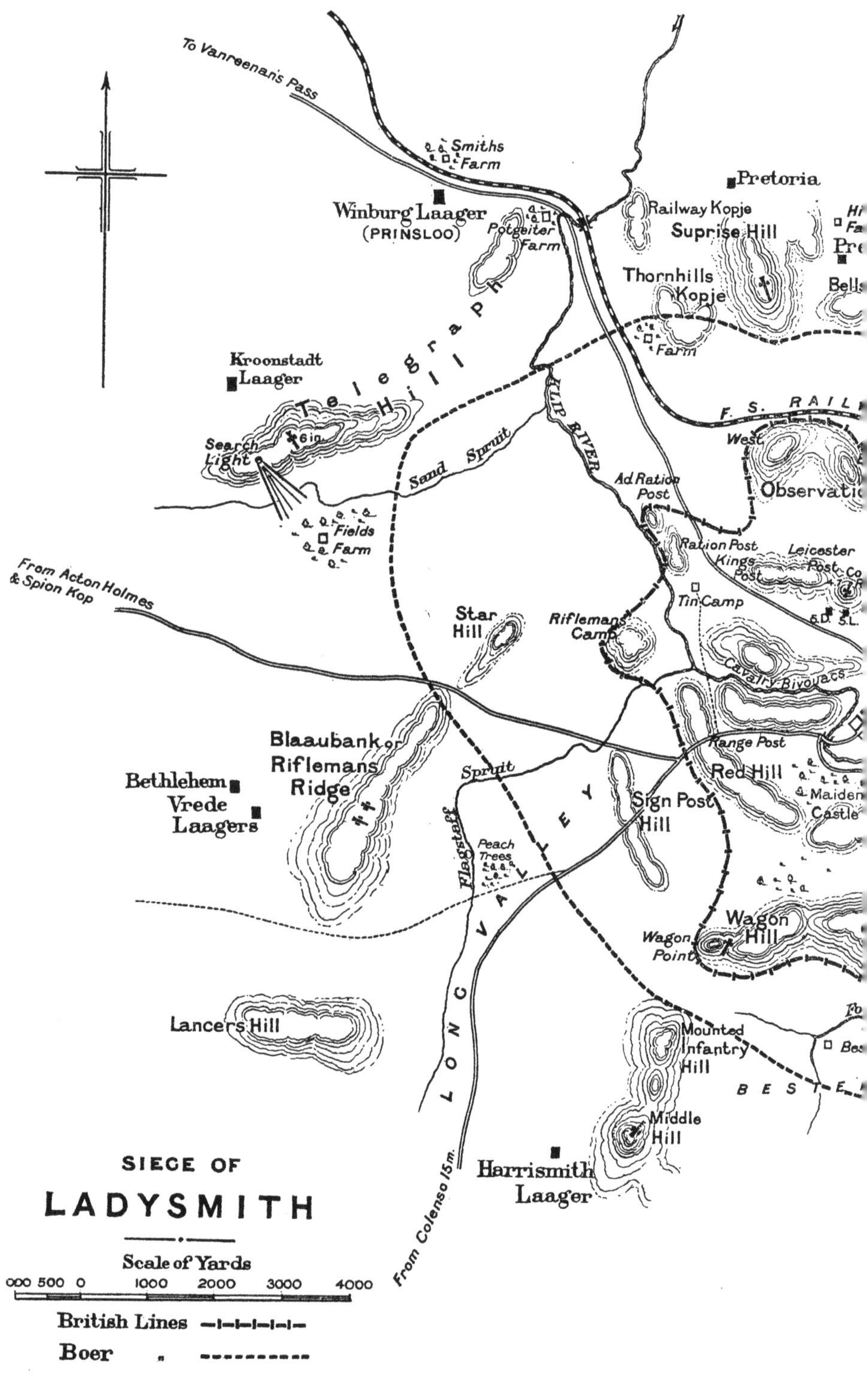
To Vanreenan's Pass
Smiths Farm
Pretoria
Winburg Laager
(PRINSLOO)
Potgeiter Farm
Railway Kopje
Suprise Hill
Thornhills Kopje
Farm
Kroonstadt Laager
Telegraph Hill
Search Light
6 in.
Klip River
F. S. RAIL
West
Observati
Sand Spruit
Ad Ration Post
Fields Farm
Ration Post
Kings Post
Leicester Post
Tin Camp
From Acton Holmes & Spion Kop
Star Hill
Riflemans Camp
Cavalry Bivouacs
Blaaubank or Riflemans Ridge
Range Post
Red Hill
Maiden Castle
Bethlehem
Vrede
Laagers
Spruit
Sign Post Hill
Flagstaff
Peach Trees
Long Valley
Wagon Hill
Wagon Point
Lancers Hill
Mounted Infantry Hill
Middle Hill
Harrismith Laager
From Colenso 15m.
SIEGE OF
LADYSMITH
Scale of Yards
000 500 0 1000 2000 3000 4000
British Lines
Boer

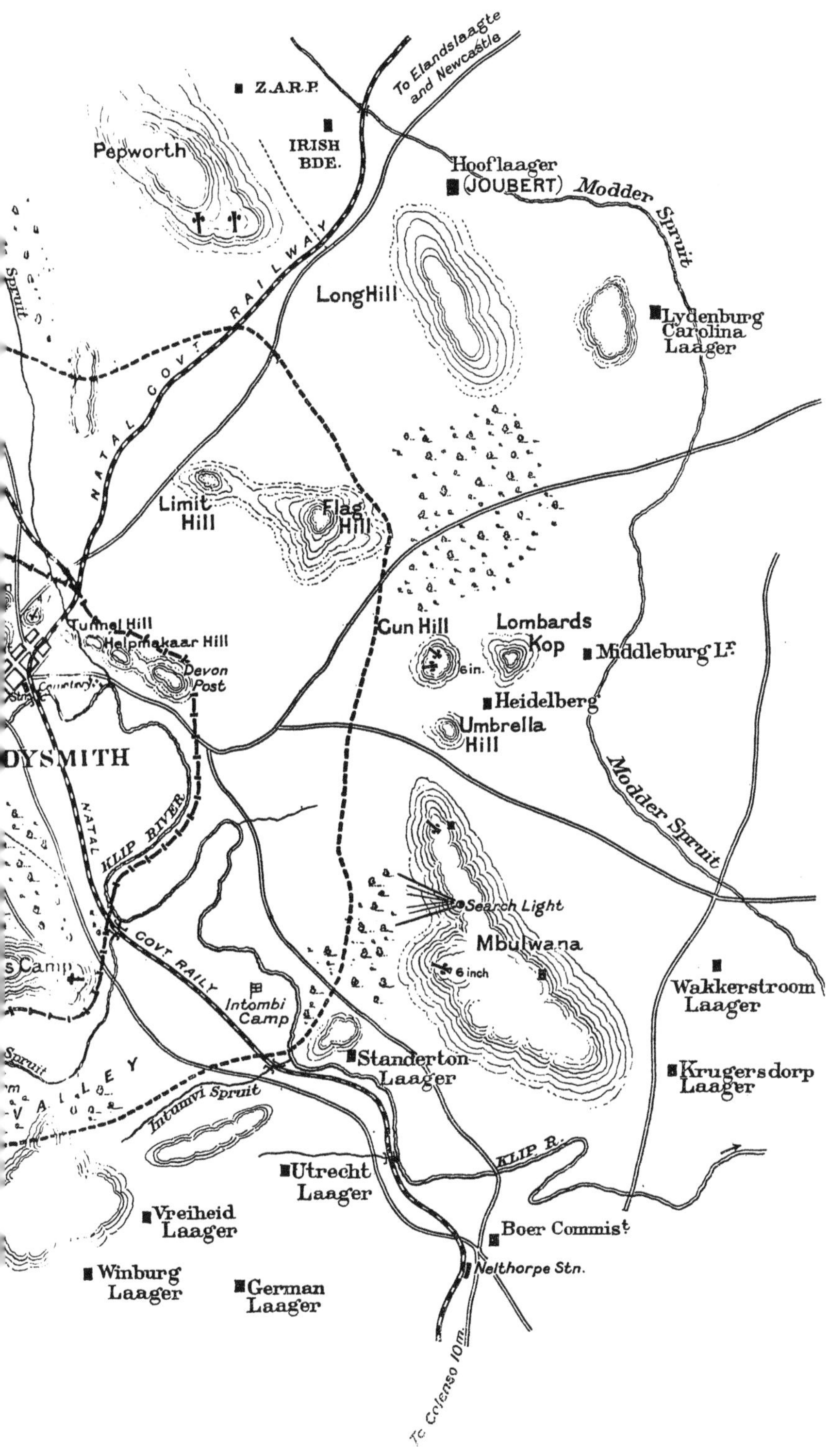

A. Doubleday & Co., Ltd.

remembered that it was the 9th of November and the Prince of Wales' birthday, in whose honour Sir George White had fired a Royal Salute, and planted twenty one successive shells in the main Boer position. The enemy again woke up, and kept the men behind the low stone wall lying very tight for some hours, and then apparently gave up all ideas of possessing the hill. As evening approached they dropped to a desultory fire; and after dark the 5th Lancer Squadrons left the hill to the Rifle Brigade, and rode back to camp.

Considering the heavy fire "A" Squadron and the troop of "C" had been subjected to from daylight to dusk, the casualties were extraordinarily small, Sergeant Saunders and Corporal Horton being wounded, and two horses killed.

The Rifle Brigade had suffered a little more. The men on Observation Hill West had had a most trying day; from daylight to dusk they had lain prone behind the eighteen inch wall in a broiling African summer sun, with no food beyond the biscuit and water some may have had in their haversacks and water bottles. About 1 p.m. a man had been seen creeping about the rocks, on the other side of the open space behind the firing line, with a huge tin can, probably containing a stew, and waiting his chance to make a dash across. His hungry comrades behind the wall watched his movements with much interest. The carrier of the stew attempted the dash, and at once attracted the attention of every Boer on Stone Wall Hill. The bullets played merrily about him, and his friends at the wall gave a yell of despair when he suddenly went a tremendous crumpler, and the tin can and its contents went flying. He carefully crawled back, but no one else attempted the feat.

Burghers from Pretoria had been the assailants on Observation Hill, while the Vreiheid commando had made an equally ineffectual attempt on Caesar's Camp from the South. The total British loss only amounted to 4 killed and 27 wounded. An enormous number of shells had been thrown into Ladysmith. The assault of this day showed that the besieging army had but little stomach for coming to close quarters. It had been a close thing on Observation Hill West before daybreak, and the Boer possession of the ridge less than 1700 yards from Cove Redoubt, and within rifle range of the Naval guns, would have seriously cramped White's defences. As a result of the affair, Observation Hill was put in a position of defence, and while the Cavalry continued to hold it by day, they were relieved by infantry at night, except at the Western end, which was permanently held by the Rifles.

The 10th to the 13th were spent by the Regiment alternately "standing to" in the river bivouac and on Observation Hill. On the 14th of November when "standing to" the 5th Lancers were suddenly ordered out about 10 a.m. with the Imperial Light Horse and the 21st and 67th Field batteries, under the command of General Brocklehurst to turn the enemy off Rifleman's (or Blaaubank) Ridge. The I.L.H. succeeded in establishing themselves on Star Hill while the rest of the force remained in support on Sign Post Hill. For some time the Boer position was heavily shelled, and a little before 1 p.m. the I.L.H. were withdrawn and the 5th Lancers covered the retirement into Ladysmith, being well shelled by the Rifleman's Ridge guns and the Mbulwana 'Long Tom,' the latter at a range of 10,000 yards. The only result of this sortie was a furious artillery duel lasting nearly two hours. At midnight, when the garrison with the exception of the sentries pursuing their solitary vigils was enjoying its well earned rest, "Boom, boom" went the Boer "Long Toms" and the startled slumberer jumped up to hear the now familiar "whizz" over his head and the final crash as a 96 lb shell buried itself in the sand. Most disconcerting was this midnight shelling at first, but with little practical result beyond the use of severe language on the part of the awakened defenders, whom custom very soon enabled to take but little notice of the disturbance. Until the 29th of November the Regiment was employed as usual on Observation Hill or "standing to" in the river bivouac; but little of interest occurred. On the 15th the Boers indulged in much sniping of Observation Hill; for a single man to show himself brought forth many mauser bullets. On the 17th "A" Squadron rigged up a dummy Lancer on Observation Hill, who took no notice of the Boer snipers but remained looking over the wall with supreme contempt, and so raised the wrath of the enemy that he drew quite a heavy fire on his devoted head every time he appeared; particularly savage against the dummy was an elderly white bearded sniper on Stone Wall Hill. Eventually the enemy gave the dummy a shell and it was withdrawn, such attention being more than was bargained for. Corporal Andrews was this day shot in the thigh. The 18th was a heavy day's shelling, and on this day a Doctor Starks, an unfortunate traveller from England shut up in Ladysmith, met his fate under somewhat curious circumstances. Every day before sunrise he pursued his way to his cave by the Klip river bed, and there remained immune from shells until after dark, when he would return to his quarters at the Royal Hotel. On the evening of the 18th the doctor had returned according to his custom. But Mbulwana fired its last shell a little later than usual, which, entering the back of the hotel, crashed right through the building and tore off both legs of the unfortunate doctor, killing him as he stood talking in the verandah at the hotel front. Another shell completely wrecked a house next to the Royal Hotel. Doctor Jamieson, Colonel Frank Rhodes and Sir John Willoughby of Jamieson raid fame were living at the hotel, and the Boers were popularly supposed to know this. If so they must have had good information of the doings in Ladysmith. About this time there were constant rumours of a Boer convoy to be attacked and captured which kept the Regiment in a constant state of suspense on "standing to" days. Instructions were given as to the best method of quickly destroying wagons, supplies, ammunition etc., and on the 22nd an Officer's patrol of 5th Lancers went out to reconnoitre a possible route towards Springfontein. The patrol returned quicker than it went out, having met with a good dusting, and reported the impracticability of such a sortie getting out "and returning" between the Boer positions on the West. On the 23rd a single shell from Mbulwana, falling in the lines of the Liverpool regiment on Junction Hill, killed 2 men and wounded 11, of whom only 5 survived. On the 24th the Boer guns on Rifleman's Ridge and Telegraph Hill cleverly captured some 228 head of cattle grazing just outside the defences by dropping shells just over them and so gradually driving them to the Boer lines. Mounted Infantry were sent after them, but too late. A rumour on this day said that a French deserter from the Boers had arrived in Ladysmith and reported the enemy as being "cock-a-hoop," but why any one should elect to desert the enemy and come into besieged Ladysmith passed the comprehension of most thinking people. The next day Joubert sent some 150 Indians under a flag of truce into Ladysmith, but Sir George White refused to take them, and the Indians were sent back, no doubt rejoicing. This was also a day of much shelling, and what was of more importance to the individual soldiers of the garrison, it saw the first reduction in the meat ration, due no doubt to the loss of the cattle the day before.

Writing in his "From Cape Town to Ladysmith," Mr G. W. Stevens, the gallant war correspondent whose fate it was to die of disease during the siege, thus describes the beleaguered town at this time :

"Deserted in its markets, repeopled in its wastes, here ripped with iron splinters, there again rising into rail roofed, rock walled caves ; trampled down in its gardens, manured where nothing can ever grow; skirts hemmed with sand bags, and bowels bored with tunnels—the Boers may not have hurt us, but they have left their mark for years on Ladysmith. They have not hurt us much, and yet the casualties mount up. Three to-day, two yesterday, four dead or dying, and seven wounded with one shell—they are nothing at all, but they mount up. I suppose we stand at about 50 now, and there will be more before we are done with it."

On the 27th the Boers mounted another 6 inch Creuzot gun on Middle Hill, and to meet this new arrival, two old obsolete pattern howitzers were posted in the nek connecting Wagon Hill with Caesar's camp, and so effectually did they tackle the new comer that the "Long Tom" was silenced on the 30th. The Boers had now moved the Pepworth Hill "Long Tom" to Gun Hill, and search lights were placed by them on Mbulwana and Telegraph Hill. In Ladysmith one night a railway engine carrying explosives was started on the Free State line and sent off to work destruction amongst the besiegers. The locomotive apparently went well for a distance, but then it left the rails, and overturning, lay gasping on its side, while Briton and Boer looked on and wondered.

On the 29th of November the Regiment was relieved from its duties on Observation Hill and formed part of a flying column to co-operate with Sir Redvers Buller and the Relieving Force. Each man was ordered to carry 2 days rations and each horse 20lbs corn. The next day while the Regiment were at "stables" in the Convent donga, a shell from Mbulwana fell into the midst of "D" Squadron and buried itself in the sand without damaging a man or a horse, in spite of the crowd of men and horses in the donga. The Boers on this day were particularly active with their guns, and the tower of the Town Hall, which was being used as an hospital, was struck. During the night Buller from South of

Colenso flashed messages by search light, but there being no search light in Ladysmith Sir George White was confined to the heliograph to Weenen Hill by day for his replies. In addition to the cavalry, White's mobile force to co-operate with Buller consisted of the Gordon Highlanders, Devons, Rifle Brigade and a half battalion of Manchesters, with three batteries of field artillery. The infantry were engaged nightly in route marching in order to keep fit, and the cavalry exercised for an hour before dawn.

In the early days of December the mobile column turned out most nights for the co-operation with Buller which never came off, and trying work the men found it, sitting for hours at their horses heads in the Ladysmith streets, waiting for they knew not what. One night as they slowly rode past the Naval brigade camp at a snail's pace, apparently for some dashing enterprise, the bluejackets whispered "good luck Tommy," and "Tommy" thanked his brother in arms and whispered he wished he knew where the devil he was going. About this time the garrison was much indebted to the War correspondents for publishing two amusing periodicals termed "The Ladysmith Lyre" and the "Bombshell." Cheering news was received from the Relief Force by helio, to the effect that Lord Methuen had defeated the Boers at Graspan, Belmont and Modder River, on his way to relieve Kimberley in the Western theatre of war.

On the 7th of December preparations were made with the utmost secrecy for a sortie against Gun Hill, so secret in fact that the garrison knew nothing of the projected enterprise. At about 10 o'clock on the night of the 7th a small force of 100 of the Imperial Light Horse and 400 Natal Volunteers were turned out under Major-General Archibald Hunter, chief of the staff to Sir George White, who, led by Henderson's guides, walked round the Helpmakaar (or Devonshire) post up the road towards Gun Hill. The force moved with the utmost quietness, light shoes being worn in place of boots. Leaving 100 men of the Border Mounted Rifles to watch his left flank, and sending the main body of volunteers to Lombard's Nek to check any counter-attack from Mbulwana, Hunter moved towards Gun Hill with the Imperial Light Horse, under Major Edwards, and the Natal Carabineers. Striking across the donga-intersected thorn scrub, Hunter arrived at the foot of Gun Hill at 2 a.m. His men deployed into line and with the I.L.H. on the left and the volunteers on the right, the little force stormed the hill. They gained the summit, the Royal Engineer Officers, Fowke and Turner, destroyed the gun, Hunter called for three cheers for the Queen, and the force retraced their steps for Ladysmith, and reached the defences by dawn with the loss of only one man killed and some eight officers and men wounded. The following description of this gallant and successful enterprise is told by an officer who took part in the affair :—" We reached the foot of Gun Hill about 2 a.m. The force was divided into two parties. The I.L.H. and the Natal Carabineers were to rush the guns, while a detachment was to establish itself on the Western side of the hill to prevent any reinforcements gaining the hill. The I.L.H. had climbed half way up the hill without meeting any sentries, when suddenly they stumbled on the Boer pickets. A sentry of a picket who must have been asleep while the assaulting party passed him, challenged from below, "Wis kom dar." "Wis kom dar," he again shouted, and a voice called out in English "shut him up." The Boer then yelled for all he was worth in Dutch, "God's truth, they are here, bring me my horse, shoot, shoot !" and, firing his rifle, disappeared, or was killed. The I.L.H. were now near the top, some rifles were fired on the crest of the hill, and Edwards called out "Fix bayonets." There was not a bayonet with the party, but the Boers on the crest of the hill did not wait to discover the fact, but firing their rifles wildly, were off down the other side. The assaulting party charged the crest and were soon up at the gun. I believe Henderson was first to the gun, which was loaded, and ranged to 8000 yards. The men ran on and ranged the far crest and kept up a good fire while the sappers blew up the gun. Bringing away the breech of the gun and a machine gun as trophies, we returned safely to Ladysmith by dawn. The Boers on Mbulwana opened a wild and useless rifle fire as we retired." Meanwhile a Squadron of the 19th Hussars, some Mounted Infantry and three companies of the Liverpool Regiment had been making a diversion during the night in the direction of Pepworth.

About 4 a.m. next day while at exercise, the Regiment was warned for a reconnaissance by the Cavalry Brigade, but it was not until nearly 7 a.m. that the Brigade turned out. The 5th Lancers were ordered to Limit Hill and Long Hill, the 18th Hussars to Bell's Spruit and towards Pepworth, while the 5th Dragoon Guards remained in reserve behind the Eastern point of Observation

Hill. However, by the time the Lancers and Hussars got away, the Boers were everywhere on the alert and rapidly reinforcing the threatened points, with the result that both regiments got a good dusting. The 5th Lancers rode out at a gallop from Ladysmith, being shelled the whole way to Limit Hill. They continued along the Newcastle road, but being checked by very severe rifle fire from the lower slopes of Pepworth, they rode in an extended column of troops parallel to the thorn bushes south of Long Hill, and then swinging round returned over Flag Hill to Limit Hill, where they remained some time being shelled by Mbulwana. Eventually a message came from Sir George White ordering the return of the Regiment, which sent them trotting quietly back to Ladysmith through the shells from Mbulwana and Surprise Hill. An Officer who witnessed this retirement from the defences thus described it: "A grand sight they were in columns of troops opened out and shells from Mbulwana and Surprise Hill "Long Toms" were bursting around them, and they trotted steadily and quietly through it all as though there were no such things as 96lb shells. A shell dropped in the leading troop of a squadron and sent up a cloud of dust and one expected to see men and horses on the ground and confusion, but no, as the dust cleared away the Lancers were still quietly trotting along." The 18th Hussars reached Ladysmith about the same time, having had a severe handling.

Casualties:— 5TH LANCERS.

Killed:—	Private Lee, 11th Hussars (attached).
	3 Horses.
Wounded:—	Lieutenant W. T. Willcox.
	L. Sergeant Austin.
	Trumpeter Goldup.
	Private New.
	9 Horses.

18TH HUSSARS.

Killed:—	2 men.
Wounded:—	1 Officer, 17 men and 32 horses.
Missing:—	1 man.

5TH DRAGOON GUARDS.

Wounded:—	1 Officer.

A couple of days after the reconnaissance the following order was published:—

"On the report of the reconnaissance of the 8th inst. rendered by the General Officer commanding the Cavalry Brigade, the Commander-in-Chief made the following note:—

'I regret the number of casualties but I was an eye witness and it was inspiriting to see the keenness and dash with which a dangerous duty was carried out by Officers and men.'"

On the 9th some Officers of the Cavalry Brigade were playing a game of polo near the Brigade camp, when a shell from Mbulwana dropping amongst them put an end to polo: most unsporting conduct on the part of the enemy. Next day, departing from their usual custom of spending a quiet Sunday, the Boers shelled somewhat severely.

On the night of the 10th of December another sortie was ordered, and this time the honour was given to the Regulars. The objective was the 4.7 inch howitzer on Surprise Hill. At 10 p.m. five companies of the Rifle Brigade moved out from King's Post, but there being bright moonlight the Riflemen lay low in a donga beyond Observation Hill. About midnight the moon disappeared and the Riflemen moved silently out of the donga towards Surprise Hill on their venture. Half a company were left to protect the left, and another the right under Bell's Kopje. By 2 a.m. the foot of the hill was reached. Here two companies commenced the ascent as the assaulting line and two others were formed outwards to support them. When almost on the brow of the hill a frantic challenge rang out followed by a burst of firing, but too late, for the Riflemen swarmed over the crest line, the battery was found, and the men thrown out in a semi-circle while the Engineers prepared the destruction of the gun. But something was wrong with the fuze, and another had to be placed. This time the gun was destroyed, and with a ringing cheer the men started down the slope. But the delay had given time for the enemy to turn out on the alarm. A cross-fire was brought to bear on the slopes of the hill from both flanks, and a party of Boers were able to get to the foot of the hill and intercept the storming party. In grim silence the Riflemen charged, and the British bayonet once again did its work. After some minutes of wild confusion, during which the enemy were shouting misleading orders in English, the men made the best of their way back to Observation Hill. It was a gallant

Outpost Duty near Lydenburg.
Fifth Lancers Transport Crossing a drift.
Fifth Lancers in the Ladysmith Trenches.

sortie and in view of the fact that the enemy were more prepared for such ventures than on the 7th the casualties were not excessive.

Killed :— 1 Officer and 14 men. Wounded :—about 50.

What news of the outside world the Staff in Ladysmith received was generally posted in orders. About this time the troops were informed of Methuen's fight at the Modder River, of the investment of Mafeking, of the intense interest with which the campaign was being followed at home, and the supposition that Methuen's column had " by now reached Kimberley. "

And so the days wore on, the Regiment " standing to " in the Klip River bivouac by day, turning out for the flying column at night, and exercising the hour before dawn. The enemy regularly shelled the town and camps. The monotony of existence in Ladysmith was somewhat broken on the 15th of December by the sound of the distant guns of the Relief Force, though what was happening South of the Tugela had to be left to conjecture. The 16th was a day of heavy bombardment, the Boers thus keeping the anniversary of their victory over the Zulus on Dingaan's Day. Meanwhile food was getting scarce, and dysentery and enteric were beginning to make themselves felt; and a plague of flies, and the intense heat made life scarcely endurable. However, the hopes of the garrison were at their highest, Buller's big guns were thundering at the Tugela heights, relief seemed shortly probable, and the troops were looking forward to an advance and settling their score with the besieging Burghers. Vain hopes! The following order issued on the 17th dispelled all illusions:

" The G.O.C. Natal Field Force regrets to have to announce that General Sir Redvers Buller failed to make good his first attack on Colenso. Reinforcements will therefore not arrive here as early as expected. Sir George White is confident that the defence of Ladysmith will be continued by the garrison in the same spirited manner as it has hitherto been conducted until the general officer Commanding-in-Chief in South Africa does relieve it." The scale of rations and forage was reduced and the garrison, at last alive to the probability of a lengthened siege, set themselves to the task with grim earnestness. The defences by now seemed practically secure against attack, but the ravages of sickness were becoming very serious. The bombardment continued,on some days being more severe than others. The first shell from Mbulwana on the 18th was an unlucky one, killing 5 men and 11 horses of the Natal Volunteers, and wounding another 9 men.

Surely, if slowly the Long Toms were, adding to their score, a soldier to the Blaaubank gun, four or five Kaffirs to a single shell from Mbulwana, a Kaffir woman by the river blown to bits, a 96 lb bursting in Ian Hamilton's kitchen, and so it goes on. The Boer search lights on Mbulwana and Telegraph Hill nightly caused merriment in their efforts to obliterate the messages flashed by the British Army south of the Tugela. In an Officer's diary of the 18th of December is an entry "Brandy, gin and stout finished, whiskey worth £5 a bottle, eggs fetching thirteen shillings a dozen and jam finished." Writing of life with the Cavalry Brigade the same officer says, "standing to every day. Appalling work this standing to. One never knows from daylight to dark but that "turn out" will be suddenly ordered for a sortie, a chase of a convoy or to capture something. Rumour very busy over Buller and his numerous victories, one day he is over the Tugela, on another he has carried the Tugela heights in a night assault and bayoneted the Boers in the trenches, while in the far distance one hears the occasional boom of his big guns. Many people cross because they think we do not know all about Buller's fight on the 15th at Colenso. One supposes it is necessary but it is galling to think that one is only told a certain amount." In the light of after events this last paragraph is interesting.

December the 22nd was an unlucky day. In the early morning Mbulwana dropped a shell into the Gloucester camp which killed 6 men and wounded 10. About 9-15 a.m., while the 5th Lancers were at "stables" and the Colonel and his staff walking round a troop of "C" Squadron, a shell from Telegraph Hill burst at the end of the line of horses of the troop, amongst the horses the men were grooming. Not a man or horse near the bursting shell was hit, but at the other end of the line, twenty yards away, 4 horses were killed outright and 3 wounded, while the following officers standing to one side of the line of horses were wounded :—

Lieutenant-Colonel J. Fawcett.
Major A. C. King.
Captain E. O Wathen.
Captain M. Oakes.
Lieutenant and Adjt. H. H. Hulse.
Sqd. Sergt.-Major Harris.

The men grooming between the horses escaped without a scratch. During the day a Mbulwana shell hit the tower of the Town Hall.

Lieutenant-Colonel Fawcett was moved to Intombi Neutral Camp, and Major King assumed command of the Regiment.

Christmas Day in Ladysmith was as cheerful a festival as the troops could make of the situation. The Natal Field Force orders published an order to the effect that "The following message has been received from Her Majesty the Queen-Empress:—"I wish you and all my brave soldiers and sailors a happy Christmas. God protect and bless you all. V.R.I." The following is an entry taken from the diary of an Officer of the Regiment:—"December 25. A merry Xmas! a dull day. Mbulwana sent us a shell containing a plum pudding. We had a great dinner at night, and the Leicester Regiment, which had saved some champagne for the occasion, sent us a bottle which was much appreciated. In addition we each had half a glass of beer, the last of our liquor. As merry a Christmas as possible was spent; I know in our mess our 'Xmas dinner if not too plentiful was jovial."

On the 27th a Mbulwana shell dropped into the Officers' Mess of the Devonshire Regiment, killing one officer and wounding nine others. For the remaining days of the year the bombardment continued merrily, and added to it was an almost nightly fusilade of rifle fire from the enemy's trenches, which though quite harmless, was disturbing to the troops in the defences. From the 29th to the 31st there were very heavy rains. The Klip river rose and washed away some horses of the mounted infantry, while from being continually wet through, men of the Regiment were to be seen riding to the bivouac naked below their cloaks in their endeavours to dry their clothes. Enteric and dysentry was rapidly increasing, and drugs and medical comforts as rapidly decreasing. The Regiment however was keeping fairly fit, and had less men sick than the majority of the other corps.

The following message from the Lord Mayor of London was flashed through by Buller:—"Kindly convey to the troops at the front hearty Christmas greetings from London, also our admiration for and sympathy with their struggles," while on the 1st of January, 1900, the search light flashed through the darkness two messages, the first from Her Majesty the Queen, "Wish you you all a bright and happy New Year. God bless you all. V.R.I.".

The second came from a far distant spot in the British Empire, "Sailors, Soldiers and Civilians Hong Kong send warm greeting to Garrison Ladysmith."

To quote again from the diary of an officer of the 5th Lancers in these early days of 1900 :—

"Grub more scarce than ever. People are cheerful considering everything, though liable to be cross over small matters. Buller's guns seem silent and consequently the conjectures are many and varied, the most general being that he is working round by Helpmakaar. Eggs fetch 23/s a dozen. How monotonous this is getting. Shelling only on the part of the Boers, for we appear to be saving our ammunition to join in Buller's relief, but at the present rate of his progression there won't be many people left to meet him. Sickness very bad. Our horses suffering from short rations. Our men are splendid, one never hears a growl, they are starved, they live in tents which are worse than useless, they get wet to the skin most nights in the week, and have to wear their wet clothes during the day ; or, as I saw one man mount his horse with no clothes on but his cavalry cloak, his helmet and his boots. The rest of his kit he hoped to dry in the bivouac if it was a fine day. They look after their horses too. Private Snoad of my squadron has a young horse looking, in spite of short rations, quite sleek, well groomed and fit. How it is managed one does not enquire, but the rider looks as if he could do with some of his mount's sleekness. They are all the same, never a grumble, but jolly and anxious for a smack at the Boers, and though somewhat haggard and thin, keeping wonderfully fit. Gallant old Regiment !"

CHAPTER XVI.

The Defence of Ladysmith.—Wagon Hill, 6th January, 1900.

The siege of Ladysmith was now entering upon its third month, and yet no signs of surrender from the garrison. The bolder spirits in the Boer laagers were clamouring for a great effort to attack and capture the place. They pointed to the steady reinforcements Buller was receiving South of the Tugela, to French's successful efforts at Colesburg, and to the evident danger of troops arriving in South Africa from every part of the British Empire. It was high time the siege was ended, and the investing commandos set free for service elsewhere. Joubert was against it, but at a general " Krygsraad " was compelled to yield to the pressure of a large majority in favour of a desperate assault on Ladysmith.

The plan adopted was the delivery of a main assault against the Platrand (the Caesar's Camp-Wagon Hill ridge). A vigorous attack was also to be made on Observation Hill, while a general demonstration along the whole perimeter of the defence would prevent the withdrawal of troops to reinforce the defence of the Platrand. The points of attack against this ridge were the Eastern and South Eastern end of Caesar's camp, and the Southern and South Western end of Wagon Hill and Wagon Point, while the supports were to concentrate in the bed of Fouries' Spruit and behind Mounted Infantry Hill. The Mbulwana and Telegraph Hill Creuzot guns would enfilade the ridge, while half a dozen guns drawn from Colenso were to be placed at various points along the southern side of Besters' valley. Finally, the attack was to be delivered before dawn, and its actual date was left to Joubert, whose order was to be published at the last minute, to prevent any chance of Ladysmith discovering the Boer intentions.

Caesar's Camp was being held by the Manchester battalion ;

three companies of the 60th Rifles were on the eastern end of Wagon Hill, and two weak squadrons of the Imperial Light Horse formed the garrison of the western portion of Wagon Hill and of Wagon Point ; and on the night of the 5th of January a 4.7 Naval gun was moved to Wagon Hill to be placed in position on Wagon Point. With the gun was a working party of a company of Gordon Highlanders, Engineers and some sailors. The gun arrived at the foot of the hill, but the attack began soon after, and so there it remained. Major-General Ian Hamilton was in command of this section of the defence.

After dark on the evening of January the 5th Joubert sent his orders for the great assault to be carried out in the early hours of the following morning, and a few hours later the commandos rode out of their laagers and began assembling at the appointed rendezvous. That same evening an officer of the I. L. H. had gone out reconnoitring towards Middle Hill, and the pickets on Wagon Hill had been if possible more than usually wide awake, waiting for his return. About 2 a.m. the working party with the Naval gun were hard at work on Wagon Point, and about 2.30 a.m. an advanced picket of the I. L. H. heard men scrambling amongst the boulders at the foot of Wagon Hill. They reported the matter, and the squadron was aroused. A quarter of an hour later there was no doubt about it, the picket could plainly hear the Boers climbing up the hill. A sentry challenged and the picket fired a volley down the slope. A crash of musketry was the answer, and the enemy scrambled up the hill and a desperate fight ensued at point blank range, until a couple of rounds from a Hotchkiss checked the Boers. Some I. L. H. on Wagon Point were moved towards the direction of the firing, and gained the nek below the Crows Nest. Meanwhile the outbreak of firing had created confusion amongst the working party on the Naval guns. The Gordons rushed to their arms, and the sappers and sailors formed up behind the gun emplacement. In the darkness and confusion the fierce fight went on, and the I. L. H. lost many officers and men. It was difficult to distinguish friend from foe, and many a gallant Light Horseman rushed headlong into the enemy. About 3.15 a.m. the Boer fire was steadily increasing, and to prevent being outflanked, the I. L. H. on the slope retired to the top of Wagon Hill, a murderous fire being poured on the men as they became silhouetted against the grey dawn on the sky line. The Boers immediately pressed forward, and at the same time a small party of Boers, endeavouring to work round the foot of Wagon Point to the British rear, were checked by the guard over the 4.7 gun still lying at the foot of the hill.

The outburst of firing on Wagon Hill had warned the troops on Caesar's Camp, and the crest line was held at double strength by the pickets and their morning reliefs. The Boers climbing its southern face were checked by the steady shooting of the Manchesters, and lying low amongst the rocks and bushes, kept up a heavy fire on the sky line. The men of Utrecht, however, had successfully climbed the Eastern slopes, and at 4.15 a.m. arrived at the crest, and after a short and desperate struggle overpowered the extreme left group of the picket there. They now turned to the left and began rolling up the thin line of the British on the South Eastern crest. To quote from the "Times" History, "The Manchesters held on to their sangas and rifle pits with the most heroic determination, and group after group had to be completely annihilated before the Boers could win a few yards of crest line. But slowly and surely the enemy, with equal gallantry, pushed their way. Re-inforced from below at day-light they were now in possession of the eastern crest." They then commenced to work along the southern crest, but supports were sent to the pickets in front of them and the Boers were checked.

Meanwhile daylight on Wagon Hill had revealed an extraordinary situation. The Free Staters were in possession of the summit of the South-western shoulder of Wagon Hill, and continuing to their right was a line of Boers cunningly lying low amongst the rocks and boulders along the crest, and to their left front were more burghers lining the eastern slopes of the nek almost to the Crows' Nest (I. L. H. Fort). Facing them in the Crows' Nest, whence they commanded the whole summit of Wagon Hill and Wagon Point, lay a handful of Imperial Light Horsemen, "grimly firing away in their unyielding resolve to cling to the position committed to their charge." Away to their left lay the 60th Rifles, some on the south eastern crest firing down into the Besters Valley, and others amongst the rocks on the plateau facing the Free Staters. Across the nek was the small devoted party of Light Horsemen who had won their point during the dark, of whom only five were destined to extricate themselves later. At the farthest end of Wagon Point and in the gun emplacement, Sappers, Highlanders and blue jackets were busy keeping the Boers down at the foot of the hill, and worrying their supports on Mounted Infantry Hill. On the actual summit of Wagon Point was a span of oxen in the wagon of the working party quietly grazing through the constant stream of bullets skimming across the plateau, and unconscious of the hell going on around them. It is interesting to note that eight of these sixteen oxen were alive the following morning. As the light strengthened the Boer supports in Besters' Valley and on Mounted Infantry Hill opened fire, and the bursting shells of the enemy's artillery added to the din on the Platrand, and many a gallant soldier discovered lying in the open in the pitiless dawn was shot as he lay.

About 4.45 a.m. the rest of the Imperial Light Horse came galloping up from their camp in Ladysmith town, and were sent up to various points along Wagon Hill and Wagon Point. The arrival of the reinforcing squadrons resulted in a renewed outburst of fire, and officers and men fell fast. The remainder of the battalion of 60th Rifles arrived a little before 7 a.m., and repeated attempts were made by the Riflemen to charge across the open at the Free Staters on the crest, whose deadly short range fire was playing such havoc. Three officers fell dead together with most of the men who followed them, as they successively attempted the rush. The fighting gradually became stationary, Briton and Boer lying under what cover there was, and separated only by a few yards of open grass on the summit, and for hours across the deadly strip of open did the hum of Mauser and Lee-enfield bullets continue. The 21st Battery, with an escort of 5 D.G. from near Range post, had meanwhile come into action against Mounted Infantry Hill.

To return to the fight on the Eastern end of the Platrand, we left the Transvaalers in possession of the Eastern crest of Caesar's Camp. Against these a small party of Gordon Highlanders advanced, but were effectually checked by the enemy. About 6 a.m. the 53rd Battery R.F.A. moved out of Ladysmith and came into action in the scrub south west of the Town, and at a range of some 2000 yards, burst shrapnel amongst the Boers on the Eastern point of Caesar's Camp, and gradually forced them back to the slopes on the south eastern end, where they remained in cover. Soon after coming into action the battery received the attention of the Mbulwana "Long Tom," who steadily fired shell after shell at it, of which, however, the British gunners took not the slightest notice, and continued their bombardment of the Caesar's Camp slope until about 9 a.m., when, having completed their task, they withdrew to cover. Of the many gallant performances that day nothing surpassed the spirited action of this battery. A sergeant of the battery who had an arm and a leg taken off by a Mbulwana shell while sitting on the trail of a gun exclaimed, "Roll me out of the way lads and buck up." Meanwhile some companies of the Rifle Brigade and the remainder of the Gordon Highlanders had arrived in reserve at Caesar's Camp, the latter regiment having lost their Colonel, who was mortally wounded soon after crossing the Klip River.

To turn to the doings of the 5th Lancers. At 2 a.m. the Regiment as usual turned out to exercise, and soon after the search lights on Mbulwana and Telegraph Hill commenced flashing, while half an hour later a sputtering of rifle fire was heard in the direction of Wagon Hill, followed by firing on Caesar's Camp and Observation Hill. The Regiment had reached the river bivouac before dawn, by which time the rifle fire along the Caesar's Camp—Wagon Hill ridge—was very heavy, and evidently more than the false alarm it was thought to be. By 6.30 a.m. the fusilade was tremendous, and Wagon Hill and Caesar's Camp apparently heavily engaged, and the horses were kept saddled up ready to turn out. As the light improved the Boer guns opened fire, their attentions being chiefly directed on Caesar's Camp and Wagon Hill, though some dropped shells all round the defences. At 9 a.m. a galloper arrived with an order for the 5th Lancers to go to Caesar's Camp below Manchester Post. The bombardment round Ladysmith by this time was tremendous, and shells were whistling from all directions. The big guns on both sides were paying each other marked attention, while Abdy's battery of Field Artillery could be seen giving the Boers on the eastern end of Caesar's Camp a very warm time, in spite of Mbulwana's angry remonstrance. Coming up out of the bivouac, the Regiment trotted down the road to Fly Kraal and Manchester Post in a long column of sections opened out to a distance of twenty yards between sections, the whole journey of over two miles being done in full view of Mbulwana, which shelled heavily but inaffectually. Arrived at Manchester Post, the squadrons awaited events in the dongas below the hill, while on the top the desperate fight was waging, and a continual stream of bullets and splinters of shell went whistling and shrieking high overhead.

By 11 a.m. Caesar's Camp was safe, and the Boer attack there had failed, though the gallant Transvaalers were still clinging on to its southern slopes. On the other end of the Platrand, however, a desperate and nearly successful effort was made by the Free Staters. A little before 1 p.m. the fight on Wagon Point had gradually subsided into a position of "check," both sides still lying in their cover of boulders and not a few of the wearied men asleep, when a small party of Boers quietly scaled the extreme western point, and were almost on the summit before they were perceived. It was nearly a Majuba over again, the burst of point blank fire, and the sudden appearance of the Burghers sending the astonished defenders headlong down the hill, but only for a minute. The panic was gallantly checked, and following their officers, the mixed troops sprang forward to the crest. A dozen Boers were before them on the summit, and all but three hung back. This gallant three rushed for the 4.7 gun pit. It was a wild race; a British Officer was in there first, and with his revolver killed the nearest Boer. Another Boer fired from outside the pit, but an officer and a corporal of Royal Engineers killed him and his companion at once. The immediate danger was over, and reinforced by the 18th Hussars the troops re-occupied the summit.

In the mean time the 5th Lancers were awaiting events in the dongas below Manchester Post. At 1 p.m. a staff officer rode down from Manchester Post and calling to the first 5th Lancer Officer he saw, he shouted, "The cavalry are to gallop to Wagon Hill at once." The word was quickly passed, and the three squadrons were mounted and away through the scrub and round a shoulder of the hill before Mbulwana could get in a shell. A mad gallop to Wagon Hill, and the Regiment pulled up below the 60th Rifle redoubt. Dismounting, the led horses were put down into some dongas for shelter from the Boer guns on Blaaubank ridge, and the men ran up on the Northern slopes of Wagon Hill. Halting, the squadrons formed in mass. Soon after some 19th Hussars and 5th Dragoon Guards arrived and formed below the 5th Lancers. Meanwhile Sir George White had ordered the hill to be cleared at all costs. His cavalry had been sent round, and Ian Hamilton was told that the Devonshire regiment were on their way.

To quote again from the "Times" History, "All day long the glowing African sun had beaten down upon the backs of the men lying with their faces close to the sweltering stones. But now the whole heavens darkened, and at four o'clock the threatening thunder storm burst with a deluge of rain rare even in South Africa. The blinding, drenching downpour was full in the faces of the Boers, who, fearing that the British would now attempt to rush them off the crest, began firing wildly into the mist. Breaking out again as if by magic, the rifle fire rolled along the crest of Wagon Hill and spread to Wagon Point and to Caesar's Camp in one unbroken deafening roar, reaching and sustaining an intensity which it had never touched before. But the exhausted troops on Wagon Hill were not yet ready for the attempt." To the thunder was added the din of bursting shrapnel. Suddenly and clear rang out "5th Lancers" and the squadrons were up and over the crest, and running like demons, heaven alone knew where, through a hail of bullets and a pandemonium of appalling sounds. The only course for an individual was to run straight and as hard as possible. On the squadrons sprinted, until, as if by instinct, they all fell flat below a line of boulders a foot and a half high, on the summit of the hill and close to the firing line of the 60th Rifles. "C" Squadron was on the right, "A" in the centre and "D" on the left, and some 19th Hussars were mixed up with the 5th. A short run and a trying one, but thanks to the storm there were few casualties. All around were dead and wounded Riflemen and Light Horsemen. The wounded had a bad time, those unable to move lying with the cruel hailstones of an African storm beating on their faces. "D" squadron on the left was soon after withdrawn. The rain continued but the thunder had passed, and in its place the Boer guns sent shell after shell bursting on the plateau. The rifle fire was intense, and sounded to the men lying prone behind the rocks as if all the bees in Africa had been let loose,—a continual buzz and hum. Two troops of "C" Squadron had rushed to some rocks on the right, and slightly in advance of the line, and were enfiladed by the Free Staters on the slopes of the Crows' Nest. The Crows' nest loomed up on the right of the Regiment, and one wondered if there were sufficient Britishers to hang on to it.

Between the Crows' nest and the right of the regiment was the open plateau running out into a salient, on the crest of which were the Free Staters. The din went on, the hum of bullets and the crack of the bursting shrapnel, when suddenly, a little before 6 p.m., a mass of men appeared over the northern crest in rear of the Regiment. Wearing their grey infantry great coats, and with bayonets fixed, they came on at a run. The tension was extreme. Here and there the Lancers saw a gallant Devonshire man rolled over like a shot rabbit. Swinging to the right they went, their Colonel turning round, as with his sword he again directed them to swing to the left, and with a ringing cheer the mass of men charged down on the Free Staters. Mad excitement! the glorious side of war! Lieutenant Hill, in command of the right troop of "C" Squadron of the Regiment, flung his helmet away and cheered, and jumping up to join in the grand charge, dropped dead with a bullet through his brain. The Boers with one last wild volley, turned and bolted down the hillside, and the Devons lined the crest the Free Staters had held so successfully throughout the day. For another half hour the firing went on, then kindly night put an end to the sixteen hours of desperate fighting, and the Platrand was still held by the Ladysmith garrison. "A" and "C" squadrons spent the night in the position they had lain in during the afternoon, while "D" Squadron was in support on the northern slopes of the hill. A cold and miserable night, every one wet through and many of the killed and wounded lying where they fell. The enemy did not renew the attack in the morning, and the first few hours were spent in collecting the killed and wounded, "C" Squadron carrying the dead of the Gordon Highlanders from Wagon Point. Dawn on this Sunday morning produced a ghastly spectacle on the Platrand. The British losses were 18 officers and 150 men killed, and 25 officers and 224 men wounded, a total of 417 casualties. It was difficult to determine the Boer losses, the accounts being so varied, but their casualties were probably much the same as those of the defenders. Fouries Spruit, which had been dry in the morning, was a raging river after the storm, and must have carried off a good many of the retreating Boers in the evening. A large percentage of wounds on both sides were in the head. On the Sunday afternoon the dead were buried on the northern slopes of Wagon Hill, the funeral being attended by the Officers of the Regiments who had lost officers and men.

One amusing story is told of a soldier who was taken prisoner in one of the early Boer rushes. His captors kept him on the southern slope of the hill, and he was put in charge of two burghers, who took it in turn to climb to the crest and " do some shooting. " In the evening the prisoner and his guard, hearing a clattering, looked up and saw the other guard, who had been shooting, come bounding down the hill like a goat in a desperate hurry. Without stopping he merely yelled, " Karki comin', " and the other guard was up and bounding down the hill after his comrade with a parting " goodbye. " The Devons were charging.

The Officers of the Regiment engaged on Wagon Hill during the action of the 6th of January were :—

Major S. C. King—Commanding.
Lt. R. C. Brown-Clayton—Acting Adjutant.

" A " Squadron.

Capt. J. B. Scriven.
2nd Lt. H. F. Fraser.

" C " Squadron.

Capt. E. O. Wathen.
Lieut. W. T. Willcox.
2nd Lieut. W. H. T. Hill.

" D " Squadron.

Capt. A. Parker.
" W. A. Adams.
Lieut. A. V. L. Wood.
" J. N. Jardine.
" C. Arkwright.

The losses sustained by the Regiment were :—

2nd Lt. W. H. T. Hill.—killed.
Died of wounds. No. 4853 Private Andrews.
Wounded—Capt. E. O. Wathen.
No. 4214 Private Horner.

The other attacks on Ladysmith on the 6th of January call for little notice as they were never pressed home.

CHAPTER XVII.

The Defence and Relief of Ladysmith.

On the night of 11th-12th of January two non-commissioned officers of the Regiment, Corporal Simkin of "D" Squadron and Corporal Trill of the Band were sent out on a patrol up the Long Valley. Dismounting beyond Sign Post Hill and leaving their horses, they crept up towards the "Peach Trees" where they discovered a strong picquet of the enemy. A very smart and plucky piece of work.

On the following night "D" Squadron were ordered out to surprise the Boer picquet, but on the report of Trill and Simkin, who were sent on to reconnoitre and had discovered the Burghers were not in position, the squadron returned to camp.

A shell from the Surprise Hill gun, fired at a working party on Cove Redoubt, came over the Kopje on the afternoon of the 16th of January and burst in the Officers' Mess tent of the 5th Lancers. There were luckily only two officers in the tent at the time, one of whom, Captain Scriven, was wounded, while the other was not touched.

Sickness was rapidly increasing, and the heat intense, the thermometer on the 21st of January registering 105° in the shade. Food was getting extremely scarce, and an officer noted in his diary that on January the 19th potatoes cost one shilling each, eggs were fetching one guinea a dozen, and a penny box of matches could be bought for a shilling. Four days later the same officer writes that a dozen of whiskey was offered for sale for £100, and eventually raffled for and sent to the hospitals, while eggs were now fetching two shillings each.

For some days the guns of the Relief Force had been heard pounding away to the West of Ladysmith, and some healthy excitement had entered into the monotonous struggle of the garrison against hunger and disease. The rumours were extraordinary and conflicting : one day Buller had defeated the Burghers ; on another he himself had been driven back over the Tugela ; and yet again that a diversion was being made by a force out to the East of Ladysmith by Waschbank. Finally on the evening of the 24th of January the excitement had reached its pitch, for from the highest points in the western defences British shells could be seen bursting on the reverse slopes of Spion Kop. A more than sanguine 5th Lancer on Cove Redoubt declared he could see English infantry on the top of that now famous mountain, while there was no doubt whatever about the long continuous stream of wagons to be seen winding their way to the north along the Van Reenans Pass road, and the naked eye was able to discover the rapidly disappearing laager at the foot of Spion Kop. Darkness brought the anxious watchers back to camp, but the Relief Force was evidently in possession of Spion Kop, and there was not much to bar its way between it and Ladysmith. Relief at last loomed in the distance, and one looked forward to the events of the morrow. Alas ! At break of day those able to do so ran up Cove Redoubt, and gazing out to the West, saw the Boer laager again nestling at the foot of Spion Kop. What did it mean ? If Sir George White knew, he was keeping it all to himself. Meanwhile all the talk and conjecture failed to support the hopes of the last evening, and the enemy's "Long Toms" treated the garrison to an extra large dose of their 96lb. shells. During the night of the 27th a terrific storm brought down every tent but two in the Regiment. The sufferings of the sick and wounded in their tents at the Intombi hospital, many of which collapsed during the storm, must have been somewhat trying. In an officer's diary there is the following note against the 28th of January:—"Whew ! Buller has failed again, and Ladysmith is still in the merry stage of siege. It appears that Spion Kop was taken by Woodgate, hence the Boer retreat we saw, and during the night the British retired from the hill and the Boers returned. Woodgate was killed."

On the 28th of January Lord Roberts telegraphed from the Cape Colony to Sir George White via Sir Redvers Buller's army, "I beg you will yourself accept and offer all those serving under your command, my warmest congratulations on heroic splendid defence you have made."

"At the end of January," an Officer writes in his diary, "extraordinary things are happening. The Cavalry are being turned into Infantry. We cannot feed our men, we cannot feed our horses, so the horses must suffer to feed the men. We have handed our lances, swords, carbines and saddles into the Ordnance Store, and drawn out rifles and bayonets in their place. We have taken our places in the defences with the infantry, and the Garrison is eating our horses. It gives one something to think about, being one of a brigade of British Cavalry suddenly turned into Infantry and ordered to eat their own horses. All except 75 horses per regiment are turned out to pick up what grass they can on the outskirts of the defences, and a number are rounded up every evening to be slaughtered for food. The poor beasts have a bad time of it, many die of exhaustion and a number are killed by Boer shells. For some days they used to stampede back to their regimental camps, especially at "feed" times, and had to be driven away by the men. The 5th Lancers horses were turned out on Observation Hill. The Regiment now joined the infantry garrison at Leicester Post and Cove Redoubt, while one troop was sent out to Advanced Ration Post at the most north westerly point of the defence. Of the remainder of the Brigade, the 5th Dragoon Guards went to Observation Hill, the 18th Hussars on to Caesar's Camp, and the 19th Hussars to Helpmakaar Post. The 75 mounted men of each regiment camped on their former day bivouac, and endeavoured to keep their horses alive. A duty that the Regiment now carries out, in turn with the 5th Dragoon Guards, is that of "Flag of Truce Duty" on the Newcastle Road. Lying on the Eastern point of Observation Hill all day with a telescope glued to his eyes, the officer in charge of the party watches the approach of a Boer flag of truce and has speedily to clamber down the Kopje, mount and gallop out to meet it, and stop the wily burgher approaching too near to the defences. Amusing are some of the conversations between the bearer of the Boer flag and the Officer who meets him. "Would you like a bottle of whiskey ?" on one occasion asked a Boer. "No thanks, we have plenty in Ladysmith" was the reply, and the burgher laughed and refused to believe the lie. On another occasion, while waiting the result of his report to Head Quarters, a British Officer was handed by his opponent a copy of a Transvaal paper. "What lies this paper of yours tells about the war," remarked the

Officer. "No more than yours," was the reply, and the Boer produced an absolutely impossible battle picture in a London illustrated paper.

The diary continues on February the 2nd, "For dinner to-day we had chevril (a horse bovril) and the haunch of a mule. No doubt one could manage it if the meat and soup were good instead of being tainted by the hot weather and the plague of flies. My diet generally runs to one egg every other day (costing 3/-) a mess of violet powder discovered in a chemist's shop made with the help of some grease into a blanc-mange, which makes one smell like a girl's school, and what ration biscuits I can get hold of. To drink, coffee made of mealies without milk. To smoke, nothing but some Kafir tobacco, sunflowers and saved up tea leaves. We are told now that the Boers are having stormy meetings, and are discussing one more determined assault on Ladysmith. All night long now the garrison stands to arms, and our men are warned that the enemy may try to get in dressed in khaki, and using English bugle calls."

The guns of the Relief Force were once more being heard in the distance, and the enemy had placed another gun on Surprise Hill, an old time howitzer throwing a solid round shot which could almost be seen coming through its course and finally bumping along the ground. Their search lights were very busy and in addition, star shells were sent up every night. In Ladysmith dummy batteries were built during a night, often causing much waste of ammunition on the part of the burghers in the morning. At the end of the first week in February the sick and wounded at Intombi had reached 2,500, and scurvy had made its appearance. Rations were again reduced to ¼ on the 11th, and Officers and men found the effort required to walk to the more distant posts a trying one. In spite of the monotony and discomforts of life, however, every one remained cheery and sanguine of relief.

The last public sale of luxuries was held on the 13th, when the following prices were reached :—

1½ lb Tin of jam £1.11.0
1 jar stewed Peaches £1.5.0
1 packet "Old Gold" Cigarettes 17/6 and £1.5.6
50 Common Cigars £9.10.0
100 Havana Cigars £18.10.0
A Vegetable marrow £1.5.0

HH

Tomatoes each—2.6
¼lb Navy Cut Tobacco £3.0.0
¾ lbs sugar—10.0
A few bottles of Brandy £7 per bottle.
" Whiskey £6 "

The sale of these two last items was prohibited, all spirits being sent to the hospitals. Eggs had now reached the figure of 4/- each.

The enemy were now seen to be attempting to dam the Klip river where it leaves the Ladysmith plain in the gorge between Mbulwana and Besters' hill, but with what object was a matter for conjecture; probably that of flooding the town. A Naval gun at the eastern point of Caesar's Camp planted an occasional shell amongst the Boer working parties.

At last on the 14th of February good news came flashing through from the army south of the Tugela. Lord Roberts had sent a message, " With a large force of troops very strong in cavalry, artillery, and mounted infantry, I have invaded the Orange Free State. Tell your troops that I hope the events of the next few days will enable me to help Ladysmith. " On the next evening a large force of Boers was observed trekking northwards in the direction of Van Reenan's Pass, and it was supposed that they were leaving to oppose Roberts' advance in the Free State. Meanwhile, judging by the gun fire heard in the direction of Colenso, the Relief Force were daily heavily engaged with the enemy opposing their crossing of the Tugela. Soon too came the news that Kimberley had been relieved by Lord Roberts' army, and shortly after from Sir Redvers Buller, " Have taken Cingola mountain, everything going splendidly. " Some welcome rains coming on cooled the atmosphere for a day or two, but the rising river caused a few casualties, amongst them being a private of "A" Squadron drowned. Disease was now rampant; out at Intombi hospital, with no medical comforts or food, men for a few days were being buried at the rate of seventeen a day.

Inside Ladysmith the Garrison sat tight, fighting disease and straining every nerve to hold out if necessary to the bitter end; what that end might be no one ever suggested. The naval guns were now silent, presumably reserving their last few rounds of ammunition, and the Boer " Long Toms " daily shelled the defences without reply. Every night the weak and sickly men in the Ladysmith trenches stood to arms getting what sleep they could, and waiting for the final assault which never came. The strain was beginning to tell, officers and men, in spite of devotedly hanging on, were gradually going under and being taken off to Intombi. Conjecture was rife, and men were laying 6 to 4 that the Relief Force would reach the outposts to-morrow night, the next night, and so on. They were evidently fighting hard, judging by the cannonade to the south. On the night of the 25th a small sortie was made, and the Boer abattis round Gun Hill burnt, producing an enormous rifle fire from the enemy's trenches. South Africans inside the town suggested the anniversary of Majuba as the day of the final attempt of the Boers to take Ladysmith by assault, and on February the 26th, after having been on half rations for a few days, the troops were once more reduced to the now familiar quarter. What had happened? The reduction to $\frac{1}{4}$ rations invariably took place after a non-success south of the Tugela, but no one believed in another reverse. " However, on a $\frac{1}{4}$ ration the garrison can hold out until the end of March and then—; that there will be no surrender every one places implicit trust in our gallant chief." The 27th of February, the anniversary of Majuba, saw an extremely severe bombardment through which the naval guns kept silent, but no general assault by the Boers.

And now comes the story of the last day of the siege:—" On the 28th of February about 12.30 p.m. when we were sitting in the mess tent about to begin our so called lunch, "boom" went Mbulwana, and sent a shell right over the town and away to the Western defences, a range of some 11,000 yards; but beyond remarking on the powers of a modern big gun, no one, as usual, took much notice of it. Only one shell, and, as we were soon to discover, the last to be fired into Ladysmith. Shortly after, however, we were effectually awakened and interested, for "crash" went the naval gun in the Princess Victoria battery just above our heads. The sailor guns had been silent now for many days, and as Princess Victoria fired again, we rushed out and watched her shell bursting on Long Tom's redoubt on Mbulwana. The sailor gun on Caesar's Camp joined in, and the blue-jackets steadily sent shell after shell at Long Tom. What did it all mean? Our glasses soon showed us, for there, sure enough, was a derrick rigged up over the redoubt, and the Boers were endeavouring to unship Long Tom. The sailors shot well; their shells burst on the redoubt, beyond it, under it and all round it, and once knocked the derrick over. We prayed the gun would not be taken away, for of all the guns round Ladysmith the Mbulwana "Long Tom" had been the most vicious. Some of us rushed up Cove Redoubt with our glasses and sat alongside Princess Victoria, unmindful of her deafening crashes. And what a sight greeted us! "Whoop!" "Garn away!" and we threw our caps in the air, for there, on every road leading to the North, were strings of wagons and Boers all over the veldt, all riding and driving for all they were worth. There was no doubt about it this time, the investing army was in full retreat. Some one raised a cheer, and then suddenly the same thought seemed to strike us all, and in silence we realised our impotence. "Oh God! where are our horses!" "However, Buller's cavalry will soon be after them," some one said. An hour or more passed and still the stream of flying Boers, but no British cavalry yet on the scene. "Where are they? Surely they are not going to let them get away!" "I'd give a thousand pounds for horses." "But what's the good of it, we've eaten our horses, greedy beggars."

In the evening dusk the roads were still blocked with Boers, of whom there appeared to be no end, and the usual thunderstorm was threatening. It is extraordinary how a thunderstorm always arrives on an eventful day. Suddenly some one pointed away to the Kopjes beyond Intombi and asked "What's that?" There was a long snake like movement of mounted men coming down the Kopjes. "Those Boers will be cut off for a certainty," remarked some one.

"No they won't for they're Khaki."

"By the God of War it's the relief column at last."

We watched them pass Intombi, and canter through the thorns towards the river, and any doubts were quickly dispelled as the distant cheers on Caesar's Camp reached us. On they came, and a couple of squadrons of Natal Volunteers and some Imperial Light Horse rode through the drift into the town. They were the first glimpse we had had of the outside world for four months.

Volunteers were called for in order to make an effort to follow up the Boers on the morrow, the qualification being the ability to walk two miles. In the morning a small force started away on the tail of the retreating enemy. The Infantry, however, got no further than Limit Hill, while the few cavalry went but little further. Two officers' patrols of 5th Lancers went as far as End Hill and up the Long Valley, and reported all clear. Through Buller's helio we heard from Lord Roberts, "Cronje and all his force have surrendered to me. I trust the Government will be pleased to hear of this success on Majuba Day." During the morning a squadron of the 13th Hussars rode in with their wallets full of tobacco, which our men much appreciated.

On the morning of the 3rd of March those of the Garrison not on duty turned out, and lined the road to welcome the Relief Force, who marched through the Town to their camps out to the West. A sickly looking lot they were, those Ladysmith soldiers, white, pinched and drawn faces, and clothes hanging like sacks. It was a hot morning, and many succumbed while waiting. At last from the town came the strains of a regimental march, and soon after an escort of Royal Dragoons rode up followed by Sir Redvers Buller's staff. Then came along the various regiments of his army. In the post of honour, and leading the infantry, came our old friends the Dublin Fusiliers, or what remained of them, and the cheers of the two regiments as they met were worth listening to. The men of the Relief Force looked wonderfully fit, but the severe fighting they had been through was easily seen in the regiments at half their strength and with few officers. The Dublin Fusiliers marched through with some four officers only. As they passed we heard of Harts Hill, Pieters Hill and of all the stubborn fighting on the Tugela. Some of the regiments were told by their officers not to smoke while passing through Ladysmith, in deference to the feelings of the garrison who were without the soothing weed. A curious trait of the soldiers is this consideration for the feelings of others at times when all such feelings are usually dead.

In the Natal Army Orders of the 6th of March, 1900, the following telegram from the Commander-in-Chief was published:—

"Please convey to all ranks under your command my appreciation as a soldier of their splendid work at Ladysmith. They fought nobly and deserved the splendid success they achieved. Congratulate also Sir George White and those under his command for the gallant manner in which they have maintained the old fighting reputation of our Army under very trying circumstances. Wolseley."

Congratulations for both the Defence and Relief Forces were received from all parts of Cape Colony and Natal, from most of the great Towns and Societies of the United Kingdom, from India, Canada, New Zealand and Australia.

On the 10th of March, 1900, Lieut. Cyril Arkwright of the 5th Lancers died of enteric fever at Intombi, and was buried there.

" In consequence of the gallant conduct of Her Irish soldiers, Her Majesty Queen Victoria orders that all Irishmen, whether serving in Irish regiments or not, shall be allowed to wear the Shamrock on St. Patrick's day. " (Natal Army Orders, 16 March 1900).

Lieut-General Dunham Massy C. R., Colonel of the 5th Lancers, wired :—"Please let all ranks know how pleased I am with all that has been done and how proud I feel in being at the head of the 5th Lancers."

The following is a list of Officers, non-commissioned Officers and men of the 5th Lancers who were killed, died of disease, and were wounded during the Defence of Ladysmith :—

2nd November	No 4458	Pte. Stapleton	Wounded
,, ,,	4808	,, Buller	,,
9th ,,	3344	Cpl. Horton	,,
,,	4231	Sergt. Saunders	,,
17th ,,	4052	Cpl. Andrew	,,
30th ,,	4757	Pte. Archer	Died of Disease
8th December	3944	Pte. Lea	Killed (11th Hussars)
		Lieut : W. T. Willcox	Wounded
	4218	Trumpeter Goldup	,,
	3407	Pte. New	,,
	4058	Lt. Serg ; Austin	,,
14th ,,	3187	Pt. Deas	,,
22nd ,,		Lieut : Col. J. M. Fawcett	,,
		Major A. C. King	,,
		Capt. E. O. Wathen	,,
		Capt. M. P. R. Oakes	,,
		Lt. : & Adjt. H. H. Hulse	,,
	3550	S. S. S. M. Harris	,,
24th ,,	4854	Pte. Burford	Died of Disease.
30th ,,	5005	Pte. Donoghue	,,
6th January		2nd Lieut : W. H. T. Hill	Killed.
	4765	Shoeing Smith Dogett,	Died of Disease.
		Capt E. O. Wathen	Wounded.
	4853	Pte. Andrews	,,
	4214	Pte. Horner	,,

9th January	4367 Pte. Clifford	Died of Disease.
12th "	4772 Pte. Cornwall	"
14th "	4879 Pte. Austin	"
15th "	4424 Pte. Pritchard	"
16th "	4505 Pte. Gadie	"
	Capt J. B. Scriven	Wounded.
	4618 Pte. Marriott	"
21st "	4766 Pte. French	Died of Disease.
22nd "	4846 Pte. Sullivan	"
	4950 Pte. Prince	"
23rd "	4853 Pte. Andrews	Died of Wounds.
	3325 Pte. Stanton	Died of Disease.
25th "	4092 Pte. Cullis	"
	4096 Pte. Hodson (11th Hussars)	"
2nd February	4058 Sergt. Austin	"
3rd "	4925 Pte. Rea	"
10th "	4746 L. Cpl. Hill	"
11th "	4495 Pte. Watson	"
13th "	2723 S. Sergt. Farrier Stewart	"
14th "	4808 Pte. Butler	"
16th "	4365 L. Cpl. Nixon	"
17th "	4389 Pte. Coughlan	"
	4601 Pte. Groocott	"
18th "	4299 Pte. Dowsett	Drowned.
20th "	4844 Pte. Hewer	Died of Disease.
21st "	4226 Pte. Rider	"
22nd "	3811 Corpl. Mearman	"
5th March	3933 Pte. Gansden	"
9th "	Lieut. C. Arkwright	"
20th "	4813 Pte. Bull	"
21st "	4109 Pte. Jeffrey	"
24th "	4440 Pte. McGrath	"
26th "	4887 Pte. Webber	"
28th "	2740 Sergt. Ryan	"
1st April	4096 Pte. Vanbeek	"
3rd "	4926 Pte. Sleigh	"
15th "	4858 Pte. Wain	"
25th "	4235 Pte. Lowley	"

The deaths in March and April were the result of disease contracted during the siege.

Before the war Ladysmith had had an unenviable reputation for enteric fever, so the ravages of that disease amongst a large force cooped up within a circumscribed area during the four months of siege is not surprising. During the last week in November there were 15 cases of enteric and 72 of dysentery in hospital. A month later the enteric cases numbered 441 and dysentery 361. The medical returns show during the siege 10,673 admissions to hospital out of the garrison of 13,497 men. Many of these, of course, were re-admissions, but the numbers are conclusive evidence of the ravages of disease during the investment, and of the strenuous work carried on by the medical staff in Ladysmith. History does not relate, neither do " returns " show the numbers of sick men who devotedly stuck to their duty when perhaps hospital would have been a wiser course. The week of the Spion Kop failure found 842 cases of enteric and 472 of dysentery in hospital.

And so ended the Siege of Ladysmith.

In spite of hardships endured, the heavy sickness and mortality, and the starvation and awful monotony experienced during the defence of Ladysmith, the discipline and cheerful endurance of the men of the Regiment was splendid.

The following passage is taken from the "Times" History of the War in South Africa :—

" From the gallant veteran in command harassed by his responsibilities, distracted by the news of disaster . . . without, shaken by fever,—down to the simple private enduring hunger and discomfort in the trenches, one and the same resolution animated the whole defence. Ladysmith was to be held at all costs ; whether relief came or not the defence was to be carried on to the last ; when all else failed the garrison would try to break its way out or perish in the attempt. It was in this spirit that the flag was kept flying, and it is for this reason that the siege of Ladysmith will live in the annals of the British Army. "

CHAPTER XVIII.

1900 *to* 1902.—*Move to the Transvaal.—Meerzicht.— Amersfort.—Twyfelaar.—Van Wyks Vlei.—Battle of Belfast.—Crocodile Valley.—Lydenburg.—Krugerspoort.—Machadadorp.—Belfast.—Carolina.—Lake Chrissie.—Piet Relief.—Middelburg.—Move to the Cape Colony.—Peace.*

After the Relief of Ladysmith the Regiment, together with the 18th and 19th Hussars moved out to Field's Farm, and on the 11th of April to Pound's Plateau, below Mbulwana, where the brigade was gradually re-horsed and brought up to strength with drafts from England. Lt.-Colonel Fawcett having been invalided home, Major King assumed temporary command of the Regiment.

During May the Regiment joined the Drakensburg Defence Force, "D" Squadron being at Jonono's Kop, "A" at Besters' Station on the Orange Free State railway, and "C" and Headquarters at Smith's crossing. On the 6th of June a patrol of "A" Squadron were in action with the enemy near Brackwall Station and had the following men wounded :—

No. 3390 Private Skinner (dangerously) died 14th June.
No. 4745 Private Nicholls.
No. 3773 Private Taylor.
and a horse killed.

Upon arriving at the level crossing over the railway near Brackwall Station, the patrol had been halted, and Corporal Smith and two men were sent across the line and over some six hundred yards of open to reconnoitre a kopje commanding the railway. At the foot of the hill the Corporal sent Private Taylor dismounted up the kopje, remaining below himself with the other man and the horses. On arriving at the summit Taylor worked along a spur and found himself within a few yards of a large party of Boers, who were as surprised as himself at the meeting. They opened a heavy fire, wounding him as he made his way down. Running to the crest, the Boers commenced firing at Corporal Smith and his comrade below them. Smith, with great coolness, sent away the man with him, and with the led horse remained dodging about under a heavy fire at sixty yards range waiting for Taylor to join him. The latter, by rolling and scrambling down the hill, managed to reach the Corporal, who picked him up and rode back with him to the patrol. For his gallant conduct Corporal Smith was awarded the Distinguished Conduct Medal. On the 22nd of June " C " Squadron moved to Modder Spruit.

On the 27th of July "D" Squadron and Headquarters entrained at Ladysmith for Volksrust, " C " Squadron marching to Elandslaagte and there joining the train, while " A " squadron remained behind at Besters' station.

On leaving the Drakensburg Defence Force, the G.O.C. wrote to the Officer Commanding expressing his sense of " the excellent work they have carried out with such zeal and intelligence. The cheerfulness and discipline shown by all ranks caused him to regret that they are now to be dissociated from him. This is all the more a subject of regret as he was associated with them during the Siege of Ladysmith, and he thus feels he is parting with old comrades, during a time which will be historic. He is confident that the Regiment will add still further to the reputation it has so gallantly maintained in the past. "

The two squadrons arrived at Volksrust, the Transvaal border town, at 8 a.m. on the 28th of July, and detraining, marched to Sandspruit. Next morning they marched to Meerzicht, some 9 miles to the North, and there rejoined Brig.-General Brocklehurst's 2nd Cavalry Brigade (18th and 19th Hussars). The Regiment arrived at Meerzicht about midday, and found the troops out engaged in a reconnaissance in force towards Amersfort. Soon after their arrival the squadrons were turned out to help in the retirement of the British force, "C" Squadron being on the left and "D" on the right. Camp was reached at 9 p.m., the only casualty being a horse of "C" Squadron hit, though some of the other regiments suffered more heavily.

General Sir Redvers Buller was now at Pardekop, some 12 miles to the west, and was there preparing a large force for a move north. At Meerzicht was the 2nd Cavalry Brigade, a brigade of Infantry, and artillery, while the railway between Standerton and Natal was strongly held. All tents were sent to the railway, and on the 7th of August the forward move of the Meerzicht and the Pardekop troops commenced. They consisted of the 2nd Cavalry Brigade, Dundonald's mounted brigade of irregulars and volunteers, of "A" battery R.H.A., the 21st, 42nd, and 53rd batteries R.F.A., a howitzer battery, two 5 inch, two 4.7, four 12 pounders and four pom-poms, with the 4th (or Ladysmith) Division of infantry, a total of 9000 men and 42 guns. The objective of this first day's march was the township of Amersfort. The whole force advanced on the Rooi Kopjes, whereupon the Boers fell back on a ridge a couple of miles S.W. of Amersfort. The infantry, supported by artillery, gradually dislodged the enemy from this ridge, while Dundonald's men worked round by the left and the Cavalry brigade by the right. Before this latter move, the Regiment sent an officer's patrol to a hill on the right, the patrol coming under a heavy fire. The 5th Lancers and a squadron of 18th Hussars now worked round the hill, and the enemy leaving, it was soon occupied by mounted infantry. The remainder of the action does not call for much comment. The Regiment rode on some five miles to the south of Amersfort, and afterwards, returning to the brigade, eventually reached Amersfort at 10 p.m. Buller's casualties this day numbered some 22 ; of the 5th Lancers one horse was killed, and Sergeant Scott, three men and some horses wounded. Owing to the non-arrival of the transport the troops spent a cold and hungry night.

Next day at 6.30 a.m. "C" Squadron rode through a dense fog to water at a spruit in the centre of the town, and just below General Buller's house. One troop had watered and were dismounted at the top end of the cutting leading to the water, while the other troops of the squadron were watering or awaiting their turn to water, when whizz came a bullet out of the fog followed by others in rapid succession. It was somewhat disconcerting to be thus fired at within a few yards of the General's quarters, in the centre of a large British force. The men, however, led their horses out of the spruit with only one horse killed, and two horses and a charger wounded. It transpired some time later that three bold Boers had penetrated through the fog and the outposts to lie in wait by the water side ; of these three men one was mortally wounded when endeavouring to pass an infantry picket. One of the Boers was an Englishman. The remainder of the day was spent by "C" Squadron on the Rooi Kopjes, covering the left of the transport column, while "D" remained in support in Amersfort.

The next day the force marched to Mooifontein, the Regiment

forming the cavalry of the advanced guard, small parties of Boers retiring before the advance. The Vaal river was crossed on the 10th, the enemy for a short time disputing the passage of the bridge, and on the 11th the force bivouacked at Klipfontein. The 5th Lancers in the advanced guard halted some distance beyond the bivouac, and spent two most uncomfortable hours in a grass fire. An officer with an advanced troop writes in his diary, "given a hill held by enemy in no great strength, and up which a troop well opened out can gallop, it is ten to one that if one goes as hard as horses can lay legs to the ground, brother Boer will scoot down the other side, only waiting to fire a hurried and useless shot ; Lances are too much for him." Another officer of a troop in the vanguard was given a most reliable and smart guide of the country, who was invariably missing when required. "Time after time I turned for my guide and found him not. "Where is he ?" "Saw him going to that farm, sir," and half an hour later up he would canter as pleased as Punch with a goose and a fowl or two hanging round his saddle. "Where in the devil have you been ?" "Oh ! just inspecting —'s farm." was the cool answer. Eventually his scent for food being of so much more use to him than his eye for country, I told him to go to blazes and not come back again." The night of August the 12th was spent at Ermelo, and the following day Botha's Rust, near the source of the Vaal was reached.

The 15th found the force at Twyfelaar, where the Boers were discovered in some force. A man of the 19th Hussars was killed from a farm flying a white flag. Here Buller got in touch with French's cavalry Division of Roberts' army, and a halt of five days was made while the supply column went in to the Pretoria-Delagoa railway to be replenished. On the 18th a patrol of "D" squadron were fired on from a hill top, up which they promptly galloped, and secured a prisoner who was too late to join his comrades. A stronger party of Boers then tried to cut off the retreat of the patrol as they retired. The prisoner attempted to escape and was shot.

On the 21st of August, the supply column having returned, Buller marched from Twyfelaar, his left being covered by French. The Boers under Louis Botha were holding a strong position north and south of the Pretoria-Delagoa railway, with commandos extending to the south and south west of Belfast opposing Buller's march. By noon the advanced guard of the Natal army had reached Van Wyks Vlei, some eight miles to the north of Twyfelaar. The 18th Hussars were in action in front when a message arrived that the mounted infantry were hotly engaged on the Frischgewaagd ridge, some three miles to the east. "D" squadron of the Regiment with one troop of "C," who had been support to the advanced guard, were at once sent off to the help of the mounted infantry, who were found on a ridge commanded by the enemy at a short range, and closely engaged. Lieutenant Wood's troop crept into the firing line of the M. I., while Lieutenant Jardine's troop, with a pom-pom, came into action on their left, and Lieutenant Willcox's troop was sent down a ravine to the right, and working round, brought a flanking fire to bear on the Boer left. For some hours a heavy fire went on, and eventually some companies of the Gordon Highlanders arrived and reinforced Wood and the mounted infantry. They were followed just before dusk by half a battalion of the Leicester regiment, who worked down the valley on the right. Under cover of the infantry the mounted infantry and the three troops of Lancers withdrew, followed soon after by the infantry. This day cost the force some 44 casualties. The next morning the Commanding Officer of the Mounted infantry regiment wrote to the O.C. 5th Lancers thanking him for the assistance given him by the three troops of Lancers. The force halted at Van Wyks Vlei, while two battalions and some irregulars and guns engaged the Boers who had caused the trouble of the day before, and drove them off the ridges with a loss of ten men on the British side.

On the 23rd Buller advanced to Geluk Farm. The Regiment was again in the advanced guard, and came in touch with the Carabineers of French's force. The Boers opposed Buller's cavalry pretty strongly, and an Officer with the 19th Hussars was killed. An Artillery duel went on for some time, but the enemy gradually gave way. On their arrival the infantry occupied an outpost line, in which operation the leading companies of the Liverpool regiment went too far, and suffered severely, with a loss of ten killed, forty-eight wounded, and some thirty missing. That night "C" squadron occupied a gap in the outpost line between Buller's infantry and French's cavalry.

From an Officer's diary:— "I had been asleep about an hour and a half when I found.... pulling at my blankets with the cheerful news that he had a message to be passed along the outpost line to French's picquets, and I was to go with him. I cursed my luck, and off we started, and on getting clear of the bivouacs put out our lights. Having posted the 5th Lancer picquets in the dark, I merely had a general idea of the direction, but on we stumbled. We walked for an hour, which I judged ample time to reach the outpost line, but still no signs of it. We had no wish to walk past our picquets and into those of the Boers, so we "tacked" to the right and then to the left and lay with our ears on the ground and listened, but devil of a sign or sound of man or horse. It was a pitch dark night, we dared not shout or show a light in case we had got beyond the outposts, We sneaked about for some time longer and suddenly right in front of us we saw a single flash of light and then darkness again, but we had marked the direction and stalked it. Sure enough in a few minutes against the sky line we saw a dark mass of horses. Were they British or Boers? We lay down for what seemed hours to await some sign. It came, the clanking of a sword scabbard as it struck something. Boers don't wear swords, they must be our fellows, but risking nothing we crept on our stomachs to the mass and "Halt! who comes there" as a sentry heard us. "Friend. Who are you?" "Picquet of the...... " came the answer, and we were safe. Sad to relate the sentry had been smoking, but the match with which he lit his pipe was a godsend to us." This story is related as it was told as being an object lesson to all sentries.

On August the 24th and 25th the Natal Army halted on the Geluk ridges, Buller intrenching himself and preparing for the coming battle; while Lord Roberts had arrived at Belfast on the railway. Buller's outposts were engaged both days, and on the evening of the 25th the camp was shelled. The Boer Government, with President Kruger, was now at Watervall Onder, east of Machadadorp, on the railway.

The next day, Sunday the 26th, Buller left Geluk and moved north towards the railway, while French operated against the Boer right, north of the railway, and between the two, Robert's guards and mounted infantry advanced. A brigade of Bullers' infantry were engaged all day, and the Regiment, with the 2nd Cavalry Brigade, (Natal) were more or less under shell fire throughout the day on the left of the Natal Army. The casualties in the force during the day were 8 killed, 34 wounded and 3 missing.

By nightfall Buller's force was overlooking the Boer position at Dalmanutha. His troops spent a cold night on the ground, with in many cases but little transport. During the night a man of "D" squadron was shot on outpost duty.

On the morning of the 27th, on the ridges he had won the previous day, Buller found himself opposite the Bergendal farm, which was really the centre of the Boer position opposing the united forces of Roberts and himself. Across an undulating valley, about two miles north of his position, and just south of the railway, lay the Bergendal farm. Close to the farm, in amongst the rocks of an apparently unapproachable kopje, Botha had placed his best troops, the "Zarps" of Johannesburg and Pretoria (the Zuid Africaanshe Republick Politie). The kopje, in the shape of a natural platform, was some two hundred yards east to west, and a hundred from north to south, where it fell into the valley dividing it from the Natal Army's position. It was a strong position, and looked like certain death to any troops who attempted to cross the glacis in front of it. Just north of the farm and kopje ran the railway. Another commando was in the farm and along an outcrop of rocks still further to the left. With the Zarps were a pom-pom and a maxim, there was also a gun or two with the commandos on their left, while two Long Toms on the heights north of the railway had the range of Buller's ridge.

South west of the Bergendal farm, and at a range of but little over 2200 yards, Buller placed the 21st and 42nd Field Batteries, four 12 prs, a howitzer battery, two 4.7 inch and two 5 inch guns. On this part of the ridge, and north, to the west of the farm, was a brigade of infantry, while running south along the ridge was another infantry brigade, and Dundonald's irregular horse with the baggage. The 2nd Cavalry Brigade, a battery of Horse, another of Field artillery, and the mounted infantry were on the northern flank of the position, just south of the railway, and north west of Bergendal.

Buller's position ran north and south along the Vogelstruis-poort ridge, while that of the Boers south of the railway ran in a south easterly direction from the Zarps' Kopje by the railway. The Bergendal position was to be taken by assault under cover of the artillery fire.

Early on the morning of the 27th "D" Squadron and one troop of "C" went with the cavalry brigade to their position on the northern flank of Buller's army, while two troops of "C" were ordered off as escort to guns.

Shortly before 11 a.m. the bombardment of the Bergendal position began. The shelling of the artillery was chiefly directed on the kopje at Bergendal farm, and for three hours did the luckless Zarps undergo the concentrated fire of Buller's thirty-eight guns. So severe was it, that while schrapnel steadily and without intermission swept the kopje from end to end and the rocks themselves were torn and rent by the explosion of the lyddite shells, the defenders were unable to leave the post, neither could reinforcements reach it.

The artillery having thoroughly prepared the way, the infantry were ordered to the attack. The honours of the assault were assigned to the 2nd Rifle Brigade and the 1st Inniskilling Fusiliers, and gallantly the two battalions faced the ordeal before them. In spite of the artillery fire the defenders of the Bergendal kopje and the commandos on their right and left opened a heavy rifle fire on the advancing companies. The Rifleman were on the left and the Fusiliers on the right. In a succession of rushes across the bare, open, rifle swept plain, which in the form of a glacis sloped gradually up to the Zarps' position, the infantry surmounted the kopje. It was a gallant assault. The rifle Brigade suffered the most, having 3 Officers and 10 men killed, and the Colonel and four more officers and 63 men wounded, while the casualties of the Inniskillings numbered 17.

The 2nd Cavalry brigade, which up till now had been lying idle on the left flank, got orders to move to the right. With the 5th Lancers leading, the brigade galloped along the Bergendal position and past the Zarps Kopje out towards Delmanutha station. The 5th continued on to the east, followed by the Chesnut troop of Horse Artillery. The Boers were everywhere in full retreat, and with the troop of "C" Squadron in front, the Regiment continued on until stopped by the retreating enemy, when the battery came into action. The pursuit was not pressed further, and the Regiment got back to the bivouac of the army at Bergendal soon after dark. During this gallop Corporal Cottingham and three men and a horse of the troop of "C" Squadron were wounded.

The Cavalry and an infantry division of Roberts' army north of the railway, had not been heavily engaged on the 27th.

The following Officers of the Regiment were present at the battle of Belfast:—

Fifth Lancers Watering, on march to Belfast with Buller.

Boers retreating after the battle of Belfast.

Ladysmith Town Hall.

To Machadadorp after Belfast.

Major A. C. King — Commanding.
Lieutenant R. C. Browne-Clayton — Adjutant.

"C" Squadron.

Captain M. P. R. Oakes.
Lieutenant W. T. Willcox.
2nd Lieutenant J. Darley.
" G. McClintock.

"D" Squadron.

Captain Collis.
Lieutenant A. V. L. Wood.
" J. B. Jardine.
2nd Lieutenant Dugdale.

Next day the force marched to Machadadorp, the Regiment forming the rear guard, and on the 29th Buller reached Helvetia, on the Lydenburg road. The 5th were still in the rear guard, spending the night on the road midway between Machadadorp and Helvetia, and reaching the latter place on the morning of the 30th, where, in addition to the Natal Column, were Roberts' Cavalry and Guards. Buller, taking half his force to the heights above Nooitgedacht, met some 2000 British prisoners just released by the Boers.

On the 1st of September the Natal Column started for Lydenburg, a troop of "C" Squadron remaining at Helvetia. For some ten miles the column climbed ridge after ridge, and moving along roads of the worst description, by midday reached Schoeman's Nek, from which place the force scrambled its way down into the wild and magnificent valley of the Crocodile river. A halt was called at Badfontein, the troops bivouacking at the kopje Aileen, an isolated sugar loaf kopje in the centre of the valley. In front of Buller a semicircle of mountains closed the valley, the Lydenburg road winding its way through the centre of the valley and up the precipitous heights of Witklip. A forward movement was ordered for the following morning, but no sooner had the advanced troops moved off, than the enemy, strongly posted on the Wilklip (Rietfontein) heights, effectually prevented any further advance. Dundonald's mounted brigade were closely engaged all day, but a direct attack on the position being too costly, the column returned to its bivouac at Badfontein in the afternoon. During the night the Cavalry Brigade were ordered back to Helvetia, and by forced marches joined a detachment under Ian Hamilton, who advanced along the Dulstroom road parallel to Buller's march on the other side of the mountains. On the morning of the 6th the Cavalry Brigade led Hamilton's advance through the difficult defiles, picketing the hill sides until the infantry came up, and gradually forcing back the enemy's snipers in front of them, so that before midday they had cleared the pass, and appearing on Botha's right flank, found him in retreat. Buller now climbed out of the valley, and sent his mounted infantry and irregular horsemen to join the Cavalry brigade, and press on into Lydenburg. The Boers retreated through the town before the cavalry, and took up a position on the Paardeplaats mountain, east of and overlooking Lydenburg town. The following day Buller and Hamilton closed in on Lydenburg, and no sooner had the troops commenced to settle down in their bivouacs in and around the town, than Botha's Long Toms opened on them, causing a few casualties.

Next morning an attack was made upon Paardeplaats. Hamilton's column, protected on his right by the cavalry Brigade, advanced against the Boer left, while Lyttleton, with Dundonald's mounted men on his left, moved against the Boer right. In the centre the whole of the artillery was in support. By noon the Long Toms withdrew, and by 3.30 p.m., after some six hours' stiff climbing, the infantry of both attacks had reached the crest. The Boers fought a resolute rear guard action, and aided by a thick mist, were able to get away their guns and transport and retire towards Spitzkop. The total British casualties of the day only amounted to 26. Meanwhile "C" Squadron during the afternoon were ordered to escort a convoy of a 100 empty wagons and some Boer prisoners to the rear. Retracing their steps they marched to Witklip and back along the Crocodile valley to Schoeman's Nek, where the whole of the Sunday night was spent in loading the wagons with supplies. The convoy got back again to Lydenburg on the night of the 11th.

Ian Hamilton was now recalled by Roberts to the railway, and Buller with half of the Natal Column moved on to Spitzkop in pursuit of the Boers, the Cavalry and an infantry brigade remaining as a garrison in Lydenburg. The Regiment remained in Lydenburg with the Cavalry Brigade, and was engaged in convoy work, outpost and reconnaissance until September the 30th, when the Cavalry Brigade, some companies of the Leicestershire

Regiment and a battery under Brigadier-General Brocklehurst marched north for Krugerspoort. In the deep rocky ravine of the Spekboom river the scouts of the Regiment were engaged with a few of the enemy ; Private Haughton of "D" Squadron was killed, and two others wounded. By midday Krugerspoort was reached, and as the leading troops of the 5th Lancers galloped into the village, the Boers galloped out to the north under the cover of their pom-poms. Some prisoners were taken in the village. Next morning about 11 o'clock Buller's column arrived from Pilgrim's Rest and Spitzkop and joined Brocklehurst's camp. The Boers opened fire with a Long Tom and two other guns during the afternoon and evening, killing an Officer of the Devons and a man of the 5th Lancers, and wounding some twelve men of the S.A.L.H. During the night 300 men from the 5th Lancers, 18th and 19th Hussars attempted to capture the Long Tom by a night surprise, but on reaching its position of the previous day, they found that it had been removed, and the Boers had gone. The next day the whole force marched back to Lydenburg, the 5th Lancers in the rear guard. For the remainder of its time in Lydenburg the Regiment was engaged in convoy duty, outposts and reconnaissance. On the 8th of October, during a reconnaissance to the north of the town, Private Trill was killed.

The Natal Field Force was now broken up, and General Sir Redvers Buller left for England, and on October the 11th the Regiment left Lydenburg for Machadadorp, picking up the troop of "C" Squadron at Helvetia on its way. The squadrons remained at Machadadorp until the 28th of October and were engaged during that time in patrolling. On the 15th and 16th "D" Squadron escorted a convoy of ammunition to General French's Cavalry division near Geluk, and had a somewhat lively march, being pressed by the enemy. However, they accomplished their mission and rejoined with a man and six horses shot. On the 28th the Regiment marched to Belfast and joined Smith-Dorrien's column, and "C" Squadron was detached to form a post at Nooitgedacht.

Early in November "A" Squadron arrived by train from the Natal border and rejoined Headquarters. Since July they had been employed chiefly in patrol work at Albertina and Harrismith in the North eastern corner of the Orange River Colony. Lieutenant-Colonel Fawcett resumed command from sick leave.

On the 15th of November the lances and carbines were taken

from the Regiment and the men were armed with the infantry rifle. Until its departure from Belfast the squadrons were constantly on the move and engaged in making night marches against neighbouring commandos of the enemy. On the 17th of November "D" Squadron relieved "C" at Nooitgedacht, the latter rejoining Head Quarters.

On the night of the 8th of January the Boers made a most determined attack on all the British posts along the railway from Pan to Machadadorp.

The night was a foggy one, and at Belfast, a little after midnight, parties of the enemy succeeded in getting amongst the British camps. An outlying post of the Belfast garrison was Monument Hill, held by the Royal Irish Rifles. The rifle fire in this direction was extremely heavy, and a small patrol of lancers brought in a Rifleman who reported that he thought the post had been captured by the Boers. A second patrol was then ordered to go to Monument Hill and endeavour to discover what had occurred there. For this patrol Sergeant Evans, Corporal Forbes and Private Aldridge volunteered. The distance to the hill was little over a mile, and the three men rode to within three hundred yards of the post. Dismounting, they arranged that while the sergeant and the private remained with the horses, Corporal Forbes should proceed on foot and endeavour to ascertain the situation. To quote Forbes' own words— " I left the road and struck across the veldt, and by running, creeping, crawling and rolling I managed to get up to the wire entanglements which encircled the post. The difficulty now was to get through the wire. I could hear shouts and groans and there was some shooting going on, but whether Briton or Boer was in possession I could not tell. I dare not go round the entanglement to the entrances as I knew they would be guarded, and so by a series of wriggles soon found myself inside the post. What was to be done now? I knew if I were seen I should be shot, whoever held the hill, so I continued to wriggle and roll on my stomach. I soon came across the effects of the fight, the dead bodies of riflemen and Boers, and the tents which had been cut down on top of the Irishmen. Some one was calling " Water, for God's sake give me water " and suddenly a dog barked a few yards to my right and I could just distinguish a man. I immediately covered him with my rifle but he apparently had not seen me. I remained where I was for some

A Fifth Lancers Patrol.
Catching a dinner.
Crossing the Spekboom River.

time and then slowly crawled back a little and worked my way to where I heard the shouts for water coming from. I soon found two of the Irishmen badly wounded and asked them in a whisper what had happened, but the only reply was a piteous appeal for water. I then crawled some 50 yards to the cook house and found a camp kettle with some water in it, and slowly wriggled back to the two wounded men, and filling my cap with water, gave them a good drink. They then told me that the Boers had rushed the sentries in the fog, cutting down the tents on their occupants and shooting and clubbing the men with their rifles as they rushed out, and although the garrison had made a gallant fight they were overpowered and the post captured. There was a lot of shouting going on by the Boers, and I quietly crawled towards it, and then there was a shot. Beyond a man standing on the monument I could see nothing, and so gradually crawled back to the wire entanglements, and as soon as I was clear of these ran back to the horses, where I found Evans and Aldridge safe, and we rode back to camp and made our report." For his conduct on this occasion No. 4216 Corporal H. N. Forbes was eventually awarded the medal for Distinguished Conduct in the Field. A troop of "D" Squadron under Lieutenant Rose acquitted itself well in a small isolated post at Wildfontein, for some hours resisting a most determined attack by the enemy.

On the 24th of January "A" and "C" Squadrons marched with a column to Wonderfontein, and the next day to Twyfelaar, where "C" Squadron got in touch with a considerable force of the enemy and was engaged up to dark. During the action Sergeant Hamilton and three men were wounded, Sergeant Hamilton and Private Morton severely. The march was continued next morning, and the enemy eventually driven through Carolina and that place occupied. The following day Carolina was evacuated, the Regiment finding the advanced guard. During the retirement the rear guard was hotly engaged, and on arrival in camp the Regiment had to assist the rear guard, who reached camp after dark. The retirement was continued the next day, and Wonderfontein reached on the 30th January.

On the 3rd of February the Regiment formed part of General Smith-Dorrien's column, which acted on the left flank of General French's drive towards the Swazi and Zululand border. Lake Chrissie was reached on the 6th, where the enemy showed in some force. "C" squadron was in advance, and coming under a close fire, drew swords and charged. The Boers, however, cleared in time. During the night a resolute attack was made on the British camp. The attack was eventually repulsed, but with considerable loss to the column. The Regiment had seven casualties, among them being Sergeant Morgan, who died of his wounds, and no less than one hundred and fifty of the horses were killed or lost in a stampede ; two horses found their way back to Wonderfontein, a distance of over thirty miles. The infantry suffered severely. In one of the trenches held by the West Yorkshire Regiment every one of the defenders was killed.

On the 9th the march was resumed, and at about 9.30 a.m. a Boer laager was seen. The Imperial Light Horse, supported by "C" Squadron, were detached to capture it. They successfully accomplished the feat, and took a large number of wagons and stock. Considerable difficulties in the shape of swollen rivers and bad roads were now experienced, and on the 15th the force reached Amsterdam. During these operations Sergeant Hyde, in charge of a small patrol of the Regiment, greatly distinguished himself. The enemy had derailed a train, which they were prevented from looting by Sergeant Hyde, who stubbornly held on to a ganger's hut with his patrol, from whence he commanded the train. For his action he was awarded the Distinguished Conduct Medal. Corporal Kirby, on another patrol, was wounded. A halt was made at Amsterdam until the 28th, during which time the troops had a trying experience. The rain was incessant, and the column without tents, in addition to which rations for man and horse were reduced to a minimum. There was plenty of fresh meat, but no groceries, biscuits or vegetables. A raid was made on a farm for provisions, during which Private Hardy was killed, and Privates Palmer and Power wounded. The column arrived at Derby on the 28th, and next day a force was detached into Swaziland and a further capture of wagons and stock effected.

On the 3rd of March a small force under Major King was ordered out to forage for supplies. During the retirement of this detachment Lieutenant Dugdale's outpost was severely handled, and he greatly distinguished himself by bringing in some wounded men of the 19th Hussars who were with him. For his gallantry on this occasion Lieutenant Dugdale was awarded the V.C. The story of Dugdale's V.C. is as follows :—

On the 3rd of March Lieutenant Dugdale, 5th Lancers, was in command of a small outpost near Derby. He was ordered to cover the retirement of Major King's force, after which retirement his own party came under a heavy fire at a range of about two hundred yards, when a sergeant, two men and a horse were hit. Pulling up, Dugdale dismounted and placed one of the wounded men on his own charger; he then caught another horse, galloped up to a wounded man and took him up behind him, and brought both men safely out of action.

Major King now assumed command of the Regiment, Lieutenant-Col. Fawcett going down country sick. The march was continued to Piet Relief which became the headquarters of the Regiment until the 14th of April, during which time the squadrons were constantly employed on convoy duty. A bad outbreak of glanders occurred while at Piet Relief. On a convoy duty Corporal Sandham was missing, his body being found some days later. From the 15th to the 27th of April the force marched back to Wonderfontein in charge of a large convoy. The difficulties of the march, owing to the bad state of the roads after the recent heavy rains, were enormous. So bad were they that on one occasion it took a whole day to move a quarter of a mile.

On the 28th the Regiment left Smith-Dorrien's column and marched to Belfast, and thence to Middelburg. "A" Squadron under Major Spurrell then left to join Colonel Spen's column near Barbeton, and "C" and "D" Squadrons joined General Bindon Blood's column and marched to Carolina, where the two squadrons were mostly employed in convoy duty to the railway.

On the 10th of July the Regiment received orders to move down into the Cape Colony, and on the following day left Carolina for the railway, the two squadrons entraining at Wonderfontein. Passing through Pretoria, Naaupoort in the Cape Colony was reached on the 16th. During the railway journey to Cape Colony the Officers were conveyed in goods vans, and the men in open trucks and flats. The cold was intense, and the men, taking every advantage of the slow rate of travelling, fastened their stirrup leathers to rings on the sides of the trucks, and catching hold of them, ran alongside the train. Remounts were taken over at Naaupoort, and on the 18th "C" and "D" squadrons marched for Middelburg, where they joined a column being organised under Colonel

Hunter-Weston. Leaving Middelburg on the 22nd, the newly formed column reached the railway at Wildfontein on the 22nd.

At Wildfontein the Regiment was deprived of its swords.

Naaupoort was reached on the 27th, where "A" Squadron rejoined, and the following day "D" Squadron under Captain Parker was detached to join Colonel Kavanagh's column. "D" Squadron did not rejoin Head Quarters until the end of the war.

Following the fortunes of "A" and "C" Squadrons, we now find them engaged in a period of constant hard work and continuous trekking without much fighting. On August the 2nd Private Long was killed near Zuurfontein, and Sergeant Sewell wounded. For the remainder of the year Head Quarters and the two squadrons were marching and countermarching with very little rest. A gallant exploit by Private Worsencroft on the 22nd of November is deserving of mention. On that date a patrol of "C" Squadron under Lieutenant Arkwright was in touch with a force of Boers. The Officer sent Worsencroft with his report to an Imperial Yeomanry post at Petersville. Worsencroft's ride was a somewhat adventurous one. He rode through the Boer force at night, and delivering his message, returned by the same route before daylight. Five times that night he came under the enemy's fire, and during his ride he covered a distance of thirty four miles.

On the 23rd a patrol under Sergeant Hamilton, accompanied by an Officer of the Imperial Yeomanry, located a considerable force of the enemy. Fixing up a heliograph, the patrol endeavoured to send back a message, but the Boers at once attacking, the message was not finished. With the exception of the Officer, who was wounded and captured, the patrol managed to get away, though the majority had to walk to Hamalfontein, their horses having been shot.

The pursuit of Kritzinger's Commando on the 16th of December was an arduous affair. From 8 a. m. until dusk, without a halt, did the Regiment, with Colonel Doran's column, endeavour to catch the Boer leader. A great part of the pursuit was carried out at a gallop. Kritzinger was eventually driven on to the blockhouse line, but the strain of the pursuit had been so severe that barely fifty horses of the Regiment finished the chase. No food or forage was available that night, but the Coldstream Guards entertained the Regiment most hospitably and made the men as comfortable as their very limited means would permit. Various Boer com-

Colonel E. H. H. ALLENBY, C.B.

Late Commanding the Regiment

mandos were pursued by the column, Latigan, Wessels and others, each being driven in turn in conjunction with other British columns. Occasionally the Boers would fight, and during these operations Lieutenant Darley, Sergeants Wilson and Saunders and Corporal Craven were severely wounded.

The new year found the Regiment still hard at work marching over a considerable area of the Cape Colony. The commander of the column was changed more than once, and still the strenuous trek went on. From the 17th of February until the 20th of May Malan's commando was being pursued about Carnavon, Victoria West, and S.E. towards Port Elizabeth.

Then followed a chase after Fouchees' commando, close on whose heels followed the British column until, on June the 2nd, came the news of the declaration of peace.

"D" Squadron meanwhile had been engaged in the same arduous work with Kavanagh's and Caldwell's columns, and latterly had embarked at Cape Town for Port Nolloth, where, with Cooper's force, it took part in the relief of Ookiep.

The want of space prevents a detailed account of the doings of the squadrons of the Regiment in the Cape Colony during the last stage of the war. That the work was hard, very hard, every officer and man who took part in the operations can testify. The continual chasing, for week after week, of an enemy who would not stop to fight, required the endurance and discipline of the best of troops. And the best of troops were there. The regiments of British Cavalry with the columns in the Cape Colony, by their splendid example, carried with them the local corps of horsemen, and together side by side, British Cavalry and Local Horse rode through the strenuous time for the good of the common cause.

Soon after peace was declared "D" Squadron rejoined Head Quarters, and the Regiment moved to East London and embarked for Home.

CHAPTER XIX.

1902 *to* 1906.—*England.—Colchester.—The War Memorial.—Aldershot.—Manœuvres.—Death of Lt.-General Massy.*

The Regiment disembarked at Southampton on the 21st of October, after an absence from England of 14 years, the last three of which had been on active service, and proceeded to Colchester, where the new commanding Officer, Lieutenant-Colonel E. H. H. Allenby C. B. joined and took over the command. A few days later the band and one squadron, under the command of Captain E. B. Wilson, took part in the Coronation procession in London of His Majesty King Edward VII.

In welcoming the Regiment home, the Colonel, Lieutenant-General Dunham Massy, wrote to the commanding Officer, " I was very glad to hear from you of the safe arrival home of the Regiment, and that their behaviour was exemplary. The Regiment has always been smart, and well conducted. " The Past Officers also invited the Present Officers of the Regiment to a Welcome Home dinner in London.

In the following August and September the Regiment took part in the manœuvres on Salisbury Plain, being brigaded with the Royal Dragoons and a Provisional Regiment of Cavalry formed from the depôts of regiments in South Africa.

The 9th Lancers was at this time affiliated with the 5th, and periodical drafts were sent to India for the 9th.

At an inspection by the G.O.C. Eastern district, that officer reported he " was much pleased with the appearance of the men on parade, and that the horses looked well and in hard condition. "

In August, 1904, the Regiment took part in the summer training of the 2nd Army Corps under Field Marshal Sir Evelyn Wood on Salisbury Plain, where it repeatedly gained the highest praise for its excellent work.

On the 26th of August, in St. Patrick's Cathedral, Dublin, the

Major-General COOKE, C.V.O.

Honorary Colonel of the Regiment

memorial erected by the officers and men of the 5th Lancers, in memory of their comrades who fell in the South African War, was unveiled by General Lord Grenfell, commanding the Forces in Ireland. The impressive ceremony, and the scene within the National Irish Cathedral, was one which those who witnessed it are unlikely to forget. The vast interior of the beautiful old Cathedral was crowded to overflowing by an enormous congregation, while outside, in all the approaches, stood a large number of Irish men and women in sympathy with the Irish Lancers in the honouring of their gallant dead.

The memorial is placed in the north choir aisle beside the grave of the Duke Schomberg, under whose command our predecessors of Wynne's Dragoons endured the horrors of the fever stricken camp of Dundalk in 1689, and marched with William of Orange to the Boyne, where the gallant old Duke met his death while rallying his Dutch and French regiments in the crossing of that historic river. Another coincidence worthy of mention, is the guard of honour inside the cathedral so kindly furnished by the 6th Inniskilling Dragoons, quartered in Dublin, whose predecessors, together with Wynne's Dragoons of Enniskillen, on the 1st of January, 1689 were first called into existence as Cunningham's Dragoons of Enniskillen. Both regiments fought side by side in all the actions of the time around that famous city ; together they served under Schomberg at Dundalk, and together took a distinguished part in the crossing of the Boyne, at Aughrim and in Flanders.

A contingent of officers, N.C.O. and men of the Regiment journeyed over to Dublin from their camp on Salisbury Plain to attend the ceremony, while Lieutenant-General Dunham Massy and many old officers were present, and the regiments quartered on the Curragh and in Dublin were all represented.

As the " Last Post " rang out from the trumpets of the Inniskilling Dragoons and echoed amongst the columns and aisles of the old Cathedral, many an eye in the vast congregation was dimmed, and many a mind must have dwelt on the concluding words of the Dean's address, " For that long roll of the dead must speak to us all, as it will speak to our children's children in the generations to come, of the awful cost of war, if it speaks, too, —as it does speak, thank God—of the courage and honour and obedience which men prize more than life. Their graves lie across the seas, but here in this Temple of Peace their memory shall remain."

The inscription on the memorial runs :—

IN MEMORY OF THE OFFICERS, N.C.O.'s, AND
MEN OF THE
FIFTH ROYAL IRISH LANCERS

Who lost their lives in South Africa
Between March 4, 1898, and September 19, 1902.

COLONEL

J. J. Scott Chisholme.

LIEUTENANTS

C. Arkwright and W. H. T. Hill.

SERGEANTS.

Staff-Sergt. Farr. H. Stewart
F. Austin
J. Hollis
G. Morgan.
S. Ryan
H. E. Sandys
F. Traill
R. Willison

CORPORALS.

E. Mearman
G. Cornwall
T. Coughlan
W. Cullis
T. Dervan
J. E. Donoghue
F. Dowsett
J. Everett
G. H. File
S. J. Fleming.

LANCE-CORPORALS.

T. Bethell
M. Blair
A. Hill
A. C. Nixon
Shoeing-Smith G. A. Doggett
A. Sandham
E. B. Smyth
H. T. Warner.

PRIVATES.

A. E. Archer
H. T. Andrews
N. Armitage
K. Howard
W. James
R. H. Jeffrey
O. T. Kinsey
W. Leach
S. Long
W. Lowley

J. M'Grath
H. O'Neill
T. Pickard
C. Prince
R. Rea
T. Skinner
J. Sullivan
W. R. Trill
P. Veitch
T. J. Watson
H. Austin
R. Barnes
H. Blunden
J. Buckley
E. Bull
W. T. Burford
M. Butler
J. Byrne
D. Byrnetes
W. Chankley
D. Clifford
S. Cooper
C. J. Oliver
A. Otway
A. H. Plume
J. Pritchard
A. W. Rider
O. Sleigh
H. D. Thompson
W. Vandepeer
J. Wain
H. C. Webber
D. Fownes
W. French
E. J. Gadie
E. E. Gatfield
A. Gausden
J. Griffin
A. G. W. Groocott
A. Hardy
W. Haughton
A. Hetherington
F. Hewer
O. C. Hobson.

From the 6th to the 13th of September the Regiment was brigaded with some squadrons of Household Cavalry and Essex Imperial Yeomanry on the defending side during the Essex manœuvres, the brigade being commanded by Colonel Allenby, 5th Lancers. The 1st Army Corps, under Sir John French, embarked at Southampton, and landing near Clacton-on-Sea, drove back the defenders under Major-General Wynne to the west of Essex, when, in accordance with a general idea, the invading force retired to the sea pursued by the defenders, and re-embarked at Clacton, being much harassed by Allenby's cavalry on their northern flank. The Regiment obtained much credit for its work during these manœuvres.

At his final inspection of the 5th, the General commanding the Eastern District bade farewell to the men on the foot parade of the Regiment, and spoke of the excellent conduct of the Lancers during their stay in Colchester.

On the 26th and 27th of September the Regiment marched by squadrons to Aldershot, where it took over the West Cavalry Barracks from the 14th Hussars, and was brigaded with the King's Dragoon Guards and 8th Hussars in the 1st Cavalry Brigade, commanded by Major-General Scobell.

The 12th Lancers were now affiliated with the 5th, and the 9th with the 21st Lancers.

On the 24th of May H. M. the King was present at field operations of the Aldershot Army Corps to the south of the Hog's Back. His Majesty witnessed the attack of the 1st Cavalry Brigade against Kettlebury Hill, and subsequently commanded that all ranks should be informed of his Majesty's satisfaction with the manner in which the field operations were carried out.

On the 8th of June the Regiment was present at the Royal Review of the Aldershot Army Corps on Laffan's Plain by His Majesty the King. The King of Spain was present at the Review. The smart appearance of the Regiment was commented upon by all.

On September the 8th the Regiment, with the 1st Cavalry Brigade, marched to the Berkshire Downs and joined the camp of the Cavalry Division, consisting of the 1st Cavalry Brigade (K.D.G., 5 L., 8 H.,), the 2nd Cavalry Brigade (7 D.G., 14 H., 21st L.) and the Household Cavalry Brigade. On the 23rd of September the Brigade took part in the manœuvres of the Aldershot Army Corps in the Thames Valley.

On the 19th of October Colonel Allenby was promoted to Brigadier-General to command the 4th Cavalry Brigade, and Major H. W. G. Graham D.S.O. was appointed to the command of the Regiment.

In May H. M. the King once more witnessed operations of the Aldershot Army Corps on the Fox Hills and Chobham Ridges; in August the 1st Cavalry Brigade was inspected by Sir John French on the Berkshire Downs, and during September the Regiment took part in the manœuvres of the Army Corps on the Sussex Downs.

On the 21st of September the Regiment lost its Colonel, Lieutenant-General Dunham Massy, who died at his residence in Tipperary. The following notice of his death was published in the Regimental Orders :—

"The Officer Commanding has to inform the Regiment with the deepest regret of the death of Lieut-General Dunham Massy, C.B., which took place at Grantstown Hall, Tipperary, at 2.30 p.m. yesterday.

General Massy for eight years, from 1871 to 1879, commanded the Regiment, and for the last ten years has been Colonel of the Regiment.

Lieutenant-Colonel H. W. G. Graham, D.S.O.

His services are too well known to need record in these orders, and offer to all an example of high courage and devotion to duty which justly deserve the imitation of every officer, N.C.O. and man in the Regiment."

Lieut.-General Massy was succeeded by Major-General Thomas Cooke C.V.O., formerly of the 17th Lancers.

APPENDICES

I.

APPENDICES

BIOGRAPHIES OF OFFICERS

BRIGADIER-GENERAL JAMES WYNNE.

James Wynne was the eldest son of Owen Wynne by Catherine, widow of James Hamilton, and daughter of Claud, 2nd Baron Strabane. Owen Wynne, who lived in Wales, migrated to Ireland about 1688.

The Brigadier's early military service was passed in the infantry. He was a captain in Colonel Stewart's Regiment of Foot, the present 9th Norfolk Regiment, and with that corps he took a prominent part in the relief of Londonderry under Major-General Kirke.

In 1689 James Wynne was one of the Officers selected by Kirke to organize the levies which had been doing such strenuous service at Enniskillen. He was appointed a Colonel of Dragoons and raised the famous Enniskillen Regiment known at the time as Wynne's Dragoons, and in later years, as the 5th Royal Irish Dragoons.

Colonel Wynne led his Regiment throughout the War in Ireland, which ended with the Peace of Limerick in 1691.

In 1694 Wynne took his Regiment to Flanders, where he commanded a brigade of dragoons under Marlborough, being appointed a Brigadier-General on the 4th of October.

In an action near Rouselaer he was badly wounded, although at the time it was not thought fatally. He died of his wounds, however, a few days later, on the 15th of July, 1695, at Ghent.

GENERAL CHARLES ROSS.

Charles Ross of Balnagowan was a son of the 11th Baron Ross.

He joined Wynne's Regiment of Enniskillen Dragoons as a captain in 1689 on the formation of the Regiment, and with it served throughout the war in Ireland.

In 1694, as Lieutenant-Colonel he accompanied the regiment to Flanders. On the death of James Wynne on the 15th of July, 1695, Marlborough appointed Ross Colonel of the Regiment.

In 1704, through the instrumentality of its Colonel, the title of the Royal Dragoons of Ireland was accorded to Ross's Dragoons.

On March the 9th, 1702, Colonel Ross was appointed a Brigadier-General of Dragoons, and on January the 1st he was promoted to Major-General. He served throughout Marlborough's campaigns in Flanders and commanded a brigade of dragoons at Blenheim, Ramillies, Oudenarde and Malplaquet.

Promotion to Lieutenant-General followed on the 1st of January, 1707. On the 1st of May, 1711, Ross was appointed Colonel-General of all the Dragoon Forces, and on January the 1st, 1712, he was promoted a General.

In September, 1713, General Ross was appointed Envoy Extraordinary to the Court of France.

He was removed from the head of the Royal Irish Dragoons by George I. on the 8th of October, 1715. He was, however, re-appointed to the Colonelcy of his old Regiment on the 1st of February, 1729, which appointment he continued to hold until his death at Bath on the 5th of August, 1732. He was buried at Fearn in Ross-shire.

LIEUTENANT-GENERAL OWEN WYNNE.

Lt.-General Owen Wynne was the third son of Owen Wynne who settled in Ireland about the year 1688, having previously lived in Wales, and was a younger brother of Brigadier-General James Wynne who raised the Regiment.

Owen Wynne in 1688 was serving in the army of James II, but being a Protestant, he transferred his allegiance to the Prince of Orange on the breaking out of the Revolution. He was with Major-Gen. Kirke's force sent from England to the relief of Londonderry, and he also took some part in the defence of Enniskillen, and served through the War in Ireland.

Owen Wynne was appointed a major in his brother James Wynne's Dragoons on the 1st of November, 1694, and served with his Regiment through the Flanders campaign of 1694 to 1697, being promoted Lieut.-Colonel in July, 1695, taking the place of Charles Ross, promoted Colonel of the Regiment on the death of James Wynne.

He served under Marlborough and was promoted Colonel in 1703, and in 1705 he raised and commanded a regiment of foot. The year 1706 saw him a Brigadier-General, and in 1709 he was promoted to Major-General.

In 1715 Major-General Wynne raised and commanded the regiment now known as the 9th Lancers. From the head of Owen Wynne's Dragoons he was transferred to the Colonelcy of the 5th Horse, now the 4th Dragoon Guards.

Promotion to Lieutenant-General followed in 1726, and in 1728

Owen Wynne was Commander-in-Chief of His Majesty's Forces in Ireland. In August, 1732, he was transferred from the 5th Horse to the Colonelcy of his old Regiment the Royal Irish Dragoons, which appointment he retained until his death in 1737.

Owen Wynne represented Ballyshannon in Parliament from 1715 to 1727, and from 1727 to 1737 was member for Sligo. He was also a Privy Councillor, and in 1736 was Governor of Londonderry. It is stated that he several times refused a peerage.

FIELD MARSHAL VISCOUNT MOLESWORTH.

Richard Molesworth was the third son of Robert Molesworth, Ambassador at one time to the Court of Denmark and raised to the Peerage in 1716.

He was educated for the Law and sent to the Temple, but preferring a more active life, he volunteered for the War of the Spanish Succession, and obtaining an ensigncy in Orkney's Regiment of Foot, now the Royal Scots, went to Flanders on the 14th of April, 1702. He served with his regiment at Blenheim, and at Ramillies he was an Aide-de-camp to Marlborough.

It was during the battle of Ramillies that he saved the life of his great chief, an incident which for some reason appears to have been kept somewhat secret at the time, but which nevertheless has been handed down in history. The Duke while leading a cavalry charge, was cannoned against when jumping a ditch, and unhorsed, his charger getting away. Molesworth, who was close at hand, came to his assistance, and dismounting, helped Marlborough to his own horse, and remained to face the enemy alone on his feet. The French, however, were so intent upon pursuing the Duke that Molesworth escaped with a few sabre cuts. He then recovered the Duke's horse, and rejoined him, when Marlborough remounted his own animal. While doing so, the equerry who was holding the Duke's stirrup, had his head taken off by a cannon ball.

There is a picture in the August 1901 number of the *Royal United Service Magazine* of a very rare medal struck in honour of the occasion, but the present owner of the medal cannot be traced. The *obverse* bears the profile of Captain Richard Molesworth equipped as a Roman warrior, with the inscription " Richard Molesworth Brit. Trib. Mil. " round the margin. On the *reverse* is a figure representing Victory leading by the hand a warrior, trampling on broken artillery with the motto "Per Ardua."

In 1708 Molesworth was present at the Relief of Brussels as Captain and Lieutenant-Colonel in the Coldstream Guards, and at Mons in the following year he was blown up by a mine, but not seriously injured.

He next commanded a regiment of foot which was disbanded after the Peace of Utrecht. In 1715 he raised a regiment of dragoons with which he fought at Preston, where he was severely wounded. His dragoons were disbanded in 1718, and in 1725 he was appointed Colonel of the 27th Foot, from which he was transferred in 1732 to the Colonelcy of the present 9th Lancers.

He succeeded to the title in 1726 as 3rd Viscount.

In 1735 Viscount Molesworth was promoted Major-General, and in 1737 appointed to the Colonelcy of the Royal Irish Dragoons, which appointment he held until his death.

Promotion to Lieutenant-General followed in 1739, and in 1741 he was Master-General of the Ordnance in Ireland, and a Field Marshal in 1757.

Viscount Molesworth was a member of the Privy Council in 1733. He was twice married. After a brilliant career he died in October, 1758.

GENERAL THE HON. JOSEPH YORKE K.B.

The Hon. Joseph Yorke was a son of Lord Hardwicke, the Lord Chancellor.

General Yorke commenced his career in the Foot Guards, with whom he served in Flanders in 1743, and was present at Fontenoy and the other actions of that campaign.

He also fought with his regiment during the Jacobite rebellion in 1745.

He was appointed Colonel of the 5th Dragoons, from Colonel of the 8th Dragoons, on November the 27th, 1760, and on April the 4th, 1787 he was again transferred to the head of the 11th Light Dragoons. He was a Major-General on the 18th of January, 1758, a Lieutenant-General on the 11th of December, 1760, and General on August the 29th 1777.

GENERAL SIR JAMES CHARLES CHATTERTON, BART., K.H.

Sir J. C. Chatterton served in the 12th Light Dragoons in Portugal, Spain, Flanders and France, from 1811 to 1818, including the affairs of Feute Gurnaldo and Aldea de Ponte, the sieges of Cuidad Rodrigo and Badajoz ; the actions of Usagre, Ilerna, Passage of the Tormes near Salamanca, and the attack of the enemy's rear guard, Heights of St. Christoval, Rendo, Castrejohn and battle of Salamanca ; affairs at Tuedela, Valadoled, Celadadel, Caminho, to the investment and siege of Burgos ; actions at Monasterio, Bridge of Bannel, Quintana Palla Venta del Pozo and Cabezon, actions on the retreat from Burgos to Salamanca, thence to the combat and passage of the Hembra, Torrequemada, and outposts to Cuidad Rodrigo, upon the advance of the army in 1813 ; the passage of the Ebro and Esla, action at Osma and battle of Vittoria, actions at Villa Franca, Tolosa ; to the siege and capture of St. Sebastian, the actions on crossing the Biadossa and carrying the enemy's fortified entrenchments.

The battle of Neville and the actions at St. Jean de Luz, at Angelet, the Mayor's House, Bidart, the battles on the Nive from the 9th to the 13th of

December, 1813, the passage of the Adour, and investment of Bayonne, occupation of Bordeaux, the passage of the Garonne, and affairs at Etaiers at the passage of the Drodogne, besides various skirmishes and minor affairs.

In 1815 the battles of Quatre Bras and Waterloo, to the advance on and capture of Paris. He had received the Cross of Knight of the Royal Hanoverian Guelphic Order, that of San Fernando of Spain, the War Medal and four clasps for the battles of Salamanca, Vittoria, Neville and Nive, and the Waterloo medal.

He received the commands of Her Majesty to attend and bear the great Banner on the occasion of the funeral of the late illustrious Duke of Wellington in consideration of his long, faithful and distinguished services.

He was appointed the Colonel of the 5th Light Dragoons, of the Lancers on February the 23rd, 1858, and was transferred to the Colonelcy of the 4th Dragoon Guards on November the 22nd, 1868. He was promoted to Major-General in 1854, Lieut.-General in 1859 and General in 1866.

LIEUTENANT-GENERAL THE HON. CHARLES WEMYSS THESIGER.

General Thesiger was appointed a Cornet in the 5th Madras Light Cavalry on the 27th of February, 1849. He was transferred to the 14th Light Dragoons in August, 1853, and was promoted Lieutenant in December of the same year. In June, 1857 he was transferred to the 6th Dragoons, and in July, 1858 was promoted Captain.

The year 1861 saw him a Major, and from December, 1868 to January, 1878 he commanded the Inniskillings, being promoted a Colonel in December, 1873. He was Major-General on the 7th of October, 1884, and Lieutenant-General in April, 1891.

From April, 1885 to March, 1890, General Thesiger was commanding the Curragh Brigade, and was Inspector-General of Cavalry in Ireland.

He was appointed to the head of the 5th Lancers on the 24th of January, 1892 which appointment he held until transferred to the Colonelcy of his old Regiment, the 14th Hussars, in 1896.

His war services were in China in 1860, when he was present at the capture of Pekin.

LIEUTENANT-GENERAL WILLIAM GODFREY DUNHAM MASSY C.B.
5TH LANCERS.

Born in 1839, General Massy was the eldest son of Major H.W. Massy of Grantstown, Tipperary. He was educated at a private school and Dublin University, and entered the army as an Ensign in October, 1854.

The following year he proceeded to the Crimea and took part in the later stages of the siege of Sebastopol, and was under fire at the battle of Tchernaya, where his gallantry attracted some notice. In the assault on the Redan, on the 8th of September, 1855, although wounded in several places, he led the Grenadiers of the 19th Regiment, and was struck in the left thigh with a ball which shattered the bone. When the assault fell back Lieutenant Massy found himself amongst the Russians, but he was thought to be dying and left on the field instead of being taken prisoner, and was subsequently rescued by the British.

His case was considered hopeless, and for six months he lay on a camp stretcher, but his fine constitution and indomitable pluck carried him through.

The Commander-in-Chief brought his gallantry to notice, and expressed great praise for his conduct and the fortitude with which he bore his sufferings. At the time "Redan" Massy was probably the best known officer in the British Army. He was promoted captain in 1857, received the medal with clasp, Knight of the Legion of Honour, and the Turkish medal.

He later became associated with the 4th Dragoon Guards and the Military Train, and was transferred to the 5th Lancers on the raising of the Regiment in 1858.

On the Regiment going to India he became an A.A.G. in that country, following which he commanded the 5th Lancers from 1871 to 1879.

He commanded a cavalry brigade in the Afghan War of 1879-80, and with his brigade took part in the action of Charasia and at the siege of Sherpore. After the occupation of Kabul, Massy's brigade was employed in the operations against Mahomed Jan, and his brigade of 9th Lancers, Bengal Cavalry and four horse artillery guns, fought a difficult action at Killa Kaza. For his services he was twice mentioned in despatches, and obtained the medal with two clasps.

In 1886 came promotion to Major-General, and in 1893 followed that to Lieutenant-General.

From 1888 to 1893 he commanded the troops in Ceylon, and it was during that period that he received the reward for distinguished and meritorious services. His work in India is memoralised at Rawal Pindi by a fine archway, which stands at the entrance of the bazaar, and also by a spacious market, built by Sirdar Surjan Singh at a cost of £20,000.

In 1894 Lieutenant-General Massy was appointed Colonel of the 4th Dragoon Guards, and in 1896 transferred to the Colonelcy of his old Regiment the 5th Lancers.

Lieutenant-General Massy died at his residence at Grantstown Hall, Tipperary, on the 21st of September 1906.

MAJOR-GENERAL THOMAS ARTHUR COOKE, C.V.O.

Major-General Cooke was born in 1842. His first services in the Army were with the 5th Foot, (Northumberland Fusiliers) which Regiment he joined in May, 1862 as Ensign, and was promoted Lieutenant three years later.

In 1866 he joined the 17th Lancers as Lieutenant, and was promoted Captain in 1870, Major in 1881, and Lieutenant-Colonel in 1882. He commanded the 17th Lancers from 1886 to 1888. He served with the 17th Lancers in the Zulu War of 1879, and was present at the battle of Ulundi and minor operations, for which he received the South African medal and clasp.

He was a Colonel in 1886 and promoted Major-General on the 23rd of May, 1898.

Major-General Cooke served on the District and Head Quarter Staff in India from 1888 to 1894. He was Sir George Luck's Chief Staff Officer at the Cavalry Camp at Alighar, India, in 1892, and commanded a Cavalry Division during the latter portion of the manœuvres. From 1894 to 1899 he was General Officer Commanding Sind District, and from May to October 1900 he commanded the troops on Salisbury Plain. He commanded the Colonial Troops in England during the Coronation Celebrations in 1902.

Major-General Cooke was appointed a C.V.O. in 1902, and on the 21st of September, 1906, was appointed to the Colonelcy of the 5th Royal Irish Lancers.

COLONEL JOHN JAMES SCOTT CHISHOLME.

John James Scott Chisholme, of the ancient Border family of Chisholmes of Stirches in Hawick, was born at Stirches on August the 1st, 1851.

His earlier services were with the 9th Lancers, with which Regiment he served in the Afghan War 1879-80. On December the 11th, 1879, two squadrons of the 9th Lancers, with the 14th Bengal Lancers and four R. H. A. guns, under the command of Brigadier-General Dunham Massy had been severely engaged with a large force of Afghans under Mahomed Jan near Kila Kazi. On the 13th, while attending the burial of those killed in the action, the 9th Lancers were suddenly ordered to turn out. With the 5th Punjab Cavalry and some of the 14th Bengal Lancers and Guides Cavalry, the 9th fell in with the enemy near the Siah Sing heights, and catching them on both flanks scattered them over the plain. In the charge of the 9th Lancers the Commanding Officer, Captain Butson, was killed, and Captain Scott Chisholme was shot through the thigh, the flash of the rifle burning his clothes, so close was the discharge. In spite of his very severe wound Scott Chisholme remained in the saddle and brought the Regiment out of action. For his gallant conduct he was promoted brevet-major.

For some years Scott Chisholme was Military Secretary to the Governor of Madras, and on May the 8th, 1889, he exchanged into the 5th Lancers. In August, 1894, he was appointed to the command of the 5th Lancers, and took his Regiment from India to Natal in 1898. After being granted an extension of a year in command, he left the Regiment in August, 1899.

On the first murmurings of the South African War in August, 1899, the town of Pietermaritzburg in Natal was alive with all sorts of rumours as to a private corps, said to be mounted infantry, being formed at the Agricultural Show yard. The rumour, however, gave place to nothing definite. So well was the secret kept that numbers of small parties of men arrived at the station to join the corps from all parts of South Africa quite unnoticed. The movement had commenced in Johannesburg where a large number of men had expressed a wish to fight for the Mother Country in the event of war. The formation of a mounted corps of Irregulars out of the loyalists on the Rand and elsewhere, was suggested, and some of the leading inhabitants selected the officers, and the officers brought the men they knew. The offer was made to the Imperial Government, and until the result was known the matter had necessarily to be kept secret. During this time of suspense, however, the originators of the movement were not idle. They and their friends supplied the funds, and the men were engaged and horses bought.

In September the assent of the Imperial Government was received, and authority granted for the formation of a corps of 500 mounted riflemen, for service in the event of hostilities. Colonel Scott Chisholme was given the task of organising and commanding the Corps, and Her Majesty the Queen granted to it the title of "The Imperial Light Horse."

In the Show yard of the Natal capital Scott Chisholme soon got his corps organised into a useful regiment of mounted riflemen, and on the outburst of hostilities the I.L.H. proceeded North with the Regular troops in the country; in its first action on the Elandslaagte hill the Regiment lost its gallant Colonel in the moment of victory. Twice wounded, he still struggled on at their head until a third bullet gave him the death he always desired. His example of courage and devotedness to duty was never forgotten by his Light Horsemen, whose doings during the long war were ever in keeping with their early training under their first Colonel.

The ideal of a "beau-sabreur," a splendid horseman, one of the best Colonels who ever commanded a cavalry regiment, and a sympathetic friend, he was beloved by the officers and men of his Regiment. The memory of "Jabber" Chisholme will live long in the 5th Lancers.

A monument has been erected on the spot where he fell on the Elandslaagte hill by his comrades in the 5th Lancers and the Imperial Light Horse, and another in his native place of Stirches at Wilton Parish Church.

He was the last in the male line of the Chisholmes of Stirches.

SUCCESSIONS.

SUCCESSIONS OF COLONELS.

James Wynne	25th December 1688
Charles Ross	16th July 1695
Thomas Sydney	8th October 1715
Charles Ross	1st February 1729
Owen Wynne	6th August 1732
Richard, Viscount Molesworth,	17th June 1737
John Mostyn	18th October 1756
The Hon. Joseph Yorke, K.B.	27th November 1760
Robert Cuninghame, Lord Rossmore,	4th April 1787
Sir James Charles Chatterton Bart, K.H.	23rd February 1858
Edward Pole	22nd November 1868
Henry Darby Griffith, C.B.	1st January 1872
The Hon. Somerset John Gough Calthorpe,	18th November 1887
The Hon. Charles Wemyss Thesiger,	24th January 1892
William Godfrey Dunham Massy, C.B.	4th October 1896
Thomas Arthur Cooke, C.V.O.	21st September 1906

SUCCESSION OF COMMANDING OFFICERS.

Lieut.-Colonel	Charles Ross	Before 1695
"	Owen Wynne	20th July 1695
"	Robert Hunter	1705
Brevet-Colonel	Hugh Caldwell	1707
Brevet-Colonel	Jno. Hill	September 1709
Lieut.-Colonel	Ric. Gore	1st November 1711
"	Thomas Sidney	Before 5th October 1715
"	Wriothy Betton	23rd September 1719
"	Alex Rose	12th September 1729
"	William Cope	10th May 1740
"	Christopher Clarges	20th February 1749
"	William Hill	25th August 1760
"	Hugh Cane	31st March 1768
"	James Stewart	7th January 1778
"	The Hon. Chas. Wm Stewart	1st January 1797

Lieut.-Colonel	Alex J. Goldie	17th February 1798
„	Geo. Aug. Filmer Sulivan	19th February 1858
„	Robert Portal	12th March 1861
Colonel	W. Godfrey Dunham Massy	31st October 1871
„	William Lloyd Browne	5th March 1879
„	W. Ward Bennitt	29th July 1885
„	Alfred Bissel Harvey	31st March 1889
Lieut.-Colonel	Cecil F. Johnstone-Douglas	1st January 1893
Colonel	John James Scott Chisholme	12th August 1894
Lieut.-Colonel	Jas. F. Malcolm Fawcett	12th August 1899
Colonel	Henry Jenner Scobell	27th July 1901
„	Edmund Henry Hynman Allenby, C.B.	2nd August 1902
Lieut.-Colonel	Herman W. Gore Graham, D.S.O.	19th October 1905

SUCCESSION OF ADJUTANTS.

	Griffith Lloyd	1st August 1694
	Mat. Watts	20th June 1696
	David Ross	August 1703
Cornet	William Ross	23rd August 1707
Lieutenant	Wm. Ross	13th June 1752
„	James Ross	9th April 1756
„	Rich. Wolfe	7th May 1757
Cornet	Thomas Bowater	18th October 1764
„	Rich. Wolfe	16th May 1766
„	Richard Vyse	18th March 1767
„	William Broom	28th November 1771
„	Thos. Tickell	17th January 1774
„	Wynne Fawcett	5th March 1783
„	George Broome	31st October 1792
„	John Taylor	12th November 1794
Lieutenant	Edward Fraser Weaver	5th March 1858
„	Frederick Sedley	28th July 1863
„	J. Dennis	25th June 1869
„	Cyril Wm. Bowdler Bell	10th July 1871
„	Thomas Fletcher	29th November 1876
„	Julian J.N. Spicer	21st June 1879
„	John Sinclair	9th May 1883
„	Malcolm McNeill	26th May 1884
„	John Robert Harvey	31st October 1888
„	Henry V. Platt	24th October 1889
Captain	Henry Vincent Bailey	9th May 1892
Lieutenant	Harold Hatton Hulse	9th May 1896

Capt.&Brevet Major	R.C. Browne-Clayton	9th May 1900
Captain	Walter Temple Willcox	12th September 1903
Lieutenant	Charles J.A. Maberly	12th September 1906

Note. — The above list is incomplete as in many of the earlier Army Lists the Adjutants are not given.

DRILL.

The exercises and drill of Dragoons at the time the Regiment was raised are interesting.

When mounted Dragoons were exercised as Horse, each squadron was drawn up in three ranks with the Officers immediately in front of their men.

The distances and intervals in the cavalry exercises were only three ; the Open Order, which was six feet between ranks or files ; the Close Order, which was three feet between ranks or files ; and the Close Order from Close Order (i. e. doubly close) which was head to crupper or knee to knee.

When dismounted, they were formed and exercised as Foot, so far as the evolutions were concerned ; for the rest the following were the words of command, the men being mounted, to act dismounted :—

Dragoons have a care (take heed.)
Sling your muskets.
Make ready your links.
Clear your right foot of your stirrup.
Dismount and stand at your horses' heads.

(The six outside men remained mounted to take charge of the horses.)

Link your horses to the left.
March clear of your horses and shoulder as you march.
Halt.

(The Dragoons were then formed up in the same way as a foot regiment.)

Have a care of the exercise.
Officers to the right-about.
Take your posts in rear.
March.
Dragoons have a care. (the men pull off their right hand gloves and stow them under their waist belts.)
Lay your right hand on your musket.
Poise your musket.
Rest your musket.
Cock and guard.
Present; fire.
Recover your arms with the cock half bent (i.e. half-cock).

Rest upon your musket.
Handle your daggers (i.e. bayonets).
Fix them in the muzzle of your muskets.
Poise your muskets.
Charge to the front.
To the right (left, right-about, left-about) Charge.
Recover your arms.
Rest upon your muskets.
Handle your bayonets.
Withdraw your bayonets.
Place (i.e. return) *your bayonets.*
Poise your muskets.
Rest your muskets.
Clean the pan (with the ball of the thumb).
Open your cartridge-box.
Handle your primer.
Sink and prime.
Return your primer.
Shut your pan (with your fore-fingers).
Blow off your loose corns (recovering arms at the same time).
Cast about to charge.
Handle your cartridge.
Take out your cartridge (and shut the box).
Open it with your teeth.
Charge with powder and ball.
Draw forth your scourers (i.e. ram-rods).
Shorten them to an inch (against your right breasts).
Put them into the muzzle of your muskets.
Ram down powder and ball.
Withdraw your scourers.
Shorten them to an inch (as before).
Place (i.e. return) *your scourers.*
Poise your muskets.
Shoulder your muskets.
Poise your muskets.
Rest your muskets.
Lay down your arms.
Quit your arms.
To the right about.
March clear of your arms and break.

(The men being thus dispersed, the drum beat and the men drawing their swords run to their arms " with a Huzza. ")

Return your swords.
Handle your arms.
Rest your arms.
Poise your muskets.

Sling your muskets.
To the right-about.
March to your horses.
Unlink your horses.
Shorten your bridles.
Put your left foot in the stirrup.
Mount.
Fasten your links.
Unstring and advance your muskets (on the right thigh).
Join your left hands to your muskets.
Cock and guard.
Rest your muskets on your bridle-hands.
Present; Fire.
Recover your arms with the cock half-bent.

Of individual or setting up drill there was, apparently, very little; the instruction of the recruit being limited to making him hold his head up, to "look lively," and not to swing his arms. In marching the men were to step off with the left foot, and to set their "feet down altogether, so that they may be heard," and were "to march very slowly."

POLO TOURNAMENTS WON BY THE 5th LANCERS

Inter-Regimental, Hurlingham.

1878 and 1879.

Team.

Capt. S. S. Benyon.
" E. G. Paley.
Lieut. J. Spicer.
2nd. Lt. A. C. Little.
" L. H. Jones.

1882.

Team.

Capt. Spicer.
" Tufton.
Lieut. Jones.
" Little.

1887.

Team.

Capt. Spicer.
Major Little.
Capt. Jones.
Lieut. Mundy.

All-Ireland Tournament.

1881.

Team.

Capt. Spicer.
" Tufton.
Lieut. Jones.
" Combe.

All-Ireland Tournament.

1884.

Team.

Capt. Spicer.
" Tufton.
Lieut. Jones.
" Combe.

Indian Inter-Regimental.

1890.

Team.

Lieut. Daniel.
Capt. Beddy.
Lieut. Collis.
" Bailey.

Beresford Cup, Johannesburg.

1899.

Team.

Lieut. Hill.
" Jardine.
" Hulse.
" Browne-Clayton.

Ladysmith Tournament.

1899.

Team.

Lieut. Fraser.
" Jardine.
" Hill.
Col. Scott Chisholme.

REGIMENTAL INSTITUTIONS.

1. The Regimental Dinner. For the Past and Present Officers. Held annually in London in June. Agents, Messrs Cox and Co. 16 Charing Cross.

2. The Old Comrades' Dinner, for Past Officers, Non-Commissioned Officers and Men, and Present Officers and Sergeants. Held annually in London in October. All members dining pay a small subscription for the dinner, and the balance required is met by subscriptions from Past and Present Officers. The Adjutant, 5th Lancers, assisted by a committee of Past and Present Non-Commissioned officers arranges the Dinner.

In addition to the above are the Children's Christmas Tree, Summer Outing, and other minor institutions which are supported by the Regiment.

LIST OF OFFICERS
WHO HAVE SERVED IN THE
REGIMENT

LIST OF OFFICERS WHO HAVE SERVED IN THE REGIMENT

NAME	DATE OF JOINING AND RANK	OUT OF REGIMENT AND RANK	WAR SERVICES AND REMARKS
ABERCROMBIE, Alex	1694. Quarter Master.	No trace after 1709. Lieut.	Flanders 1694-97. Blenheim, Ramillies, Oudenarde, Malplaquet.
ABERCROMBIE, John	13th April 1782.	1787. Cornet.	
ADAM, William Augustus	12th Sept. 1894. Captain.	Still serving. Major.	S. Africa 1899-02. Defence of Ladysmith.
ADDY, John	8th March 1858 Quarter Master.	1860. Quarter Master.	Crimea, Balaclava, Inkerman, Sebastopol.
AKERMAN, Hercules	17th April 1867. Lieut.	1870. Lieut.	
ALLEN, James	5th Oct. 1776. Captain.	1791. Major.	
ALLENBY, Edmund Henry Hynman	2nd Aug. 1902. Lt. Col. Commanding (B. Col.)	1905. Lt. Col. Commanding (Colonel) to command 4th Cavalry Brigade.	Bechuanaland 1884-85. Zululand 1888. S. Africa 1899-02. Relief of Kimberley, Paardeburg, Diamond Hill, etc. Commanded a column. Brevets of Lt. Col. and Colonel C.B.
ANDERSON, John	16th Dec. 1775. Cornet.	1778. Cornet.	

NAME	DATE OF JOINING AND RANK	OUT OF REGIMENT AND RANK	WAR SERVICES AND REMARKS
ANDREW, Henry Patrick	7th April 1863. Cornet.	1868. Cornet.	
ANDREWS, Rob. Fleetwood	12th May 1863. Surgeon	1873. Surgeon. Major.	
ANKETELL-JONES, John H.	14th March 1900. 2nd Lieut.	1903. Lieutenant.	S. Africa 1900-02. Wounded.
ARCHER, Nicholas	3rd May 1786. Surgeon.	1787. Surgeon.	
ARKWRIGHT, Cyril	10th Oct. 1894. 2nd Lieut.	1900. Lieutenant.	S. Africa 1899-00. Died in Defence of Ladysmith.
ARKWRIGHT, Esme F.W.	29th May 1901. Lieut.	1902. Lieutenant.	S. Africa 1901-02.
ARON, Eugene, F.S.	23rd Dec. 1893. 2nd Lieut.	1894. 2nd Lieut.	
ATKINSON, Frank Buddle	9th May 1885. Lieut.	1888. Lieutenant.	
ATKINSON, John	28th May 1858. Asst. Surgeon.	1877. Surgeon. Major.	
AYLMER, Gerald	28th May 1794. Cornet.	1799. Cornet.	Disbandment of 5th Dragoons.
AYRTON, Freder.	2nd Aug. 1882. 2nd Lieut.	1891. Captain.	Nile Expedition 1885.
BABE, Simon (Samuel)	1705 Chaplain.	No record after 1709.	Blenheim, Ramillies, Oudenarde, Malplaquet.
BAILEY, Henry Vincent	10th Oct. 1888. Lieut.	1903. Major.	S. Africa 1900.
BAINES, Charles Harman	15th Sept. 1869. Cornet.	1873. Lieutenant.	

NAME	DATE OF JOINING AND RANK	OUT OF REGIMENT AND RANK	WAR SERVICES AND REMARKS
BALDWIN G.C. Kynnersley	28th July 1863. Cornet.	1868. Cornet.	
BALDWIN, Thos.	17th Nov. 1721. Cornet.	1740. Cornet.	
BALFOUR, Jeremiah	1708. Quarter Master.	Still a Cornet. 1730.	Malplaquet.
BALL, William	8th Jan. 1740. Lieutenant.	No further trace.	
BAMFORD, John	1st July 1798. Cornet.	1799. Cornet.	Disbandment of 5th Dragoons.
BARKER, Robert	24th Feb. 1708. Cornet.	No trace after 1709.	Malplaquet.
BARROW, Luke	19th Feb. 1858. Surgeon.	1858. Surgeon.	M.D.
BEATTY, Charles	20th July 1695. Lieutenant.	1710. Captain-Lieutenant.	Flanders 1695-7. Blenheim, Malplaquet.
BEAUMONT, Montmorency	23rd April 1873. Sub. Lieut.	1875. Sub. Lieut.	
BEDDY, Francis L.	6th Feb. 1889. Captain.	1890. Captain.	
BEDFORD, Charles	17th March 1863. Qr. Master.	1865. Qr. Mr.	Died.
BELL, C.W. Bowdler	23rd Dec. 1864. Cornet.	1876. Lieut.	
BELL, Henry Urmston	21st Sept. 1892. 2nd Lieut.	1896. Lieut.	
BELLINGHAM, Henry	14th Feb. 1765. Cornet.	1773. Lieut.	
BENNITT, William Ward	29th July 1885. Lieut.-Col. Comm.	1889. Lieut.-Col. Commandg (Col.)	South Africa 1881, Transvaal Campgn.

NAME	DATE OF JOINING AND RANK	OUT OF REGIMENT AND RANK	WAR SERVICES AND REMARKS
BENYON, Joseph Spencer	31st Oct. 1871. Captain.	1884. Major.	
BERNARD, North Ludlow	1st May 1724. Lieutenant.	1754. Major.	Died.
BERRYMAN, John	19th May 1880. Quarter Master.	1883. Qr. Mr.	Crimea 1854-5, Alma, Balaclava, (Light Brigade), Inkerman, Tchernaya, Sebastopol. V.C. Indian Mutiny '58. Zulu War '79.
BETTON, Wriothy	23rd Sept. 1719. Lieut. Col.	1730. Lieut. Col.	
BIGGS, Jno.	5th May 1863. Paymaster.	1868. Paymaster.	
BINGHAM, The Hon. Arthur Maurice	18th Oct. 1899. 2nd Lieutenant.	Still serving Captain.	S. Africa 1900-02. Relief of Ladysmith.
BIRCH, Rich. Jac. Wyrley	17th April 1858. Cornet.	1863. Lieut.	
BLAKE, Dennis	14th Jan. 1775. Cornet.	1777. Cornet.	
BLAKE, Martin Pierce	23rd March 1860. Cornet.	1877. Major.	
BLAKE, Walter	14th Nov. 1775. Cornet.	1781. Lieut.	
BLAKENEY, William	10th July 1722. Cornet.	No trace after 1736. Cornet.	
BLIGH, Robert	28th Feb. 1787. Cornet.	1791. Lieut.	
BOISRAGON, Danl. Chevalau de	1695. Lieut.	Before 1709. Lieut.	Flanders 1695-7. Blenheim.

NAME	DATE OF JOINING AND RANK	OUT OF REGIMENT AND RANK	WAR SERVICES AND REMARKS
BOGGES (or BOGGEST, Fraser.	27th Jan. 1707. Cornet.	(?) 1711 Capt. serving in Evans' Dragoons 1728.	Of Halwey, Suffolk. Malplaquet.
BOLTON, Jno.	14th March 1772. Cornet.	1789. Captain.	
BOSCAWEN, Geo. Evelyn	4th Nov. 1877. Captain.	No further trace.	
BOWATER, Thos.	31st Dec. 1759. Cornet.	1776. Lieut.	
BOYCE, Hugh W.	22nd Nov. 1888. Captain.	1889. Captain.	
BOYD, Edward	4th Sept. 1860. Captain.	1864. Captain.	
BRADY-BROWNE, T.H.	26th March 1902. 2nd Lieut.	1903. 2nd Lieut.	
BRIGHT, Geo. Money	17th March 1858. Lieut.	1860. Lieut.	
BROCKLEHURST, Henry Dent	29th Nov. 1876. Lieut.	1880. Lieut.	
BROOM, Wm.	13th Feb. 1762. Cornet.	1776. Lieut.	
BROOME, George	31st May 1792. Cornet.	1799. Cornet.	Disbandment of 5th Dragoons.
BROWN, Amyatt, E.	13th May 1859. Captain.	1861. Capt.	
BROWN, the Hon: Henry	20th Jan. 1764. Captain.	1772. Capt.	
BROWE —	March 1705. Qr. Master.	No further trace.	Blenheim.
BROWNE, Dennis	27th Oct. 1774. Cornet.	1784. Cornet.	

NAME	DATE OF JOINING AND RANK	OUT OF REGIMENT AND RANK	WAR SERVICES AND REMARKS
BROWNE, Redmond	31st March 1793. Major.	1799. Major.	Disbandment of 5th Dragoons.
BROWNE, Wm. Lloyd	10th July 1860. Lieut.	1884 Lt.Col. Commanding (Colonel).	
BROWNLOW, Wm.	28th March 1775. Cornet.	1777. Cornet.	
BROWNE-CLAYTON, Robert C.	24th Dec. 1890. 2nd Lieut.	Still serving. Capt. and Brevet Major.	South Africa 1899-1902. Elandslaagte, Defence of Ladysmith, Belfast. Brevet of Major.
BRUCE, John	5st Aug. 1896. 2nd Lieut.	1898. 2nd Lieut.	
BULLOCK, Jno.	15th Oct. 1757. Cornet.	1761. Cornet.	
BURLTON, Ferdinand	16th Dec. 1775. Cornet.	1779. Cornet.	
BURLTON, Jno. Phillip	5th June 1771. Cornet.	1771. Cornet.	
BURROWS, Robert	1st Augt. 1720. Cornet.	No trace after 1730. Cornet.	
BUTLER, David Gerald	9th Oct. 1860. Cornet.	1871. Cornet.	
BUTLER, The Hon: Henry	30th Nov. 1791. Cornet.	1793. Cornet.	
BUTLER, The Hon: Sam. Rich.	31st Augt. 1793. Cornet.	1747. Lieut.	
BUTTER, John	23rd Jan. 1746. Captain.	Later a Major in 3rd Horse Carabineers.	

NAME	DATE OF JOINING AND RANK	OUT OF REGIMENT AND RANK	WAR SERVICES AND REMARKS
CAILLARD, W. Maurice C. du Q.	10th Oct. 1894. 2nd Lieut.	1898. Lieut.	S. Africa 1900-01. Diamond Hill, Belfast. Johannesburg.
CALDWELL, Hugh	9th May 1690. Captain.	1710. Lieut. Col. Commanding.	Defended Donegal Castle against 2000 Dragoons under Duke of Berwick in May 1689. Flanders 1694-7. Blenheim, (wounded) Malplaquet. Killed at siege of Douay 1710. Son of Sir Jas. Caldwell of Castle Caldwell, County Fermanagh.
CALTHORPE, The Hon: Somerset John Gough	18th Nov. 1887. Colonel (Major General).	1892. Colonel. Transferred to 5th Dragoon Guards.	Crimea 1854-5. Alma, Balaclava, Inkerman. Sebastopol.
CANE, Hugh	12th March 1754. Captain.	1777. Lieut. Col. Commanding.	
CARANDINI, Frank J.	1st March 1893. Major.	1895. Major.	Afghan War 1879-80.
CARDEN, Sir Frederick Walter	26th March 1858. Lieut.	1870. Major.	
CARDEN, John	9th May 1794. Cornet.	1799. Lieut.	Disbandment of 5th Dragoons.
CARMICHAEL, Ludovic Montefiore.	15th Nov. 1861. Cornet.	1885. Major.	Nile Expedition 1884-5. Killed Abu Klea Wells.
CARTER, Henry Boyle	2nd May 1742. Cornet.	No trace after 1745. Cornet.	

NAME	DATE OF JOINING AND RANK	OUT OF REGIMENT AND RANK	WAR SERVICES AND REMARKS
CARTER, John	22nd Aug. 1794. Cornet.	1799. Captain.	Disbandment of 5th Dragoons.
CARTER, Thomas	27th June 1739. Cornet.	No further trace.	
CAULFIELD, Rich.	24th Feb. 1711. Lieut.	No trace after 1730. Lieut.	
CAULFIELD, Wm.	17th Dec. 1757. Cornet.	1772. Lieut.	
CHAFFEY, John	21st May 1858. Cornet.	1877. Major.	
CHAIGNEAU, Jno. Clement	12th March 1774. Chaplain.	1777. Chaplain.	
CHANCE, Oswald Kesteven	18th Oct. 1899. 2nd Lieut.	Still serving. Captain.	S. Africa 1899-02. Relief of Ladysmith
CHAMBERS, Richard	20th July 1867. Lieut.	1875. Lieut.	
CHAPMAN, Art. Gerard	11th Sept. 1876. Sub-Lieut.	1880. Lieut.	
CHATTERON, Sir James Charles Baronet K.H.	23rd Feb. 1858. Colonel (Major General.)	1868. Colonel (General) Transferred to 4th D.G.	Portugal, Spain and France 1811-18, and Waterloo.
CHICHESTER, Cornwallis H.	22nd Feb. 1864. Captain.	1885. Lieut. Colonel.	Crimea 1850. Tchernaya, Sebastopol.
CHISHOLME, John James Scott.	8th May 1889. Major.	1899. Lieut-Col. Commanding.	Afghanistan 1879-80. Brevet of Major. South Africa 1899. Killed at Elandslaagte in command of the Imperial Light Horse. The last in the male line of the Chisholmes of Stirches.

NAME	DATE OF JOINING AND RANK	OUT OF REGIMENT AND RANK	WAR SERVICES AND REMARKS
CLARGES, Christopher	15th April 1749. ? Major.	1760. Lieut-Colonel. Commanding.	
CLARGES, George	22nd June 1757. Captain.	Before 1765. Captain.	
CLARGES, Gould	8th Sept. 1725. Cornet.	No trace after 1730. Cornet.	
CLERKE, S. Wm.	25th Oct. 1715. Cornet.	No trace after 1730. Cornet.	
COCKBURNE, George	30th Nov. 1789. Captain.	1791. Captain.	
COCKING, Ralph	21st Oct. 1729. Chaplain.	1776. Chaplain.	
COCKSEDGE, Wm.	1694. Surgeon.	No trace after 1709. Cornet.	Resigned appointment of Surgeon and appointed Cornet. Blenheim, Ramillies, Oudenaerde, Malplaquet.
COLLIS, Wilfred E. Russell	12th Nov. 1884. Lieut.	1903. Major.	S. Africa 1900-02. Relief of Ladysmith. Belfast.
COMBE, Christian	5th Oct. 1878. 2nd Lieut.	1880. Lieut.	
CONGREVE, Wm.	5th April 1724. Lieut.	No trace after 1730. Lieut.	
CONINGHAM, Henry	1st Dec. 1725. Captain.	No trace after 1736. Captain.	
CONINGHAM, Sim. L.	1786. Lieut.	1792. Capt.-Lieut.	
CONWAY, Henry	27th July 1737. Lieut.	No further trace. Lieut.	
COOKE, Richard	17th Jan. 1774. Cornet.	1796 Captain.	

NAME	DATE OF JOINING AND RANK	OUT OF REGIMENT AND RANK	WAR SERVICES AND REMARKS
COOKE, Thomas Arthur	21st Sept. 1906. Colonel (Major General).	Still at the head of the Regiment.	Zulu War 1879. Ulundi. C.V.O.
COOPER, Henry A.	21st Aug. 1901. Lieut.	Still serving Lieut.	S. Africa 1901-02.
COPE, Anthony	1st Feb. 1714. Cornet.	1741. Captain.	
COPE, William	1st Feb. 1714. Captain.	1742. Lieut. Colonel. Commanding.	
COPPINGER, Robert.	9th July 1745. Cornet.	1752. Lieut.	Cashiered.
CORNWALLIS—	1696. Captain.	No other trace.	Flanders 1696.
CORR, Francis	29th May 1796. Cornet.	1799. Cornet.	Disbandment of 5th Dragoons.
COSTELLO, H.	24th Jan. 1883. Lieut.	1885. Lieut.	Nile Expedition 1884-5. Wounded Abu Klea Wells. Died of enteric fever at Abu Fatmeh 27th May.
COTTON, Geo. W. Vernon	22nd May 1858. Cornet.	1860. Cornet.	
COURTENAY, Joseph	1799. Cornet.	1799. Cornet.	Disbandment of 5th Dragoons.
COWAN, Joseph Henry	17th March 1858. Lieut.	1860. Lieut.	Died.
COX —	29th Feb. 1796. Cornet.	1797. Cornet.	
CRAMER, Marmaduke	1799. Lieut.	1799. Lieut.	
CREIGHTON, The Hon. Abra^m^.	31st Dec. 1782. Cornet.	1784. Cornet.	Disbandment of 5th Dragoons.

NAME	DATE OF JOINING AND RANK	OUT OF REGIMENT AND RANK	WAR SERVICES AND REMARKS
CREIGHTON, John	9th Feb. 1750. Cornet.	Before 1767. Captain.	
CREIGHTON, The Hon. John	17th April 1784. Cornet.	1791. Cornet.	
CROFTON, Edward Hugh R.	5th Jan. 1876. Lieut.	1877. Lieut.	
CROMMELIN, Alec.	28th Aug. 1753. Surgeon.	1761. Surgeon.	
CROWDY, John	3rd Feb. 1875. Riding-Master.	1880. Rdg. Master.	
CUNARD, Alick	26th Aug. 1905. Lieut.	Still serving Lieut.	
CUNINGHAME, Robert	See Rossmore.		
CURTIS, Robt. Foulkes	25th June 1784. Cornet.	1786. Cornet.	
CURTIS, William	1st Oct. 1784. Lieut.	1787. Lieut.	
DANIEL, Arthur William	14th May 1884. Lieut.	1894. Captain.	
D'ARCEY, William	29th Feb. 1796. Cornet.	1799. Captain.	Disbandment of 5th Dragoons.
DARLEY, John E.C.	15th Nov. 1899. 2nd Lieut.	1901. Lieut.	S. Africa 1900. Relief of Ladysmith. Belfast. Wounded Cape Colony.
DAVIS, Simon Farthing	28th Nov. 1767. Cornet.	1777. Lieut.	
DAWSON, Henry	29th Feb. 1793. Cornet.	1794. Cornet.	

NAME	DATE OF JOINING AND RANK	OUT OF REGIMENT AND RANK	WAR SERVICES AND REMARKS
DAWSON, Thos.	6th May 1760. Surgeon.	1770. Surgeon.	
DEANE, John	8th March 1780. Cornet.	1789. Lieut.	
DENNIS, James	24th March 1863. Cornet.	1875. Lieut.	
DESPARD Rich.	20th Sept. 1789. Chaplain.	1792. Chaplain.	
DE SATGE DE THOREN, Baron L.A.	27th Sept. 1905. 2nd Lieut.	Still serving Lieut.	
DILKES, —	1760. Cornet.	1760. Cornet.	Irish War 1689-91. Flanders 1694-7. Blenheim, Ramillies, Oudenarde, Malplaquet.
DODWELL, Michael	1st Oct. 1797. Cornet.	1798. Cornet.	
DOYNE, Mordaunt B.	30th Jan. 1884. Lieut.	1797. Captain.	Suakin 1855.
DRURY (Drewry), Robert	20th July 1689. Lieut.	No trace after 1711. Brevet Maj.	
DRURY, Robert	28th May 1794. Cornet.	1796. Cornet.	
DUGDALE, Frederick Brooks	18th Oct. 1899. 2nd Lieut.	1902. Lieut.	S. Africa 1900-02. Relief of Ladysmith Belfast. V.C. Killed hunting 1902.
DUNBAR, John	1704. Cornet.	17— Capt.-Lieut.	Blenheim. A Major. in Owen Wynne's Dragoons (9th Lcrs) in 1715.

NAME	DATE OF JOINING AND RANK	OUT OF REGIMENT AND RANK	WAR SERVICES AND REMARKS
DUNBAR, Ric.	1704. Qr. Master.	No trace after 1709.	Blenheim, Ramillies, Oudenarde, Malplaquet.
DUNN, Geo. Carr	7th Dec. 1867. Asst. Surgeon.	1875. Asst. Surgeon.	
DYER, Jno. Akin	15th March 1858. Paymaster.	1863. Paymaster.	
DYNON, John	26th Feb. 1858. Captain.	1861. Captain.	
EARLE, Henry H.	12th Oct. 1901. 2nd Lieut.	1903. 2nd Lieut.	South Africa 1901.
EDGEWORTH, William	14th May 1858. Lieut.	1863. Captain.	
EDMONDS, Wm. Blake	3rd Aug. 1870. Vet. Surgeon.	1875. Vet. Surgeon.	
ELLIS, Robert	1799. Lieut.	1799. Lieut.	Disbandment of 5th Dragoons.
ERROLL, George Earl of	26th Aug. 1786. Captain.	1791. Captain.	
ERSKINE, William	14th Nov. 1787. Lieut.	1787. Lieut.	
EVANS, John	1709. Quarter Master.	No futher trace.	Malplaquet.
EWING, Alex.	31st Dec. 1858. Cornet.	1863. Lieut.	
EWING, Alex.	13th Feb. 1863. Cornet.	1864. Cornet.	
FARRER, Jno.	1st Jan. 1723. Lieut.	No trace after 1730. Lieut.	
FAUDEL-PHILLIPS, Henry F.	26th March 1902. 2nd Lieut.	Still serving. Lieut.	

NAME	DATE OF JOINING AND RANK	OUT OF REGIMENT AND RANK	WAR SERVICES AND REMARKS
FAWCETT, Benjamin	31st Dec. 1795. Cornet.	1799. Lieut.	
FAWCETT, Jas. F. Malcolm	1st Dec. 1880. 2nd Lieut.	1901. Lieut-Col. (Commanding.)	S. Africa 1899-01. Defence of Ladysmith.
FAWCETT, (?FANFITT) Wynne	5th March 1783. Adjutant.	1793. Cornet.	
FIFE, Sir George Aubone	23rd Oct. 1875. Lieut.	1879. Lieut.	Canada 1866. Afghan War 1879-1880. Burmah 1886-87.
FIRMAN, Rich. Flood	6th Nov. 1772. Cornet.	1785. Lieut.	
FITZGERALD, G.	5th Oct. 1892. 2nd Lieut.	1895. 2nd Lieut.	
FITZGERALD, Maurice	31st Oct. 1792. Captain.	1794. Captain.	
FLEMING (Jas.)	29th Sept. 1696. Chaplain.	1704. Chaplain.	Flanders 1696-7.
FLEMING, Thom.	15th Dec. 1771. Cornet.	1788. Capt-Lieut.	
FLETCHER, Richard	1st Oct. 1784. Surgeon.	1786. Surgeon.	
FLETCHER, Thomas	6th Dec. 1873. Riding Master.	1893. Major.	
FOLLIOTT, M. Jno.	6th June 1694. Captain.	1694. Captain.	Flanders 1694-7.
FORSTER, W. Ludlow	31st March 1863. Cornet.	1865. Cornet.	
FORTESCUE, Mathew.	28th Oct. 1737. Cornet.	No trace after 1745. Cornet.	

NAME	DATE OF JOINING AND RANK	OUT OF REGIMENT AND RANK	WAR SERVICES AND REMARKS
FOX, J.G. Hubert	19th Dec. 1860. Cornet.	1868. Lieut.	
FRANKLIN, Terence	27th Aug. 1737. Cornet.	No trace after 1745. Cornet.	
FRASER, Francis	30th Sept. 1787. Surgeon.	1792. Surgeon.	
FRASER, Henry Francis	7th Dec. 1895. 2nd Lieut.	1901. Lieut.	S. Africa 1899-00. Elandslaagte, Defence of Ladysmith. West Africa. Aro Ex. 1901-2. Wounded.
FREER, G.	9th Sept. 1882. Lieut.	1883. Lieut.	
FRENCH, Arthur	24th Feb. 1775. Cornet.	1778. Cornet.	
FRENCH, Henry	21st Nov. 1747. Cornet.	1753. Cornet.	
FRENCH, John	15th March 1768. Cornet.	1778. Capt-Lieut.	
FULLERTON, Grey D'E. H.	10th Nov. 1888. 2nd Lieut.	1890. 2nd Lieut.	
GALBRAITH, Arthur	1695. Lieutenant.	No trace after 1702.	Flanders 1695-7.
GALBRAITH, Hugh	1689. Captain.	1698. Major.	Irish War 1689-91 Flanders 1694-7. Of the Tyrone Family of Galbraith
GAMWELL, Andrew	26th Feb. 1858. Captain.	1860. Captain.	Crimea. Sebastopol.
GARBETT, C.H. Vincent	28th Oct. 1871. Lieut.	1875. Lieut.	

NAME	DATE OF JOINING AND RANK	OUT OF REGIMENT AND RANK	WAR SERVICES AND REMARKS
GETHIN, Percy	1689. Captain.	Before 1702. Captain.	Irish War 1689-91. Flanders 1694-7. Married Anna, widow of Sir Fras. Gore of Artaman, co. Sligo, & daughter of Robt. Parke of Newtown, co. Leitrim.
GILBORNE, Ed. Chas. Wm.	29th Nov. 1876. Sub-Lieut.	1887. Captain.	Suakin 1885.
GLEADSTEANE, George	1st Jan. 1760. Cornet.	Before 1765.	
GOLDIE, Alexander	6th Feb. 1788. Cornet.	1791. Cornet.	
GOLDIE, Alex. J.	11th May 1791. Lieut.	1799. Lieutenant-Colonel.	Disbandment of 5th Dragoons.
GOLDIE, Thomas	17th Feb. 1798. Major.	1799. Major.	Disbandment of 5th Dragoons.
GOODAIR, Wm. Henry	27th Jan. 1883. Lieut.	1894. Captain.	Suakin 1885.
GORE, Ric.	1st April 1695. Captain.	No trace after 1711. Brevet Lieut.-Cononel.	Flanders, 1695-7. Blenheim, Ramillies, Oudenarde, Malplaquet. 9th Son of Sir Fras. Gore of Artaman, co. Sligo.
GORING, John	15th March 1798. Surgeon.	1799. Surgeon.	Disbandment of 5th Dragoons.
GORGES, Richard	27th April 1756. Captain.	1761. Captain.	
GOUGH, John	24th Feb. 1708. Cornet.	No trace after 1709. Cornet.	Malplaquet.

NAME	DATE OF JOINING AND RANK	OUT OF REGIMENT AND RANK	WAR SERVICES AND REMARKS
GRAHAM, H.W. Gore.	23rd April 1887. Lieut.	Still serving. Lieut-Colonel. Commanding.	West Africa 1889. Ashanti 1895-6. N. W. Frontier, Tirah, 1897-8. S. Africa 1899-02 D.S.O.
GRANT, Fra. Rich.	16th March 1858. Captain.	1868. Major.	
GREAVES, George	30th April 1788. Cornet.	1791. Cornet.	
GREEN, Joseph Geo.	31st Aug. 1795. Cornet.	1799. Lieut.	Disbandment of 5th Dragoons.
GREEN, Nuttall	31st Aug. 1791. Lieut.	1799. Captain.	Disbandment of 5th Dragoons.
GREEN, Phillip	27th Feb. 1875. Captain.	1877. Captain.	
GRIDLEY, H.A. Adams	28th Feb. 1874. Sub-Lieut.	1875. Sub-Lieut.	
GRIFFITH, George	3rd June 1859. Quarter Master.	1863. Quarter Master.	
GRIFFITH, Henry Darby	1st Jan. 1872. Colonel (Major General).	1877. Colonel. (General).	Crimea 1854-55. In command of Scots Greys Balaclava,(Wounded by a pistol ball in the head), Inkerman, Tchernaya, Sebastopol. C.B.
GRIFFITH, Lewis	8th Sept. 1725. Lieut.	No trace after 1745. Lieut.	
GROGAN, John	5th March 1777.	1786. Lieut.	
GWYTHER, James Edwin	25th May 1860. Cornet.	187-. Lieut.	

NAME	DATE OF JOINING AND RANK	OUT OF REGIMENT AND RANK	WAR SERVICES AND REMARKS
GWYNN, Reginald P.J.	31st May 1890. 2nd Lieut.	1900. Lieut.	S. Africa 1899-01. Elandslaagte, Defence of Ladysmith, (After 1900 with 7th Fusiliers.)
GWYNN, Jno.	10th July 1695. Cornet.	1709. Cornet.	Flanders 1695-7.
HAM (?), Edward	1704. Quarter Master.	No further trace.	
HAMILTON, Edward	1694. Quarter Master.	No trace after 1730. Lieut.	Flanders 1694-7. Blenheim, (Wounded), Ramillies, Oudenarde, Malplaquet.
HAMILTON, Fras. Henry	23rd Jan. 1863. Lieut.	1869. Lieut.	
HAMILTON, Hans.	31st Aug. 1783. Captain.	1790. Captain.	
HAMILTON, Gustav.	17th Aug. 1715. Captain.	No trace after 1730. Captain.	
HAMILTON, James	10th April 1704. Cornet.	No trace after 1709. Lieut.	Blenheim, Ramillies, Oudenarde, Malplaquet.
HANBURY, Charles	14th April 1863. Cornet.	1875. Lieut.	
HARCOURT, Henry	31st May 1790. Captain.	1792. Captain.	
HARENC, Chas. Edward	15th Oct. 1861. Cornet.	187-. Lieut.	
HARRISON, Holton	1st May 1797. Asst. Surgeon.	1798. Asst. Surgeon.	
HARVEY, Alfred Bissel	18th Oct. 1864. Cornet.	1893. Lt. Colonel (Col.) Commanding.	Suakim '85, (Wounded).

NAME	DATE OF JOINING AND RANK	OUT OF REGIMENT AND RANK	WAR SERVICES AND REMARKS
HARVEY, John Robert	3rd June 1885. Lieut.	1890. Lieut.	
HATTON, William	13th Jan. 1777. Cornet.	1786. Lieut.	
HAWORTH, Cyril H.	30th May 1891. 2nd Lieut.	1896. Lieut.	
HAY, Henry	10th March 1730. Cornet.	1737. Cornet.	
HEATH, John	24th Jan. 1733. Captain.	No trace after 1737 Captain.	
HEATLY, Henry	12th July 1777. Cornet.	1784. Lieut.	
HENDERSON, Andrew	9th May 1794. Cornet.	1796. Cornet.	
HENNIKER, B. Trecothic	30th April 1791. Major.	1793. Major.	
HERDMAN, Alfred	6th July 1867. Cornet.	1873. Lieut.	
HIGGINS, William	11th July 1722. Capt.-Lieut.	1739. Captain.	
HIGGINS, William	1745. Captain Lieutenant.	1754. Captain Lieut.	
HILL, Edward	9th June 1704. Cornet.	No trace after 1709.	Malplaquet.
HILL, John	6th Nov. 1694. Captain.	1709. Lieutenant Colonel. (B-Col).	Flanders 1694-7. Blenheim, (Wounded) Ramillies, Oudenarde, Malplaquet.
HILL, William	1st Aug. 1741. Captain.	1759. Lt. Colonel. Commanding.	

NAME	DATE OF JOINING AND RANK	OUT OF REGIMENT AND RANK	WAR SERVICES AND REMARKS
HILL, William H.T.	7th Dec. 1895. 2nd Lieut.	1900. Lieut.	S. Africa 1899-00. Elandslaagte. Killed in the battle on Wagon Hill 6th January 1900, during the Defence of Ladysmith.
HILLIER, Geo. H.	26th Feb. 1858. Captain.	1861. Brevet-Maj.	Maharajpore 1843. Moodkee.
HOBART, Robert	17th Nov. 1780. Captain.	1783. Captain.	
HODGKINSON, John	18th June 1892. 2nd Lieut.	1897. Lieut.	
HOLDERNESS, Walter	16th Dec. 1903. Riding Master and Hon: Lieut.	Still serving in that rank.	S. Africa 1896.
HOLMES, Cyril	12th Sept. 1903. 2nd Lieut.	1905. 2nd Lieut.	
HOOPER, Richard Grenside	7th Dec. 1895. 2nd Lieut.	1901. Lieut.	S. Africa 1899-00. Defence of Ladysmith.
HORWOOD, William T.F.	10th Nov. 1888. 2nd Lieutenant.	1891. 2nd Lieut.	
HUGHES-ONSLOW, A.	10th May 1882. Lieut.	1882. Lieut.	
HUISH, Henry	26th Oct. 1858. Surgeon.	1865. Surgeon.	M.D.
HULSE, Harold H.	30th Jan. 1889. 2nd Lieut.	1904. Captain.	S. Africa 1899-00. Elandslaagte. Defence of Ladysmith (Wounded).
HULTON-HARROP, D. de L.	27th July 1901. 2nd Lieut.	1904. Lieut.	S. Africa 1901.

NAME	DATE OF JOINING AND RANK	OUT OF REGIMENT AND RANK	WAR SERVICES AND REMARKS
HUNT, Arthur C.	24th Dec. 1870. Lieut.	1875. Lieut.	New Zealand 1863-66.
HUNT, John	1st Oct. 1781. Captain.	1787. Captain.	
HUNTER, John	1694. Quarter Master.	No trace after 1709. Lieut.	Flanders 1694-7. Blenheim (Wounded) Ramillies. Oudenarde, Malplaquet.
HUNTER, Robert	13th April 1698. Major.	About 1707. Brevet Lieut. Colonel.	Son of James Hunter of the Hunterston family. Is said to have had a command at the siege of Londonderry 1689. Blenheim. Governor of Virginia 1708. Taken prisoner by the French on his way to America, but soon afterwards exchanged for the Bishop of Quebec, then a prisoner in the hands of the English. Governor of New York 1709. Brig. General 1711, Governor of Jamaica 1729. Major General 1729. Died in Jamaica 31st March 1734.
HUTCHINSON, J. Haller	20th Feb. 1867. Cornet.	1870. Cornet.	
HUTCHINSON, Norton	25th June 1789. Cornet.	1795. Lieut.	

NAME	DATE OF JOINING AND RANK	OUT OF REGIMENT AND RANK	WAR SERVICES AND REMARKS
IRELAND, Courcy	1694. Quarter Master.	No trace after 1695. Qr. Mr.	Flanders 1694-5.
IRVINE, William	13th Jan. 1777. Cornet.	1799. Captain.	Disbandment of 5th Dragoons.
JARDINE, James Bruce	12th March 1890. 2nd Lieut.	Still serving. Captain.	S. Africa 1899-02. Elandslaagte Defence of Ladysmith. Belfast. An attaché with Japanese Army in Manchurian War 1904-5. D. S. O.
JENINGS, Herbert Creagh	11th June 1902. 2nd Lieut.	Still serving Lieutenant.	
JENNEY, Brabazon	7th Oct. 1758. Cornet.	No trace after 1763. Lieut.	
JENNEY, Henry	12th March 1754. Cornet.	Before 1765. Lieut.	
JESSOP, Thomas	3rd Feb. 1776. Chaplain.	1789. Chaplain.	
JOCELYN, The Hon: George	17th April 1776. Cornet.	1786. Lieut.	
JOHNSTON, Allen	1st May 1734. Captain.	No trace after 1745. Capt.	
JOHNSTON, Jno.	8th Dec. 1692. Lieut.	No trace after 1709. Capt.	Flanders 1694-7. Blenheim, Ramillies, Oudenarde, Malplaquet.
JOHNSTON, Ric.	1704. Quarter Master.	No trace after 1709. Q. M.	Blenheim, Ramillies, Oudenarde, Malplaquet.
JOHNSTONE, Fitzroy A.B.	15th Aug. 1900. 2nd Lieut.	1903. Lieut.	

NAME	DATE OF JOINING AND RANK	OUT OF REGIMENT AND RANK	WAR SERVICES AND REMARKS
JOHNSTONE-DOUGLAS, Cecil F.	30th May 1888. Major.	1894. Lieut-Col. Commanding.	Died near Simla, India, August 1894.
JONES, Charles	29th Feb. 1796. Cornet.	1799. Lieut.	Disbandment of 5th Dragoons.
JONES, Llewellyn H.	10th Nov. 1877. 2nd Lieut.	1887. Captain.	Suakim 1885.
JONES, Thomas	19th June 1798. Cornet.	1799. Cornet.	Disbandment of 5th Dragoons.
JONES, William	28th Feb. 1788. Cornet.	1796. Lieut.	
KEILY, Richard	4th March 1769. Cornet.	1776. Captain.	
KELLY, Phillip	11th Feb. 1880. Riding Master	1888. Riding Master and Hony. Captain.	
KENNEDY, Geo. E. de V.	11th Oct. 1879. 2nd Lieut.	1884. Lieut.	
KENRICK, Thomas	8th Jan. 1740. Cornet.	No trace after 1745. Cornet.	
KING, Augustus C.	21st Aug. 1889. Captain.	1901. Major.	Burmah 1886-7. S. Africa 1899-01 Elandslaagte. Defence of Ladysmith (wounded) Belfast.
KING, Gilbert	16th May 1766. Captain.	1783. Major.	
KIRWAN, Andrew	25th Aug. 1760. Cornet.	1781. Captain.	
KNIGHT, Hy. Raleigh	31st Dec. 1790. Captain.	1792. Captain.	

NAME	DATE OF JOINING AND RANK	OUT OF REGIMENT AND RANK	WAR SERVICES AND REMARKS
KNOWLDEN, R.	2nd Sept. 1905. Quarter Master and Hon: Captain.	Still serving in that rank.	
KNOX, James	1709. Quarter Master.	1753. Cornet.	Malplaquet. Died 1753.
KNOX, John	8th Nov. 1737. Cornet.	1756. Captain.	
KYNAIRD, Thomas	14th May 1735. Cornet.	No trace after 1739. Lieut.	
LADAVEZE, John	20th Jan. 1764. Captain.	Before 1767. Captain.	
LALOR, Thomas	30th May 1794. Cornet.	1795. Cornet.	
LAMBERT, Cyril E.	14th Sept. 1901. 2nd Lieut.	1903. 2nd Lieut.	S. Africa 1900-02.
LEDWILL, William	31st Dec. 1791. Lieut.	1799. Captain.	Disbandment of 5th Dragoons.
LEESON, The Hon: William	25th June 1789. Cornet.	1791. Cornet.	
LENNOX, Wills.	8th June 1749 Cornet.	1755. Cornet.	
LESLIE, Cecil	30th June 1796. Cornet.	1798. Cornet.	
LITTLE, Archd. Cosmo	15th Aug. 1877. 2nd Lieut. 11th May 1900. Brevet Major.	1892. Brevet Major. 1901. Brevet Major.	Suakim 1885. (Staff) S. Africa 1900-02. Returned to the Regiment from Reserve of officers for the War.
LLOYD, Griffith	1st Aug. 1694. Adjutant.	1702. Cornet.	Flanders 1694-7.

NAME	DATE OF JOINING AND RANK	OUT OF REGIMENT AND RANK	WAR SERVICES AND REMARKS
LLOYD, Jno.	8th May 1694. Chaplain.	1696. Chaplain.	Flanders 1694-6.
LOCKHART, Alex.	1702. Quarter Master.	No other trace.	
LORD, Thomas	20th Jan. 1764. Cornet.	1766. Cornet.	
LORD, William C.	16th March 1858. Vet. Surgeon.	1862. Veterinary Surgeon.	
LUCAS, Benj.	8th March 1757. Cornet.	1770. Lieut.	
LUMB, E.J.M.	17th June 1885. Lieut.	1887. Lieut.	
LUSHINGTON, William	6th June 1741.	No other trace.	
MABERLY, Charles J.A.	27th Sept. 1905. Lieut.	Still serving. Lieut.	S. Africa 1902.
M'CAUSLAND, John	29th June 1793. Cornet.	1799. Lieut.	
McCLINTOCK, A. George	15th Nov. 1899. 2nd Lieut.	Still serving.	S. Africa 1900-01. Belfast.
McDONNELL, Edward	9th May 1794. Captain.	1795. Captain.	
McFARLAN, Jno. Warden	20th May 1868. Paymaster.	1870. Paymaster.	
McGREGOR, Alex. Edgar	26th Feb. 1858. Captain.	1859. Captain.	Central India.
McKEANE, George	1702. Quarter Master.	No trace after 1709. Cornet.	Blenheim, Ramillies, Oudenarde, Malplaquet.
MACKENZIE, Kenneth D.	4th July 1896. 2nd Lieut.	1898. Lieut.	S. Africa 1900-02. Diamond Hill, etc.

NAME	DATE OF JOINING AND RANK	OUT OF REGIMENT AND RANK	WAR SERVICES AND REMARKS
MACKENZIE, R.S.	4th April 1900. 2nd Lieut.	1901. 2nd Lieut.	
McLAUGHLIN, Hubert J.	30th May 1888. Captain.	1899. Major.	South Africa, Transvaal 1881. Nile Exp: 1884-5. S. Africa 1899-02. (Remount Department) D.S.O.
McLEAN, John A.	8th June 1889. 2nd Lieut.	1896. Lieut.	
MACMILLAN-SCOTT, A.F.	12th Nov. 1874. Lieut.	1880. Lieut.	
McNAIR, Jno. M.	10th May 1871. Captain.	1876. Captain.	Crimea 1855. Sebastopol. Indian Mutiny. Lucknow.
MACNAMARA, Alex.	30th June 1865. Cornet.	187-. Cornet.	
McNEILL, Malcolm	5th Dec. 1883. Lieut.	1888. Lieut.	Suakim 1885.
MACOLMSON, Alex.	14th May 1858. Cornet.	1861. Lieut.	
McTAGGART, Maxwell F.	20th Feb. 1895. 2nd Lieut.	Still serving. Capt.	N. W. Frontier 1897-8. Tirah 1897-8. S. Africa 1899-00. Elandslaagte. Defence of Ladysmith.
M'TAGGART, Isaac	31th May 1792. Surgeon	1798. Lieut.	Transferred to combatant rank.
MADAN, Rich.	2nd April 1724. Captain.	No trace after 1730. Captain.	

NAME	DATE OF JOINING AND RANK	OUT OF REGIMENT AND RANK	WAR SERVICES AND REMARKS
MAGILL, George (?)	28th Dec. 1784. Cornet.	1792. Lieut.	
MAHER, John	25th July 1795. Cornet.	1798. Lieut.	
MAHON, M. Hartland	24th Dec. 1874. Paymaster.	1881. Paymaster.	
MAHON, Stephen	30th Nov. 1791. Captain.	1794. Captain.	
MALAM, Jno. William	1st Feb. 1868. Cornet.	187-. Cornet.	
MANGIN, Sam. Hen.	18th April 1766. Captain.	1782. Major.	
MANN, Jno.	16th Dec. 1695. Cornet.	No trace after 1709. Lieut.	Flanders 1695-7. Ramillies, Oudenarde, Malplaquet.
MANNING, Thomas	30th June 1795. Capt-Lieut.	1799. Capt-Lieut.	Disbandment of 5th Dragoons.
MARELLA, John	4th Oct. 1777. Lieut.	No further trace.	
MARSILLY, Peter	10th Dec. 1710. Cornet.	1739. Cornet.	
MARTIN, Edward N.M.	27th July 1901. 2nd Lieut.	1903. 2nd Lieut.	S. Africa 1900-01.
MARYON-WILSON, F.P.M.	12th Nov. 1884. 2nd Lieut.	1885. 2nd Lieut.	
MASON, Henry Mark	Oct. 1745 Cornet.	Before 1765. Captain.	
MASSINGBERD, Francis, Burell	16th Oct. 1867. Captain.	1870. Captain.	
MASSEY, Fra. Hugh.	17th Jan. 1774. Lieut.	1791. Captain.	

NAME	DATE OF JOINING AND RANK	OUT OF REGIMENT AND RANK	WAR SERVICES AND REMARKS
MASSY, W. Godfrey Dunham ("Redan")	26th Feb. 1858. Captain. 4th Oct. 1896. Colonel (Lieut. General).	1879. Lieut. Col. commanding. (Bt. Col.) 1906 Colonel (Lieut. General).	Crimea, Sebastopol, Wounded at the Redan. Died 21st Sept. 1906. C.B.
MASTER, A.C. Chester	28th Feb. 1874. Lieut.	1877. Lieutenant.	
MASTERS, Alex.	28th Oct. 1871. Lieut.	1871. Lieutenant.	
MATHER, Robert	16th April 1858. Cornet.	1860. Cornet.	
MATHEW, Charles T.	11th March 1891. 2nd Lieut.	1895. 2nd Lieut.	
MERCER, Rich.	17th March 1761. Captain.	Before 1767. Captain.	
METAXA, The Count R.B.P. Warmingham	26th July 1876. Lieut.	1877. Lieutenant.	
METGE, Jno.	30th April 1771. Cornet.	1780. Lieut.	
MICHELL, St. John Fancourt	28th Oct. 1871. Lieut.	1875. Lieut.	
MILLEFONT, David	29th June 1780. Cornet.	1786. Cornet.	
MILLER, Humphrey	25th March 1705. Captain.	No other trace.	
MITCHELL, Nathaniel	10th March 1753. Cornet.	1761. Lieutenant.	
MOLESWORTH, Richard, Viscount	27th June 1727. Colonel (Major General).	1758. Colonel (Field Marshal).	Blenheim, Ramillies, Oudenarde, Malplaquet.

NAME	DATE OF JOINING AND RANK	OUT OF REGIMENT AND RANK	WAR SERVICES AND REMARKS
MOLESWORTH, Robert	28th Oct. 1745. Cornet.	1770. Lieut.	
MOLYNEUX, Thomas	1794. Capt-Lieut.	1795. Capt-Lieut.	
MONK, Harry Percy	22th Jan. 1755. Cornet.	1767. Lieutenant.	
MONTEITH, E.V. Peshall	28th Oct. 1871. Lieut.	1875. Lieut.	
MONTROSE, Douglas Beresford Malise Ronald, Duke of	27th Feb. 1875. Sub-Lieut.	1877. Lieut.	
MOORE, Lorenzo	9th May 1794. Lieut.	1795. Lieut.	
MOORE, Lorenzo Henry	15th Dec. 1794. Cornet.	1796. Cornet.	
MOORE, Ponsonby	4th March 1745. Cornet.	1752. Lieut.	
MOORE, Richard St. Leger	19th Jan. 1876. Captain	1881. Captain.	S. Africa 1901.
MOORE, Thomas	16th May 1766. Lieut.	1776. Captain.	
MOORE, William	16th March 1775. Cornet.	1778. Cornet.	
MORLAND, Henry Courtenay	11th Feb. 1875. Lieut.	1889. Major.	
MORRIS, Charles	28th Nov. 1771. Cornet.	1773. Cornet.	
MORRISON, Richard F.	5th June 1875. Captain	1881. Captain.	Crimea 1884-5. Alma, Sebastopol.

NAME	DATE OF JOINING AND RANK	OUT OF REGIMENT AND RANK	WAR SERVICES AND REMARKS
MOSTYN, Jno.	18th Oct. 1758. Colonel (Major General.)	1760. Colonel. (Lieut-General).	
MOWAT, James S.	26th June 1901. 2nd Lieut.	1903. 2nd Lieut.	S. Africa 1900-01.
MULLER, Robert	5th Oct. 1892. 2nd Lieut.	1893. 2nd Lieut.	
MUMMERY, W.E.	23rd July 1902. Riding Master. (Hon. Lieut.)	1903. Riding Master (Hon. Lieut.)	South Africa 1902.
MUNDY, Basil St. John	23rd Aug. 1882. Lieut.	1890. Captain.	Suakim 1885.
MURPHY, Adolphus	17th March 1863. Lieut.	1863. Lieut.	Crimea.
MURRAY, Arthur	16th March 1858. Lieut.	1863. Captain.	Crimea, Sebastopol. Indian Mutiny.
NESBIT, Francis	15th April 1749. Cornet.	1757. Lieut.	
NESBITT, Thos.	17th Sept. 1761. Cornet.	1765. Cornet.	
NETTLES, William	15th Aug. 1766. Cornet.	No other trace	

NAME	DATE OF JOINING AND RANK	OUT OF REGIMENT AND RANK	WAR SERVICES AND REMARKS
NEWCOMEN, Beverley	16th Dec. 1695. Cornet.	1709. Lieut.	Flanders 1696-7. Malplaquet. 5th Son of Sir Thomas Newcomen, of Kenagh who was killed at the Siege of Enniskillen in 1689. Beverley Newcomen was "probably a child" (Dalton) when gazetted to the 5th Dragoons. No trace of him in the 5th Dragoons after 1709. In 1720 he was Lieut. Colonel of Clement Neville's Dragoons. Died 1731. A Beverley Newcomen had held a commission as Ensign of Foot in King James' Army in 1687. (Dalton's King James' Army List).
NEWCOMEN, Charles	1689. Captain.	No trace after 1702. Captain.	IrishWar 1689-91. Flanders 1694-7. Son of Sir Thos. Newcomen of Kenagh Co. Longford.
NICHOLLS, John	4th Jan. 1749. Cornet.	1772. Lieut.	
NORCOTT, Arthur	13th March 1767. Cornet.	1772. Cornet.	
NORTON, Cecil William	28th Oct. 1871. Lieut.	1882. Captain.	

NAME	DATE OF JOINING AND RANK	OUT OF REGIMENT AND RANK	WAR SERVICES AND REMARKS
OAKES, Montague Percy R.	16th Nov. 1887. 2nd Lieut.	1905. Captain.	N. W. Frontier 1897-1898. Tirah 1897-8. S. Africa 1899-01. Elandslaagte, Defence of Ladysmith. (Wounded.) Belfast.
O'BRIEN, W.W. Baird	16th Dec. 1867. Cornet.	187-. Cornet.	
OLIVER, Robert	13th Oct. 1798. Cornet.	1799. Cornet.	Disbandment of 5th Dragoons.
OLLIVER, Groome	3rd March 1863. Cornet.	1867. Cornet.	
ONGE, Samuel	8th July 1794. Cornet.	1799. Captain.	Disbandment of 5th Dragoons.
ORDE, Chas. Reginald	26th Feb. 1876. Sub. Lieut.	1879. Lieut.	
O'REILLY, James	31st Augt. 1796. Cornet.	1799. Cornet.	Disbandment of 5th Dragoons.
ORMSBY, Phillip	1745. Cornet.	1776. Captain	
OWEN, R.E.	1881. Paymaster (Hon. Capt.)	1884. Paymaster.	
PALEY, Edward Groves	11th Sept. 1875. Captain	1888. Major.	
PALISER, Richard	11th March 1769. Cornet.	1786. Captain.	
PARKER, Arthur	16th Nov. 1887. 2nd Lieut.	Still serving Captain.	
PARKER, Michael	1st April 1695. Cornet.	No trace after 1709.	Flanders, 1695-7. Oudenarde, Malplaquet.

NAME	DATE OF JOINING AND RANK	OUT OF REGIMENT AND RANK	WAR SERVICES AND REMARKS
PAYNE, Henry	18th July 1888. Riding Master, Hon. Lieut.	1901. Riding Master, Hon. Captain.	S. Africa 1899-00.
PERRIN, John	8th June 1769. Surgeon.	1776. Surgeon.	
PHILLIPS, John	25th Feb. 1741. Surgeon.	1753. Surgeon.	
PIRIE, Duncan Vernon	16th Dec. 1888. Captain.	1888. Captain.	Egypt 1882. Kassanin. Soudan '84 (Staff) El Teb, Tamai Nile 1884-1885 (Staff).
PLATT, Henry V.	11th July 1888. Lieut.	1893. Captain.	
POE, Francis	3rd Sept. 1763. Cornet.	1766. Cornet.	
POÉ, James	1694. Qr. Master.	1728. Lieut.	Flanders 1694-7. Blenheim, Ramillies, Oudenarde, Malplaquet. 3rd son of Emanuel Poé of Co. Tipperary. Died 1728.
POE, James	17th March 1761. Cornet.	Before 1765. Cornet.	
POLE, Edward	22nd Nov. 1868. Colonel. (Major-General).	1872. Colonel (Lieut-General) Transferred to 12th Lancers.	Kaffir War 1851-1853; in command of 12th Lancers and of the Cavalry and artillery in the '51 expedition over the Kei. Crimea 1855, Tchernaya, Sebastopol.
POMEROY, James	10th Oct. 1770. Cornet.	1776. Cornet.	

NAME	DATE OF JOINING AND RANK	OUT OF REGIMENT AND RANK	WAR SERVICES AND REMARKS
POOLE, G. Roland Ruscombe	31st Oct. 1871. Captain.	1876. Captain.	
PORTAL, Robert	19th Feb. 1858. Major.	1863. Lieut.-Col. Commanding.	Crimea. Alma, Balaclava, Inkerman, Sebastopol.
PRESTON, John	30th July 1796. Surgeon.	1798. Surgeon.	
PRICE, James	28th May 1794. Cornet.	1796. Cornet.	
PRICE, Wm. V. Rose	15th Feb. 1902. 2nd Lieut.	1905. Lieut.	S. Africa 1901-02.
PURCELL, Peter V.	19th Feb. 1881. 2nd Lieut.	1884. Lieut.	
PURDON, Bartholomew	28th May 1756. Cornet.	Before 1765. Lieut.	
PYLE, Sir Seymour	26th Aug. 1737. Captain.	About 1746. Captain.	
PYM, Charles Evelyn	5th Oct. 1901. Lieut.	Still serving. Lieut.	S. Africa 1901-02.
RADCLIFFE, J.L.P.	11th Aug. 1900. 2nd Lieut.	1903. Lieut.	South Africa 1902.
RANT, William	6th Aug. 1858. Riding Master.	1873. Riding Master.	Crimea 1884-5. Balaclava, Inkerman, Tchernaya, Sebastopol.
RENDALL, Jno. King.	30th Sept. 1859. Captain.	1861. Captain.	
RENNIE, John H.W.	25th Aug. 1883. Lieutenant.	1889. Captain.	Suakim 1885. South Africa 1901.
REYNELL, John	13th Feb. 1765. Cornet.	1776. Cornet.	

NAME	DATE OF JOINING AND RANK	OUT OF REGIMENT AND RANK	WAR SERVICES AND REMARKS
RHODES, Gilbert H.	13th Aug. 1902. 2nd Lieut.	1905. 2nd Lieut.	
RICE, James	20th June 1753. Cornet.	No other trace.	
RICHARDSON, J.B.	10th March 1883. Lieut.	1885. Lieut.	Suakim 1885. Killed in action.
RIDGE, J.S.	6th Sept. 1794. Cornet.	1799. Captain.	Disbandment of 5th Dragoons.
RIDLEY, Rich. Parnham	3rd Feb. 1860.	1861.	
ROBINSON, Brian W.	29th Nov. 1905. 2nd Lieut.	Still serving. 2nd Lieut.	
ROBINSON, Godfrey	15th Aug. 1883. Quarter Master. (Hon. Lieut.)	1888. Quarter Master. (Hon. Lieut.)	
ROBINSON, Jocelyn	23rd Feb. 1742. Captain.	About 1752. Captain.	
ROBINSON, Jos.	26th March 1737. Cornet.	1742. Cornet.	
ROSE, Alex.	12th Sept. 1729. Lieut-Colonel.	Given command of a regiment in 1740.	
ROSE, Hugh C.	15th Aug. 1900. 2nd Lieut.	1904. Lieutenant.	S. Africa 1901-02
ROSS, Charles	1689. Captain.		Ireland 1689-91. Flanders 1694-7.
	16th July 1695. Colonel.	1715. Colonel (General).	Blenheim, Ramillies, Oudenarde, Malplaquet. Died
	1st Feb. 1729. Col. (General).	1732. Colonel (General).	Aug. 1732. Son of the 11th Baron Ross of Balnagowan.

NAME	DATE OF JOINING AND RANK	OUT OF REGIMENT AND RANK	WAR SERVICES AND REMARKS
ROSS, David	1702. Quarter Master.	1707. Adjutant.	Blenheim (wounded).
ROSS, James	16th Dec. 1695. Lieut.	1702. Lieut.	Flanders 1696-7.
ROSS, James	9th April 1756. Adjutant.	1757. Adjutant.	
ROSS, William	20th July 1695. Captain.	Before 1704. Captain.	Flanders 1696-7.
ROSS, William	20th July 1695. Captain-Lieut.	1705. Capt-Lieut.	Flanders 1696-7.
ROSS, William	1702. Quarter Master.	About 1737. Lieut.	Malplaquet.
ROSS, William	30th Aug. 1739. Cornet.	1761. Captain.	
ROSSMORE, Robert, Lord.	4th April 1787. Colonel (Lieut. General).	1799. Colonel. (General).	Disbandment of 5th Dragoons. Formerly Robert Cuninghame, succeeded to title of Rossmore in 1797.
ROSTRON, Jno. Ashworth	19th April 1876. Vet. Surgeon.	1879. Veterinary Surgeon.	
ROTHNEY, Alex. Erskine	19th Oct. 1872. Sub.-Lieut.	1873. Sub-Lieut.	
ROWLEY, Clotworthy	13th Feb. 1762. Captain.	1776. Major.	
RUSSELL, J. Fredk. Love	12th Feb. 1873. Sub-Lieut.	1881. Lieut.	
SADLIER, R. Moore	18th Dec. 1860. Cornet.	1864. Lieut.	
ST. AUBYN, Wm. John	15th Feb. 1861.	1868. Lieut.	

NAME	DATE OF JOINING AND RANK	OUT OF REGIMENT AND RANK	WAR SERVICES AND REMARKS
ST. LEGER, Richard	9th Oct. 1775. Cornet.	1777. Cornet.	
SALIS, Henry Norman	17th March 1858. Lieut.	1861. Lieut.	Kaffir War 1851-53.
SANDFORD, Joseph	26th Sept. 1772. Cornet.	1779. Lieut.	
SANDFORD, Robert	1744. Cornet.	1756. Lieut.	
SAUNDERS, Edward Henry	2nd July 1861. Cornet.	1876. Captain	
SAUNDERS, Geo. Robt.	26th Nov. 1861. Cornet.	1864. Lieut.	
SAVARY, William	14th Feb. 1776. Cornet.	1779. Cornet.	
SAWER, Thomas	26th April 1677. Lieut-Colonel.	1678. Lieut. Col.	
SCHRIEBER, S.W. Smith	5th May 1863. Cornet.	1864. Cornet.	
SCOBELL, Henry Jenner	27th July 1901. Lieut-Colonel.	1902. Lieut.-Col. Transferred to Scots Greys.	S. Africa 1899-02. Relief of Kimberley, Paardeburg etc. commanding a column C.B. commanding 1st Cav. Brigade 1903. Maj. Gen. 1903.
SCOTT, James	25th March 1705. Surgeon.	About 1710. Surgeon.	Malplaquet.
SCOTT, James	27th May 1717. Cornet.	About 1745. Lieut.	
SCRIVEN, John B.	3rd Oct. 1888. 2nd Lieut.	1906. Captain and Brevet Major.	S. Africa 1899-02. Elandslaagte, Defence of Ladysmith.

NAME	DATE OF JOINING AND RANK	OUT OF REGIMENT AND RANK	WAR SERVICES AND REMARKS
SCROW, James	4th March 1735. Cornet.	1739. Cornet.	
SEBAG-MONTE-FIORE, William	29th Nov. 1905. 2nd Lieut.	Still serving. 2nd Lieut.	
SECKHAM, Guy L.T.	19th July 1905. Captain.	Still serving. Captain.	
SEDLEY, Frederick, The Marquis	6th May 1862. Lieut.	1871. Captain.	China 1860. Taku Forts.
SERGISON-SMITH, Hyde	2nd Sept. 1862. Captain.	1865. Captain.	
SHAWE, Douglas	4th Feb. 1871. Quarter Master.	1880. Quarter Master.	
SIDNEY, Thomas	5th Oct. 1715. Colonel.	1729. Colonel.	Died 1729.
SINCLAIR, John	13th Aug. 1879. 2nd Lieut.	1887. Captain.	Suakim 1855
SKAEN, William	24th July 1746. Cornet.	About 1753. Lieut.	
SKEFFINGTON, The Hon. Henry	2nd Sept. 1763. Cornet.	1791(?). Cornet.	
SKEF(F)INGTON, The Hon. W. John	22nd Sept. 1769. Cornet.	1772. Cornet.	
SKEL(S)TON, John	1704. Quarter Master.	No trace after 1709. Cornet.	Blenheim, Ramillies, Oudenarde, Malplaquet.
SKENE, And. Philip	31st July 1788. Captain.	1791. Captain.	
SLADE, William H.	26th Feb. 1858.	1871. Lieut-Col.	Crimea. Sebastopol.
SLEIGH, Edgar H.	8th Nov. 1905. 2nd Lieutenant.	Still serving. 2nd Lieut.	

NAME	DATE OF JOINING AND RANK	OUT OF REGIMENT AND RANK	WAR SERVICES AND REMARKS
SMITH, Cuthbert	20th June 1739. Cornet.	1772. Captain.	
SMITH, John	27th Aug. 1760. Cornet.	Before 1765. Cornet.	
SMITH, Jno. O. Gowan	5th Oct. 1858. Cornet.	1860. Cornet.	
SMITH-	7th Feb. 1759. Cornet.	1761. Cornet.	
SMITHWICK, Robert	1st Jan. 1774. Surgeon.	1784. Surgeon.	
SMYTH, James	17th April 1784. Cornet.	1789. Cornet.	
SOMERVILLE, Warburton	17th Dec. 1762. Cornet.	Apparently 1763. Cornet.	
SPENCER, A. Campbell	12th Nov. 1873. Lieut.	1888. Major.	Nile Exp: 1885.
SPICER, Julian Jno. Newton	10th Dec. 1873. Sub. Lieut.	1888. Captain.	
SPICER, Thomas	31st Jan. 1791. Cornet.	1794. Cornet.	
SPURRELL, Robert John	2nd July 1892. Captain.	1905. Major.	Relief of Hazara 1891. N. W. Frontier 1897-8. Tirah 1897-8. S. Africa 1900-02. Ladysmith.
STANHOPE, Edward	3rd Aug. 1723. Cornet.	1737. Cornet.	
STANLEY, Edward	24th Jan. 1862. Vet. Surgeon.	1870. Vet. Surg.	
STANLEY, John	16th Nov. 1785. Captain.	1791. Captain.	

NAME	DATE OF JOINING AND RANK	OUT OF REGIMENT AND RANK	WAR SERVICES AND REMARKS
STEPHENS, Daniel	30th June 1787 Cornet.	1799. Lieut.	Disbandment of 5th Dragoons.
STEPHENSON, St. George	1st Aug. 1741. Cornet.	Before 1765. Captain.	
STEUART, Charles	1st Feb. 1728. Cornet.	About 1746. Lieut.	
STEVENSON, George	1695. Lieutenant.	No trace after 1702. Lieut.	Flanders 1695-7.
STEVENSON, Phillip L.	23rd July 1890. Captain.	1892. Captain.	
STEVENSON, William	8th May 1694. Cornet.	1704. Cornet.	Flanders 1694-7.
STEWART, The Hon. Charles W.	1st Jan. 1797. Lieut-Colonel.	1799. Lieutenant-Col. Command.	Disbandment of 5th Dragoons.
STEWART, Francis	18th June 1766. Captain.	1766. Captain.	
STEWART, James	7th Jan. 1778. Lieut-Colonel.	1796. Lieutenant-Colonel.	
STRANGWAYS, The Hon. S. Digby	5th Aug. 1767. Cornet.	1770.	
STRATTON, John	3rd April 1779. Cornet.	1786. Lieut.	
STYLE(S), Charles	21st Dec. 1733. Cornet.	1757. Captain.	
SULIVAN, Geo. Aug. Filmer	19th Feb. 1858. Lt. Colonel.	1861. Lt. Colonel (Colonel).	Crimea. Balaclava, Inkerman, Sebastopol.
SUTHERLAND, Robert	22nd May 1863. Asst. Surgeon.	1867. Assistant Surgeon.	
TAYLOR, John	7th Sept. 1771. Cornet.	1785. Lieut.	

NAME	DATE OF JOINING AND RANK	OUT OF REGIMENT AND RANK	WAR SERVICES AND REMARKS
TAYLOR, John	2nd Nov. 1794. Adjutant.	1799. Adjutant.	Disbandment of 5th Dragoons.
THACKWELL, Fra. Jno. Roche	9th March 1860. Cornet.	1869. Captain.	Killed by a tiger in India.
THACKWELL, Joseph E. Lucas	10th Oct. 1874. Lieut.	1875. Lieut.	
THESIGER, The Hon. Chas. Wemyss	24th Jan. 1892. Colonel (Lieut.-General).	1896. Colonel. (Lieut-General). transferred to 14th Hussars.	China War 1860. Capture of Pekin.
TICKELL, Thomas	14th March 1771. Cornet.	1787. Capt-Lieut.	
TIDY, John	1694. Quarter Master.	1702. Qr. Mr.	Flanders 1694-7.
TIMSON, Henry	26th Feb. 1858. Captain.	1860. Captain.	Crimea.Sebastopol. (6 D)
TOLER, George	16th Feb. 1756. Cornet.	1767. Lieut.	
TOWNSEND, Phillip	30th April 1742. Lieut.	Apparently in 1743.	
TUFTON, Geo. Richard	26th Feb. 1876. Sub. Lieut.	1889. Captain.	
TUITE, James	15th April 1749. Cornet.	1756. Cornet.	
TYRRELL, George, G.M.	15th Nov. 1899. 2nd Lieut.	Still serving. Lieutenant.	S. Africa 1900-02. Ladysmith Relief Force.
USHER, John	1704. Captain.	No trace after 1709. Capt.	Blenheim, Malplaquet.
VALLANCE, Thos. Wm.	17th March 1850. Lieut.	1863. Captain.	Crimea. Sebastopol.

NAME	DATE OF JOINING AND RANK	OUT OF REGIMENT AND RANK	WAR SERVICES AND REMARKS
VALLANCE, Vane de V.M.	12th March 1904. 2nd Lieut.	Still serving. Lieut.	
VANDELEUR, Boyle	10th Sept. 1858. Cornet.	1885. Lt. Colonel.	
VANDELEUR, David Roche	4th July 1865. Captain.	187-. Captain.	
VEREKER, Amos	20th Jan. 1764.	1778. Lieut.	
VERNON, Edward	27th Nov. 1752. Cornet.	1753. Lieut.	
VESEY, Thomas	16th Feb. 1756. Cornet.	1760. Cornet.	
VINCENT-	26th Aug. 1760. Cornet.	Before 1765. Cornet.	
VYSE, Rich.	13th Feb. 1762. Cornet.	1778. Captain.	
WALSH, James	1709. Quarter Master.	About 1745. Lieut.	Malplaquet.
WARBURTON, Charles	30th Jan. 1718. Captain.	About 1745. Captain.	
WARDLAW, Charles	30th Jan. 1717. Captain.	About 1745. Captain.	
WARRE, John	10th April 1704. Captain.	Before 1714. Bt. Major.	Ramillies, Oudenarde, Malplaquet. Only son of Sir Fraser Warre, Bart. Died at Ghent before 1714.
WATERFALL, John Henry	31st July 1860. Lieut.	1863. Lieut.	
WATERMAN, George	15th Feb. 1888. Quarter Master. (Hon. Lieut.)	1905. Quarter Master. Hon. Major.	S. Africa 1899-00. Defence of Ladysmith.

NAME	DATE OF JOINING AND RANK	OUT OF REGIMENT AND RANK	WAR SERVICES AND REMARKS
WATHEN, Edward, O.	5th April 1893. Captain.	Still serving Major.	Burmah 1886-7. South Africa 1899-1902. Defence of Ladysmith. (wounded.)
WATTS, Mat.	20th July 1689. Cornet.	No trace after 1709. Lieut.	Irish War 1689-91. Flanders 1694-7. Blenheim.
WEAVER, Edward Fra.	17 March 1858. Lieut.	1864. Captain.	Crimea. Balaclava, Inkerman, Sebastopol.
WEBB,	1695. Lieut.	No other trace.	Flanders 1695.
WELCH, James	1709. Quarter Master.	No trace after 1728. Lieut.	Malplaquet.
WELLS, Mat.	20th July 1689. Cornet.	No trace after 1702. Cornet.	Irish War 1689-91. Flanders 1694-7.
WEST, Ernest, E.	10th Nov. 1888. 2nd Lieut.	1905. Captain.	S. Africa 1899-01. Diamond Hill, etc.
WEST, John	1st May 1797. Asst. Surgeon.	1799. Asst. Surgeon.	Disbandment of 5th Dragoons.
WESTENRA-	1795. Cornet.	same year.	
WESTENRA(Z), Warren	5th Feb. 1787. Cornet.	1791. Lieut.	
WESTON, Alfred	13th June 1888. Major.	1892. Major.	Afghan War 1879-80.
WHITBY, William	28th Nov. 1771. Captain.	1776. Captain.	
WHITCHURCH, Percy	1694. Quarter Master.	1702. Quarter Master.	Flanders 1694-7.
WHITE, Mor.	1709. Quarter Master.	No other trace.	Malplaquet.

NAME	DATE OF JOINING AND RANK	OUT OF REGIMENT AND RANK	WAR SERVICES AND REMARKS
WHITEHEAD, Jas. Andus	30th June 1886. Lieut.	1888. Lieut.	
WILBRAHAM, J. Watkin	15th Aug. 1783. Captain.	1786. Captain.	
WILKINSON, William	6th May 1760. Cornet.	Before 1765. Cornet.	
WILLCOX, Walter Temple	8th Nov. 1893. 2nd Lieut.	Still serving. Captain.	S. Africa 1899-02. Defence of Ladysmith. (wounded). Belfast.
WILLEY, Edward	31st March 1768. Cornet.	1772. Cornet.	
WILLIAMS, Sidney S.	7th Dec. 1895. 2nd Lieut.	1903. Lieut.	S. Africa 1901-02.
WILLIAMS, William	6 Oct. 1798. Lieut.	1799. Lieut.	Disbandment of 5th Dragoons.
WILSON, Edwin Bryce	16 Nov. 1887 2nd Lieut.	Still serving. Captain.	
WILSON, Thomas	10th July 1716. Cornet.	1756. Captain-Lieut.	
WITHERINGTON, Edward	3rd April 1786. Cornet.	1791. Lieut.	
WOLFE, Rich.	16th Dec. 1752. Cornet.	1776. Lieut.	
WOOD, Alex Vaughan Leipzig	3rd Oct. 1888. 2nd Lieut.	Still serving. Major.	S. Africa 1899-02. Elandslaagte, Defence of Ladysmith, Belfast. D.S.O.

NAME	DATE OF JOINING AND RANK	OUT OF REGIMENT AND RANK	WAR SERVICES AND REMARKS
WOODEN, Charles	21st March 1865. Quarter Master.	1871. Quarter. Master.	Crimea 1854-55. Alma, Balaclava, (Light Brigade) Inkerman, Sebastopol. V. C. Indian Mutiny 1857.
WYATT, Charles Edwyn	22nd June 1860. Captain.	1868. Captain.	Persia 1857.
WYNNE, Fawcett (Faufitt)	5th March 1783. Adjt.	1792. Adjt.	
WYNNE, James	20th June 1689. Colonel.	1695. Colonel. (Brigadier Gen.)	Irish War 1689-91. Flanders 1694-5. Died of wounds received in action near Rouselaer, 15th July 1695-7. He raised the Regiment.
WYNNE, James	20th July 1695. Cornet.	About 1705. Lieut.	Flanders 1695-7. Apparently a son of Brigadier James Wynne.
WYNNE, John	10th May 1740. Major.	No other trace.	
WYNNE, Owen	1st Nov. 1694. Major.	1705. Lieut.-Col.	Irish War 1689-91. Flanders 1694-7. Blenheim.
	August 1732. Colonel (Lieut.-Gen.)	1737. Colonel. Lieutenant-Gen.	Died 1737.
WYNNE, Owen	2nd May 1742. Cornet.	No other trace.	
WYNNE, Robert	7th Jan. 1778. Cornet.	No other trace.	

NAME	DATE OF JOINING AND RANK	OUT OF REGIMENT AND RANK	WAR SERVICES AND REMARKS
YELVERTON, W. Charles	24th Dec. 1779. Cornet.	1787. Cornet.	
YORKE, The Hon. Joseph	27th Nov. 1760. Colonel (Major General.) From Colonel of 8th Dragoons.	1787. Colonel (General). Transferred to 11th Light Dragoons.	Flanders 1743. K.B. Son of Lord Hardwicke, Lord Chancellor of England.
YOUNG (YONGE) Edward	7th Feb. 1745. Cornet.	1761. Lieut.	

NOTE:—There is no list of Officers who served in the Irish War 1689-91 or in Flanders 1694-7, but it is presumed that the Officers on the Roll of the Regiment during those periods served in the campaigns.

A.S.C.
ARTHUR: DOUBLEDAY & Co

www.ingramcontent.com/pod-product-compliance
Ingram Content Group UK Ltd.
Pitfield, Milton Keynes, MK11 3LW, UK
UKHW050146280726
14058UKWH00007B/853